Other volumes in this series:

T0286386

IET RADAR, SONAR, NAVIGATION AND AVIONICS SI

Series Editors: Dr
Prof.

Radar Tech
Using Array Ar

Radar Techniques Using Array Antennas

Wulf-Dieter Wirth

The Institution of Engineering and Technology

Published by The Institution of Engineering and Technology, London, United Kingdom

First edition © 2001 The Institution of Electrical Engineers
Reprint with new cover © 2008 The Institution of Engineering and Technology

First published 2001
Reprinted 2008

The Institution of Engineering and Technology
Michael Faraday House
Six Hills Way, Stevenage
Herts, SG1 2AY, United Kingdom

www.theiet.org

British Library Cataloguing in Publication Data
Radar techniques using array antennas
  (Radar, sonar, navigation and avionics series no 10)
  1. Radar 2. Antenna arrays
  I. Wirth, W.
  621.3′ 8483

ISBN (10 digit) 0 885296 798 5
ISBN (13 digit) 978-0-85296-798-0

Typeset by Newgen Imaging Systems, India
Printed in the UK by Anthony Rowe Ltd., Chippenham
Reprinted in the UK by Lightning Source UK Ltd, Milton Keynes

To Christiane

Contents

Preface

The intention of this book is to give the reader an introduction into the area of modern radar techniques based on active array antennas, either for use in development and research or as a decision and planning aid for the authorities within government or industry. In particular, many new possibilities are discussed which should find consideration for future radar systems. I have assumed the general basics of radar technology to be familiar to the reader; introductory literature is widely available.

I have tried to present the techniques, procedures and concepts, which can mostly be derived from signal-theoretical views and which are described with approaches and equations, from an engineering point of view, with a minimum of mathematical treatment. As far as is possible, a descriptive and functional explanation is given for each procedure.

Despite its already long history, radar technology will continue to receive attention for many years to come. The importance of radar results from its ability to provide reconnaissance, especially target detection, location and imaging, in all weather conditions and from short to long ranges. This ability results from the relatively small propagation attenuation of the electromagnetic waves with wavelengths in the centimetre to metre range. These wavelengths also permit sufficient location accuracies and resolutions with reasonable antenna dimensions. Radar works actively, illuminating the scene to be observed with its own transmitter. From this results its ability to measure target range and its independence from daylight, in contrast to visual optics.

One can thus expect future lively interesting and importance for radar technology. This statement is among other aspects confirmed by a well-established regular series of international radar conferences organised by respected and famous professional institutions such as IEE (UK), IEEE (USA), DGON/ITG (GE), SEE (F).

The idea of writing this book was first developed by contributions and the scientific coordination of the training course 'Future radar systems' of the Carl Cranz society in Germany. Here in this book the material is presented in much more detail than is possible in a short course. Beyond this, the book is essentially based on experiences, implementations and new concepts which my colleagues and I have developed and collected in the course of more than 30 years of scientific work in the radar field at the Forschungsinstitut für Funk und Mathematik (FFM, which translates into Research Institute for Radio Sensors and Mathematics) of the Forschungs-Gesellschaft für Angewandte Naturwissenschaften (FGAN, which translates into Research Association for Applied Natural Sciences).

In the introduction a general representation of the substantial future requirements for radar systems is given. Afterwards, the future importance of multifunction systems on the basis of electronically-steered array antennas is discussed. In the chapters 2, 3 and 4 there follows a short introduction into representation of signals, into the statistical signal theory and into the substantial characteristics, architectures and relationships of array antennas. These array antennas may be of linear, planar, cylindrical or volume type. These chapters shall be used as a basis for the following chapters. In chapter 5 the receiver beamforming is discussed. Chapter 6 deals with the time-discrete signal sampling and derivation of the orthogonal signal components. Chapter 7 is dedicated to pulse compression with the use of polyphase codes, with special compression procedures for sidelobe reduction and an improvement of the range resolution. In chapter 8 we treat problems and possibilities for target detection from signal series, i.e. incoherent and coherent integration, Doppler filtering, adaptive clutter suppression and the coherent processing of very long pulse series with a special test function. Chapter 9 presents sequential detection for radar application and its advantages, together with a solution for the problem of simultaneously testing multiple range bins.

Chapters 10 to 15 are dedicated to concepts, procedures and applications of array signal processing. In chapters 10, 11 and 12 adaptive jammer suppression, correction of monopulse under jamming conditions and angular superresolution are discussed. In chapters 13 and 14 the additional possibilities and special problems of radar on flying platforms are presented, especially with respect to the detection of slowly moving targets, that is space-time adaptive processing (STAP) procedures and synthetic aperture radar (SAR) with moving target indication (MTI) and target imaging. In chapters 15 the principles of inverse synthetic aperture radar (ISAR) are introduced as a way of generating radar images for detected moving targets. In chapter 16 fluctuation and spectral target characteristics, especially by jet engine modulation, are described. Both of these last chapters demonstrate the contributions to target classification which can be achieved with a multifunction radar.

In chapter 17 a survey of the experimental phased array radar system ELRA which has been developed at FFM is given. In chapter 18 follows the description of a special system concept with omnidirectional or floodlight transmission and a multibeam receiving system for radar operation with protection against antiradar missiles (ARMs). An additional application of this concept is the detection and classification of hovering helicopters. The corresponding experimental system OLPI following these concept ideas is described. Finally, in chapter 19 some remarks are made on system parameter relations and their choice for a multifunction radar system which is based on a phased-array antenna.

I would like to thank all my colleagues at the department EL (electronics) of FGAN-FFM for my long, very fruitful and enjoyable time as department head. Especially I am grateful for many years of scientific cooperation, with numerous stimulating discussions, with W. Bühring, J. Ender, I. Gröger, E. Hanle, R. Klemm,

U. Nickel, W. Sander, K. V. Schlachta and H. Wilden. Their contributions to our work are also documented by the references at the end of each chapter.

I would also like to thank Dr. Robin Mellors-Bourne, Diana Levy and Dr. Roland Harwood who, as the Book Publishing Department at the IEE, provided much support. The excellent co-operation that I received is very much appreciated. I also owe thanks to the editors of the IEE radar series, Professor E. D. R. Shearman and Professor P. Bradsell, and to the anonymous reviewers for their numerous corrections, creative comments and suggestions.

Finally, I thank the director of the FFM Dr. Jürgen Grosche and the head of the responsible department section within the German Ministry of Defence (BMVg) Hartmuth Wolff for the promotion of my work over many years in the exciting area of modern radar techniques and for the encouragement to write this book.

Wulf-Dieter Wirth
October 2000

Chapter 1

Introduction

The most demanding requirements for radar systems have so far nearly always resulted from military objectives; civil applications then benefit from the achieved results. Developments in the available technology, particularly within the areas of highly integrated semiconductor technology for digital signal processors, miniaturised microwave integrated circuits (MMIC) and efficient microwave computer-aided design methods, have enabled ever more demanding procedures and concepts to be achieved. The theoretical fundamentals and evaluation possibilities have improved simultaneously, especially with the development of computers and software. The use of powerful personal computers with program systems such as MATLAB or Mathematica is now very common.

Processing of the received signals is performed after an analogue-to-digital conversion in suitable digital signal processors or computers. Performance improvement is characterised by the measurement of the received signals with increasing precision and sampling rate. The necessary signal-processing operations are achieved with ever higher data rates and computing power. So it is possible to achieve real-time operation for more and more sophisticated signal and data-processing algorithms. From the received target signals the target location, parameters, features and even images for classification are then derived.

The extended requirements of today's and future radar systems are the following:

- Detection of targets with small radar cross section (stealth targets) and at long ranges with high detection probability and low false-alarm probability.
- Detection of targets under extreme conditions:
    objects flying at very high or low altitude
    objects emerging with high or very low speed
    targets flying or moving in shadow
    targets emerging for a short time from shadow (e.g. helicopters)
    targets with extreme and random mobility.
- Target acquisition should be followed by reliable and safe tracking, capable and effective for multiple targets, accomplished with high location accuracy in three dimensions.

- Suppression of unwanted echo signals (clutter), e.g. from ground and fixed targets, meteorological phenomena and sea swell.
- High resolution capability in the angular and range dimensions, effective also for several closely-spaced targets.
- Classification and as far as possible identification of targets using the radar echoes.
- Imaging of the ground scene with high resolution.
- Detection, location and imaging of moving objects in the clear and also above the background of the ground scene.
- Maintenance of the radar functions also with the irradiation of multiple opposing or coincidental electronic reverberation signals and all kinds of countermeasure (jamming, ECM).
- Low vulnerability against threat by antiradar missiles (ARM).

Radar systems are installed either on the ground or on ships, aircraft, missiles or satellites. When onboard moving platforms, the movement may be used for target imaging by forming a synthetic aperture antenna for high cross-range resolution.

In multifunction systems, several functions for the fulfilment of different requirements are put together. The combination of requirements or functions is determined by the application profile of the system: long-range reconnaissance, target reconnaissance, control of weapon delivery, navigation, air or maritime traffic control, observation of the weather, agriculture and environmental monitoring.

For civil applications radar systems also have to meet demanding requirements. In principle, unlimited improvement in the security of target tracking for controlling air and maritime traffic is required. The traffic must be observed and controlled under extreme conditions with as high a detection probability and accuracy as possible. Secondary radar, additionally introduced for air traffic control, can fail in principle, since it is dependent on the cooperation of all users with their transponders. Primary radar is independent of this cooperation and can thus track all traffic. Therefore, a radar system that is as efficient as possible is required for air traffic control; limitations on the requirements are provided, as is also the case for military systems, by cost.

Central to future radar technology will be the application of the array antennas, and this book is dedicated to the application of these antennas which offer great flexibility for system design in combination with suitable control and signal processing. The design work can be imagined roughly in the following way: on the basis of a required radar function we look for a signal and data-processing concept with the aid of statistical signal theory or filter theory. For a developed concept the efficiency is examined analytically or by simulation studies. The antenna concept, suitable to the required function, can then be developed using basic knowledge about the antenna array.

Within this book we will find this implicit procedure at several points. Statistical signal theory forms an important fundamental tool, and is presented as simply as possible and limited to the most important aspects. In addition, we need some basic relations for the most important characteristics of array antennas. Then we have to choose between the available array-antenna architectures.

As a disadvantage of the planar array antenna, their limited scanning coverage and thus limited field of view has to be noted: the antenna beam can be steered only to approximately 60° off broadside. For complete azimuth coverage three or four planar arrays are required. A cheaper solution is one rotating planar array antenna, but with certain limitations in the scanning performance. An alternative solution for full azimuth coverage is a volume array (crow's nest antenna). In this case all antenna elements are active for all directions at the same time.

Conformal antenna arrays may be built on the curved surface of the carrier, e.g. the fuselage of the aircraft. Then a wider field of view may also be achieved.

Since each array antenna must be equipped with 1000–5000 transmit/receive modules, the total cost is determined essentially by the price of these modules. Substantial technological progress has already been achieved, and it is hoped that the module cost will become low enough, especially by improved design methods and large-scale manufacturing.

Military systems will further play the role of pioneers, because in some instances only with the application of a phased-array radar can the required functions be achieved and therefore the high cost is acceptable. At present several radar systems with active antenna arrays are under development as demonstrator or operational systems. Some of these are now being developed by European industry (alone or in collaboration). Examples are the AMSAR, APAR, COBRA and SAMPSON (formerly MESAR) systems.

Chapter 2

Signal representation and mathematical tools

In this book we will consider antenna arrays which consist of many individual antenna elements, and therefore a large number of signals, one for each element, has to be processed at the same time. The signals can be assigned to locations on the antenna, forming discrete sampling of the spatial wave field for transmitting or receiving. Signals transmitted or received with a radar system also have to be represented as a function of time. Signal samples are formed in the spacial and temporal dimension for digital signal processing with signal processors or computers and for recording for later analysis.

These samples z are measured generally with the complex components x and y, i.e. they are to be written as complex values. The description of z by the amplitude a and the phase φ is equivalent. An individual signal is thus given by:

$$z = x + jy = a \exp(j\varphi) = a \cos(\varphi) + ja \sin(\varphi) \tag{2.1}$$

The components x and y are usually called I and Q components for inphase and quadrature phase. The sequences of samples can be produced in such a way as to be equivalent to the original time-continuous signal. More precise information on sampling will follow in chapter 6.

2.1 Vectors, matrices

We will write matrices with bold-face letters. Two-dimensional matrices will be written with bold-face capital letters. One-dimensional matrices (rows or columns) are also named vectors.

A temporal sequence of signals must be numbered for identification. We write for the sequence z_n, with $n = 1, \ldots, N$, N is thus the length of the signal sequence

$z_1, \ldots, z_N$. We can write this sequence more briefly as a column matrix or vector $\mathbf{z}$:

$$\mathbf{z} = \begin{bmatrix} z_1 \\ \vdots \\ z_N \end{bmatrix} \tag{2.2}$$

The transpose of $\mathbf{z}$ is then the row matrix $\mathbf{z}^{\mathrm{T}} = [z_1 \cdots z_N]$. In our applications we generally need the transpose and conjugate complex form:

$$\mathbf{z}^* = \lfloor z_1^* \cdots z_N^* \rfloor = [x_1 - jy_1 \cdots x_N - jy_N] \tag{2.3}$$

This one-dimensional sequence could also be derived at one time instant from a set of antenna elements by a purely spatial sampling.

In many applications signals are regarded in terms of both temporal and spatial sampling. These are then summarised in a rectangular matrix:

$$\mathbf{Z} = \begin{bmatrix} z_{11} & \cdots & z_{M1} \\ \vdots & \ddots & \vdots \\ z_{1N} & \cdots & z_{MN} \end{bmatrix} = [\mathbf{z}_1 \cdots \mathbf{z}_M] \tag{2.4}$$

Here, the temporal sequences n are again written as a single column for one antenna element m. The signals from the M spatially separate antenna elements at one time instant form a single row. For an indication of the matrix size it is helpful to write $\mathbf{Z}[N, M]$, thus the matrix $\mathbf{Z}$ has N rows and M columns.

2.2 Computing with matrices

For completeness the basic tools for working with matrices are given in the following. More detailed explanations can be found in mathematical textbooks and e.g. in the user's guides of the programming systems MATLAB and Mathematica.

2.2.1 Addition and subtraction

Two matrices $\mathbf{A}$ and $\mathbf{B}$ of the same size, and therefore the same dimensions $[M, N]$, are added or subtracted element by element. That means for:

$$\mathbf{C} = \mathbf{A} + \mathbf{B}$$

we have to take:

$$c_{mn} = a_{mn} + b_{mn} \tag{2.5}$$

2.2.2 *Multiplication*

Multiplication assumes that elements can always be multiplied in pairs. It applies the principle: a result element is the sum of the products of a row with a column. This may be seen most simply for a pair of row and column matrices. If **a** and **b** matrices are of the same dimension N we have:

$$\mathbf{a}^{\mathrm{T}} = \lfloor a_1 \cdots a_N \rfloor$$

and

$$\mathbf{b} = \begin{bmatrix} b_1 \\ \vdots \\ b_N \end{bmatrix}$$

and can form:

$$\mathbf{a}^{\mathrm{T}}\mathbf{b} = a_1 b_1 + a_2 b_2 + \cdots + a_N b_N$$

('T' indicates the transpose) or for complex elements according to equation 2.3:

$$\mathbf{a}^*\mathbf{b} = a_1^* b_1 + a_2^* b_2 + \cdots + a_N^* b_N \tag{2.6}$$

(astrisk indicates the complex conjugate transpose) which is the scalar, dot or inner product of the matrices **a** and **b**.

As a special case for a column matrix **a** the expression:

$$\|\mathbf{a}\| = a^*a = a_1 a_1^* + \cdots + a_N a_N^* \tag{2.7}$$

is the so-called 2- or Euclidian norm. It is also explained as the square of the length of vector **a**.

A geometrical interpretation is sometimes given to the scalar product: after normalisation of the vectors with respect to their length it is the cosine of the angle φ between the two N-dimensional vectors **a** and **b**:

$$\cos \varphi = \frac{\mathbf{a}^*\mathbf{b}}{\sqrt{\|\mathbf{a}\| \, \|\mathbf{b}\|}} \tag{2.8}$$

If the value of $\varphi = 90°$ then $\cos \varphi = 0$. As in the two-dimensional case, we say that the two vectors **a** and **b** are orthogonal.

Now two rectangular matrices are to be multiplied; the matrix product is defined in the following way:

$$A[M, N] \, B [N, L] = C[M, L]$$

with

$$c_{ik} = a_{i1} b_{1k} + a_{i2} b_{2k} + \cdots + a_{iN} b_{Nk} \tag{2.9}$$

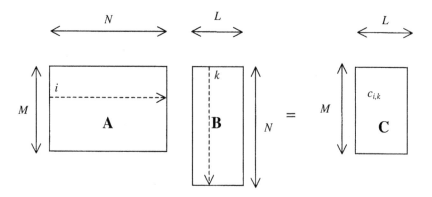

Figure 2.1   Multiplication of two rectangular matrices

The result element c_{ik} is thus formed by the scalar product of the ith row of $\mathbf{A}$ with the kth column of $\mathbf{B}$. This is illustrated in Figure 2.1.

The number of columns of the left matrix, $\mathbf{A}$, must be equal to the number of rows of the right matrix, $\mathbf{B}$; the values of M and L may be arbitrary. The multiplication of a matrix $\mathbf{A}[M, N]$ with $\mathbf{B}[N, L]$ results in a matrix $\mathbf{C}[M, L]$.

We can multiply a column by a row according to this definition, for example, as a dyadic product:

$$\mathbf{Q}_{xy} = \begin{bmatrix} x_1 \\ \vdots \\ x_N \end{bmatrix} \begin{bmatrix} y_1^* & \cdots & y_N^* \end{bmatrix} = \begin{bmatrix} x_1 y_1^* & \cdots & x_1 y_N^* \\ \vdots & \ddots & \vdots \\ x_N y_1 & \cdots & x_N y_N^* \end{bmatrix} \quad (2.10)$$

The sequence of the multiplication cannot be exchanged, without changing the result. The transpose of a product is given by:

$$(\mathbf{AB})^* = \mathbf{B}^* \mathbf{A}^* \quad (2.11)$$

2.2.3   Identity matrix

The identity matrix $\mathbf{I}[N, N]$ corresponds to the role of 'one' in scalar algebra:

$$\mathbf{I} = \begin{bmatrix} 1 & 0 & \cdots & 0 & 0 \\ 0 & 1 & \ddots & \vdots & 0 \\ \vdots & 0 & \ddots & 0 & \vdots \\ 0 & \vdots & \ddots & 1 & 0 \\ 0 & 0 & \cdots & 0 & 1 \end{bmatrix} \quad (2.12)$$

It is a diagonal matrix, containing elements, in this case the value 1, only on the main diagonal. In general the following applies:

$$\mathbf{A}[M, N]\, \mathbf{I}[N, N] = \mathbf{A}[M, N]$$
$$\mathbf{I}[M, M]\, \mathbf{A}[M, N] = \mathbf{A}[M, N]$$

(2.13)

2.2.4 Inverse matrix

Particularly important is the so-called inverse matrix $\mathbf{A}^{-1}$, defined for a square matrix $\mathbf{A}(N, N)$. Multiplication of $\mathbf{A}^{-1}$ with $\mathbf{A}$ yields the unit matrix $\mathbf{I}$.

$$\mathbf{A}^{-1}\mathbf{A} = \mathbf{I}$$

(2.14)

or equivalently:

$$\mathbf{A}\mathbf{A}^{-1} = \mathbf{I}$$

With the inverse matrix $\mathbf{A}^{-1}$ one can solve the following general problem, with $\mathbf{X}$ as the unknown matrix:

$$\mathbf{A}\mathbf{X} = \mathbf{B}$$

(2.15)

By:

$$\mathbf{A}^{-1}\mathbf{A}\mathbf{X} = \mathbf{A}^{-1}\mathbf{B}$$

we can solve for $\mathbf{X}$:

$$\mathbf{X} = \mathbf{A}^{-1}\mathbf{B}$$

(2.16)

The computation of the inverse matrix is accomplished using special algorithms, e.g. the Gauss elimination method, which are available in program systems such as MATLAB or Mathematica. To be inverted the matrix $\mathbf{A}$ must fulfil certain conditions, so that the inverse matrix exists: it must be square and non-singular.

A useful inversion formula has been derived [2]:

$$(\mathbf{A} + \mathbf{B}\mathbf{C}\mathbf{B}^*)^{-1} = \mathbf{A}^{-1} - \mathbf{A}^{-1}\mathbf{B}(\mathbf{B}^*\mathbf{A}^{-1}\mathbf{B} + \mathbf{C}^{-1})^{-1}\mathbf{B}^*\mathbf{A}^{-1}$$

(2.17)

2.2.5 Eigenvalue decomposition

To a matrix $\mathbf{A}$ applies the equation [1]:

$$\mathbf{A}\,\mathbf{x} = \lambda\,\mathbf{x}$$

The column matrix $\mathbf{x}$ is named eigenvector and the scalar number λ the eigenvalue. To a matrix of dimension $[N, N]$ there exists generally N eigenvalues, λ, which may be combined in the diagonal matrix **D**, and N eigenvectors, **x**, combined as columns in the eigenvector matrix **X**. One can thus also write:

$$\mathbf{A\,X = X\,D} \tag{2.18}$$

The column vectors **x** in **X** are mutually orthogonal, i.e. $\mathbf{X^* \, X = I}$. It follows:

$$\mathbf{X^* A X = D}$$

The matrix **A** is thus transformed into the diagonal matrix **D**.

The computation of **D** and **X** to a matrix **A** may be performed again with special procedures available in program systems, e.g. MATLAB or Mathematica.

2.2.6   QR decomposition

A matrix **A** can be decomposed into a product [1,2]:

$$\mathbf{A = QR} \tag{2.19}$$

Q is an orthonormalised matrix, i.e. it contains column vectors with the norm, given by equation 2.7, equal to 1. **R** is an upper triangle matrix, below the main diagonal it contains only elements equal to zero.

2.3   Fourier transform

To a finite time signal series s can be assigned with the Fourier transform a corresponding set of frequency values. The discrete Fourier transform, or DFT, is particularly significant for signal processing. It is defined by the following equation:

$$S_k = \sum_{n=0}^{N-1} s_n \exp\left(-\frac{2\pi j}{N} kn\right), \quad k = 0, \ldots, (N-1) \tag{2.20}$$

If the signal $s(t)$ is sampled with the regular time period T at $t = nT$ to give the signal samples s_n then S_k corresponds to the frequency spectral part of the signal s at the frequency ω_k:

$$\omega_k = \frac{2\pi}{NT} k \tag{2.21}$$

The frequency lines appear thus with the mutual distance $2\pi/NT$. A DC component of the signal corresponds to $k = 0$.

With a growing N we get a finer frequency resolution.

The spectrum, the totality of the frequency lines, is unambiguous only up to the frequency ω_{N-1}. For values of k outside the interval $0, \ldots, (N-1)$ we may write

$k = k' + pN$, with p any integer and k' within the interval $0, \ldots, (N - 1)$. Then we have:

$$\exp\left(-\frac{2\pi j}{N} kn\right) = \exp\left(-\frac{2\pi j}{N}(k' + pN)n\right)$$

$$= \exp\left(-\frac{2\pi j}{N} k'n - 2\pi jnp\right) = \exp\left(-\frac{2\pi j}{N} k'n\right)$$

The spectrum repeats itself with period N because of the cyclic exp function.

By the Fourier transform coefficients S_k according to equation 2.20 the signal s_n is completely described. The time signal series can be recovered exactly by an inverse Fourier transform:

$$s_n = \frac{1}{N} \sum_{k=0}^{N-1} S_k \exp\left(\frac{2\pi j}{N} nk\right) \tag{2.22}$$

As an illustration and short exercise we will now prove equation 2.22. By using equation 2.20 we get:

$$s_n = \frac{1}{N} \sum_{k=0}^{N-1} \left(\sum_{m=0}^{N-1} s_m \exp\left(-\frac{2\pi j}{N} km\right)\right) \exp\left(\frac{2\pi j}{N} nk\right) \tag{2.22a}$$

and by exchanging the order of summation:

$$s_n = \frac{1}{N} \sum_{m=0}^{N-1} \left(s_m \sum_{k=0}^{N-1} \exp\left(-\frac{2\pi j}{N}(m - n)k\right)\right) \tag{2.22b}$$

For $m \neq n$ there follows for the sum expression for index k:

$$X = \sum_{k=0}^{N-1} q^k$$

with

$$q = \exp\left(-\frac{2\pi j}{N}(m - n)\right)$$

Now we recall the well-known sum formula for a geometrical series:

$$X = \frac{(1 - q^N)}{1 - q}$$

Since:

$$q^N = \exp\left(-\frac{2\pi j}{N}(m - n)N\right) = 1 \tag{2.22c}$$

it follows $X = 0$.

One also says that the rows $\exp(-(2\pi j/N)nk)$ and $\exp(-(2\pi j/N)mk)$ are orthogonal to each other.

The vectors formed by both rows are orthogonal, as defined in equation 2.8.

When $m = n$ each of the exp expressions is equal to 1 and therefore the sum becomes N. The entire expression remains only s_n. This had to be proven.

The Fourier transform of equations 2.20 and 2.21 describes mathematically the spectral lines S_k, which are discrete sinusoidal signals at frequencies ω_k existing for an infinite time. This results from the implicit assumption of a periodic repetition of the signal s. In radar practice generally the signal is time limited and one can imagine the existence of the signals S_k only during the duration of the temporal signal s_n. A time-limited signal does not produce a line spectrum but, instead of each discrete S_k, a continuous spectrum with a functional form $\sin \omega/\omega$ around ω_k. However, since these spectra have mutual zeros, the orthogonality remains and the inverse transform with equation 2.22 is valid.

2.3.1   Fast Fourier transform, FFT

For computation of the DFT from equation 2.20 or 2.22 one uses the so-called fast Fourier transform FFT [3] as a very efficient numerical algorithm. Instead of N^2 complex multiplications one needs only $(N/2)\log_2 N$ complex multiplications (with $\log_2$ for the binary logarithm). This saving is achieved by skilful combinations of intermediate results. The value N must be a binary number 2^p (p integer). In modern programming systems such as MATLAB or Mathematica the FFT is of course available. For example, MATLAB performs the transform according to equation 2.20 by:

$$S_{\bar{k}} = \sum_{n=1}^{N} s_n \exp\left(-\frac{2\pi j}{N}(n-1)(\bar{k}-1)\right), \quad \bar{k} = 1, \ldots, N$$

So with $k = \bar{k} - 1$, the DC component results for $\bar{k} = 1$. The cyclic repetition explained above has to be considered.

2.4   Filter in the frequency and time domain

The transformation of signals into the frequency domain is important for the application and treatment of filters. A filter is usually characterised by its frequency response $H(\omega)$. To the signal frequencies ω_k are assigned the filter values H_k. To a signal s at the filter input we get the filter output signal y. The spectral values of the signal and the filter simply have to be multiplied, resulting in the output spectrum:

$$Y_k = H_k S_k$$

So we must first compute the signal spectrum S_k using equation 2.20, multiply by H_k and then apply the inverse Fourier transformation to come back into the time domain by equation 2.22. The result is the temporal output signal y_n:

$$y_n = \frac{1}{N} \sum_{k=0}^{N-1} H_k S_k \exp\left(\frac{2\pi j}{N} nk\right) \tag{2.23}$$

or by substituting the time-domain definition of the signal by equation 2.20:

$$y_n = \frac{1}{N} \sum_{k=0}^{N-1} H_k \sum_{l=0}^{N-1} s_l \exp\left(\frac{2\pi j}{N}(kn - kl)\right)$$

and by changing the order of summation:

$$y_n = \sum_{l=0}^{N-1} s_l \frac{1}{N} \sum_{k=0}^{N-1} H_k \exp\left(\frac{2\pi j}{N} k(n - l)\right)$$

By comparison with equation 2.22 we recognise the second sum as the time-domain representation h of the filter function H and we may write:

$$h_{n-l} = \frac{1}{N} \sum_{k=0}^{N-1} H_k \exp\left(\frac{2\pi j}{N} k(n - l)\right) \tag{2.24}$$

and finally we get:

$$y_n = \sum_{l=0}^{N-1} s_l h_{n-l} \tag{2.25}$$

This expression represents the filtering of a signal sequence s_n with a filter function h_n which is defined in the time domain. This kind of filtering is also called convolution between s and h.

Let us consider a filter for performing the convolution by means of a shift register, by which the signal s is shifted with the clock period T. At the individual shift stages the signal s_l is weighted with the factor h_{n-l}. Afterwards the summation of the weighted signals follows. The number n represents the shift between s and h, growing with the clock pulses. Such a filter is also called a transversal filter.

In Figure 2.2 we visualise the shift state $n = 2$:

$$y_2 = s_0 h_2 + s_1 h_1 + s_2 h_0$$

For $n = N - 1$ the set of s_n coincides completely with the set of h_n. In this case the signal is weighted by the temporal mirror image of h.

Later we will discuss the optimal or matched filter which satisfies the condition:

$$h_l = s_{N-1-l}^* \tag{2.26}$$

which means that all signals are weighted by their complex conjugate values. Therefore all products of the sum of equation 2.25 are real valued. This is the matched state

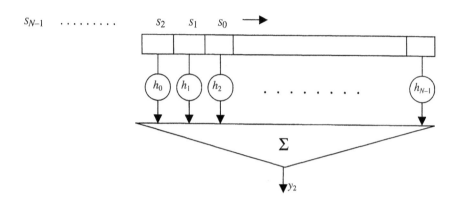

Figure 2.2   Convolution of a signal with a filter function by means of a shift register

of signal filtering resulting in a maximum output signal y. All the other shift states result in inevitable time sidelobes.

With all other shift steps n the elements of h are initially not all defined. Since h results by equation 2.22 from a given H, a periodic repetition of the values of h exists with the period N. These values are applied automatically by using equation 2.23 for the convolution. Therefore, by applying equation 2.25 the value $(n-l)$ as the index of h is always taken modulo N. If $n-l$ does not lie within the interval $0, \ldots, (N-1)$, suitable multiples of N are added or subtracted. Equation 2.25 becomes with this statement:

$$y_n = \sum_{l=0}^{N-1} s_l h_{(n-l) \bmod N} \tag{2.27}$$

Because of the periodicity, the process of filtering after equation 2.27 is known as the cyclic convolution.

If the signal s consists of a single impulse with the value $s_0 = 1$ for $l = 0$, and all the other $s_n = 0$, then $y_n = h_n$. The filter function may be produced in such a way and is therefore also called the impulse response.

The importance of equations 2.23 and 2.24 consists in the fact that on the basis of a temporal filter function h the filter effect can be computed after equation 2.23 with the FT of s and h. With larger values N this computation is then much more economical by application of the FFT compared to a direct convolution of s and h in the time domain.

Often the cyclic convolution after equation 2.27 is not adequate, since h is not periodic, but is a limited function in time. For the computation of the so-called aperiodic or linear convolution it must be ensured that the mutually shifted sequences s and h do not meet periodic repetitions. This is achieved on the basis of s and h in the time domain by filling up the time series with zeros (zero padding). The entire sequences are then, for example, twice as long.

In practice it is often the case that a very long signal sequence s has to be filtered with a very much shorter filter function h. In this case the signal sequence is appropriately decomposed into short subsequences and then filtered aperiodically. From the overlapping output sequences the actual long output sequence y_n has then to be built up again.

2.5   Correlation

We assume two signal series x_n and y_n $(n = 1, \ldots, N$, with a mean equal to zero) and we want to express their relationship or similarity. The expression:

$$q = \frac{1}{N} \sum_{n=1}^{N} x_n y_n \tag{2.28}$$

may be used for this purpose and is known as correlation. Imagine that the signals x_n and y_n are changing their amplitude and polarity randomly and independently; the products will then cancel to zero if N goes to ∞. But if both signals are the same then q will show a maximum value.

We may introduce additionally a time lag k (sample periods) and define the crosscorrelation function:

$$q(k) = \frac{1}{N} \sum_{n=1}^{N} x_n y_{n+k} \tag{2.29}$$

With this correlation function we will get a high value if x_n and y_n are similar but shifted in time by k sample periods.

If we want to indicate or detect for a signal x_n similarities or relationships in itself over time we may take the autocorrelation function:

$$\rho(k) = \frac{1}{N} \sum_{n=1}^{N} x_n x_{n+k} \tag{2.30}$$

For $k = 0$ we obtain the variance:

$$\sigma_x^2 = \frac{1}{N} \sum_{n=1}^{N} x_n^2 \tag{2.31}$$

For complex-valued signals the autocorrelation function is defined by:

$$\rho(k) = \frac{1}{N} \sum_{n=1}^{N} x_n x_{n+k}^* \tag{2.32}$$

If some equal signals have a certain phase, then this will be rotated back to zero within the product and result in a maximum contribution to the sum.

2.6   Wiener Khintchine theorem

The Wiener Khintchine theorem gives a relation between the power spectrum and the autocorrelation function for random noise signals. We may compute the amplitude spectrum for a noise signal by equation 2.20. From this follows the power spectrum:

$$P_k = S_k S_k^* \quad \text{or} \quad P_k = |S_k|^2 \tag{2.33}$$

Because we have in mind a random signal x for an infinite time we may divide the entire process into time segments with N samples each and after that take the mean value or expectation of P_k (in chapter 3 we will discuss the basics of statistics).

We compute the DFT of P_k with equation 2.20 and use equation 2.22. First we have:

$$P_k = \sum_{m=0}^{N-1} s_m \exp\left(-\frac{2\pi j}{N} km\right) \sum_{n=0}^{N-1} s_n^* \exp\left(\frac{2\pi j}{N} kn\right)$$

With this we compute:

$$\frac{1}{N}\sum_{k=0}^{N-1} P_k \exp\left(\frac{2\pi j}{N} pk\right) = \frac{1}{N}\sum_{m=0}^{N-1}\sum_{n=0}^{N-1} s_m s_n^* \sum_{k=0}^{N-1} \exp\left(\frac{2\pi j}{N}(-m+n+p)k\right)$$

The exp expression on the right-hand side is zero for $(-m+n+p) \neq 0$, according to our little exercise to prove equation 2.22. Only for:

$$m = n + p$$

we get for the sum expression the value N. So finally the result is:

$$\frac{1}{N}\sum_{k=0}^{N-1} P_k \exp\left(\frac{2\pi j}{N} pk\right) = \sum_{m=0}^{N-1} s_m^* s_{m+p}$$

and with equation 2.32:

$$\frac{1}{N^2}\sum_{k=0}^{N-1} P_k \exp\left(\frac{2\pi j}{N} pk\right) = \rho(k) \tag{2.34}$$

This relation between the power spectrum and the autocorrelation function given by the Fourier transform is named the Wiener Khintchine theorem. Sometimes we need the reverse relation of equation 2.34. This follows with equation 2.20:

$$P_k = N\sum_{n=0}^{N-1} \rho(n) \exp\left(-\frac{2\pi j}{N} kn\right) \tag{2.35}$$

2.7   References

1 BELLMANN, R.: 'Introduction to matrix analysis' (McGraw-Hill Book Company, Inc., New York, Toronto, London, 1960)
2 BODEWIG, E.: 'Matrix calculus' (North-Holland Publishing Company, Amsterdam, 1959)
3 HAGER, W. W.: 'Applied numerical linear algebra' (Prentice Hall, London, 1988)
4 COOLEY, J. W., LEWIS, P., and WELCH, P. D.: 'Historical notes on the Fast Fourier Transform', *IEEE Trans. Audio Electroacoust.*, 1967, **15** (2), pp. 76–79
5 BIALLY, T.: 'The Fast Fourier Transform and its application to radar signals'. International conference on *Radar*, Paris, France, 1978, pp. 51–59
6 BLAHUT, R. E.: 'Fast algorithms for digital signal processing' (Addison-Wesley Publishing Company, Inc., New York, 1985)
7 McCLELLAN, J. H., and RADER, C. M.: 'Number theory in digital signal processing' (Prentice Hall Inc., Englewood Cliffs, New Jersey, 1997)

Chapter 3

Statistical signal theory

3.1 The general tasks of signal processing

For the implementation of a particular radar procedure the first step is to control the transmitting beam. This can be made inertialess with electronically steered radar, using digitally controllable phase shifters and thus made completely arbitrary with respect to time and direction. As a physical fundamental condition, the delay time of the target echoes is naturally given by the speed of light and must be considered for the design of the individual radar procedures.

The receiving beam has to collect the echo signals from the same direction as that illuminated by the transmit beam. The received echo signals must then be processed. With the concept of active array antennas many degrees of freedom are available for the organisation of new radar procedures. These are generally not achievable with conventional radars using mechanical rotating reflector antennas.

A sequence of task orders is generated by the radar system's control computer according to the requirements for the radar. The function and efficiency of an individual radar procedure fulfilling the ordered task is determined to a large extent by the method of signal processing. In principle, the following basic radar tasks or functions are to be achieved within a multifunction radar (MFR). They may be mixed or combined in a time multiplex operation according to the actual priority.

Search in different range and angular areas:
- remote areas without ground clutter, possibly with weather clutter
- medium range with and without ground or weather clutter
- short range with mainlobe or sidelobe clutter

Target acquisition by detection and location orders for different target parameters such as range, cross section and Doppler frequency.

Tracking by location orders, with different target parameters and movements.

Target classification by signal analyses from signal sequences of acquired and tracked targets.

To all these primary tasks is added if necessary:

- suppression of irradiated jammer signals by adaptive antenna beamforming (ECCM)
- suppression of clutter echoes
- pulse compression
- detection procedures with CFAR (constant false-alarm rate)
- determination of the Doppler frequency of the target
- exact target location by monopulse
- correction of the location with monopulse against main beam jamming
- angular superresolution of several targets
- correction of location errors produced by multipath effects
- defensive measures against special threats such as antiradiation missile ARM, electronic support measures (ESM)
- detection of special critical targets (hovering helicopters, stealth targets)

After the transmitter has illuminated the target scenario in accordance with the required radar function, the target echoes are then picked up by the receiving antenna array and offered in an electronically suitable form as a set of spatial and temporally distributed signal samples. Unfortunately, there are also inevitable disturbance signals, at least in form of receiving noise, and sometimes additional signals from jammers and clutter reflections (ground or weather echoes).

The function of signal processing is to detect the targets from the received signals as effectively as possible and in particular to extract afterwards for these targets further information, especially parameters such as amplitude, Doppler frequency shift, Doppler spectrum, polarisation and exact position. In some applications even a target image can be extracted.

3.2   Introduction to basics of statistics

The received radar signals from targets are always superposed with the receiver noise and other disturbing signals, such as signals received from jammers, hostile noise transmitters trying to blind our radar, and echoes from clutter reflectors, that is from ground or fixed targets, trees, vehicles or meteorological scatterers. These disturbing signals are always randomly fluctuating by the nature of their origin. Radar targets, such as airplanes or missiles, are generally very complex and composed of multiple single reflection centres. The targets' movement results in varying phase relationships of the partial echoes from the multiple complex reflection elements. The superposition of all these partial echoes yields therefore a fluctuating resultant target echo.

The main task of a radar system is the detection of targets. In other words, for each resolution cell in space we have to decide 'target present' or 'no target present'. Because of the fluctuating nature of the signals this task cannot be performed with absolute certainty, but only with a limited percentage or probability of correct decisions. Of course we want to make as many correct decisions as possible according to the underlying situation. That is, we want to detect a target if it is really present, and we want to deny a target if that target is really absent and there is only noise. So we are led quite naturally to the field of statistics or, more precisely, to the need for a statistical description of radar signals and the assessment of our decisions by probabilities. Then it turns out that the way to come to a decision, that is the rule for signal processing, can be derived from the statistical description of the signal and the requirement to minimise the probability of false decisions.

3.2.1 Probabilities for discrete random variables

We are all familiar with the experiment of throwing a dice or a coin. These are the usual examples of a random regular process used in textbooks [1] for introducing the topic of statistics. Rolling a dice or flipping a coin are examples of a statistical process. Each single trial has an outcome which we may observe and note. These exclusive outcomes may be denoted by the variable X. In the case of rolling a dice we have $X = 1, 2, 3, 4, 5, 6$ as possible values for the outcomes. Now we repeat the experiment and observe the occurrence $X = 6$. If we make N experiments and observe $n(6)$ occurrences with $X = 6$, then we can define the relative frequency of occurrence $n(6)/N$. The relative occurrence of X then is $n(X)/N$. Now we can imagine repeating the experiment with a growing N. We will observe a converging value for the ratio $n(X)/N$ if we spend more and more time on the experiment, that is $n(6)/N$ will tend to 1/6 with more and more accuracy. We now define a limit for an infinite number of experiments N as the probability $P(X)$:

$$P(X) = \lim_{N \to \infty} \frac{n(X)}{N} \qquad (3.1)$$

This is the mathematical definition of probability. In practice, we would never have enough time to perform an infinite number of experiments, but the probability exactly describes the tendency of our experiment and is the only possible way of quantifying the underlying process. It gives, in the case of the dice, no forecast for the outcome of the next roll. But it forecasts the relative frequency of a specific outcome for a growing number of experiments better and better (the law of large numbers).

Computational rules for probabilities have been developed, the most important of which are summarised in the following. We have to remember the basic rule for all considerations to determine probabilities for certain events: *determine the number of possible events with the feature X in question in relation to the total possible number of events.*

An event which is certain has $P = 1$, an impossible event $P = 0$.

If we ask if event X_1 or event X_2 occurs, e.g. for a dice as the result of one throw the number 1 or 4, then the probabilities add:

$$P(X_1 \text{ or } X_2) = P(X_1 + X_2) = P(X_1) + P(X_2) \qquad (3.2)$$

If there are K possible events X_k, such as the six numbers of a dice with $K = 6$, then any one out of the possible results occurs with certainty and we have:

$$\sum_{k=1}^{K} P(X_k) = P(X_1) + P(X_2) + \cdots + P(X_K) = 1 \qquad (3.3)$$

We may observe two processes in parallel, e.g. throwing two dice and observing the results. The probability for the joint event X_i at the first dice and X_j at the second dice is given by:

$$P(X_i \text{ and } X_j) = P(X_i, X_j) = P(X_i)P(X_j) \qquad (3.4)$$

This simple product rule holds only for independent processes. There must be no influence between them, as in the case of two dice, if they are not connected somehow by a thread. The probability $P(X_i, X_j)$ is called the joint probability.

We may also consider two processes and ask, what is the probability for X_i under the condition that with the other process the event X_j occurs? The events X_i are counted only if X_j occurs at the same time:

$$P(X_i/X_j) = \lim_{N \to \infty} \frac{n(X_i, X_j)}{n(X_j)} = \lim_{N \to \infty} \frac{n(X_i, X_j)/N}{n(X_j)/N} = \frac{P(X_i, X_j)}{P(X_j)} \qquad (3.5)$$

If both processes are independent then from equations 3.4 and 3.5 follows

$$P(X_i/X_j) = \frac{P(X_i)P(X_j)}{P(X_j)} = P(X_i) \qquad (3.6)$$

Example 1: binary signal series

As an illustrative and important example we now consider the following problem:

There is a series of length M with signals X_m, that is $m = 1, \ldots, M$. Each signal may have the value $X_m = 0$ with probability q and $X_m = 1$ with probability p. The question is, what is the probability for n 1s within the series of M signals? (Figure 3.1).

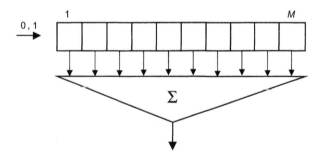

Figure 3.1   Binary signal series and summation

Example: $0\,0\,1\,0\,1\,0\,0\,1\,0\,1\,1\,1\,0\,1\,0\,0$ for $M = 16$.

Of course $p + q = 1$, that is one of the two possible signal values occurs with certainty.

For the occurrence of a certain arrangement of n 1s and $(M - n)$ 0s, resulting in the sum n for the signals, first we have the probability:

$$P_1 = p^n q^{M-n}$$

The next question is, how many different arrangements leading to the sum equal to n are possible?

We assume as an intermediate step for our development M different signals. Because we have M places for the arrangement of signals it follows for the number of possibilities: the first signal has M possible places, leaving for the second signal $(M - 1)$ places, the third $(M - 2)$ places and so forth. So we have $M(M - 1)(M - 2) \cdots 1 = M!$ possible arrangements for M different signals.

But we have as signals only 0s or 1s. All different arrangements of 1s with a certain pattern are not distinguishable and we have $n!$ possible repetitions. The same holds for the 0s: there we have $(M - n)!$ repetitions. We have to divide $M!$ by the number of repetitions to come to the number of possible arrangements of 1s and 0s. The result then is for the probability of n 1s or for the sum equal to n:

$$P(n) = \frac{M!}{n!(M - n)!} p^n q^{M-n} = \binom{M}{n} p^n q^{M-n} \tag{3.7}$$

The expression:

$$(p + q)^M = \sum_{n=0}^{M} \binom{M}{n} p^n q^{M-n} \tag{3.8}$$

where $\binom{M}{n}$ is the number of combinations of M elements taken n at a time, also written $^M C_n$, is well known as the binomial law. Equation 3.7 is therefore known as the binomial distribution. It applies to many problems with alternative single events.

Because $p + q = 1$ we see that the sum expression fulfils the condition given by equation 3.3:

$$\sum_{n=0}^{M} P(n) = 1$$

3.2.2 Continuous random variables

Radar signals received by the antenna and then amplified and mixed down to a suitable frequency band are analogue or continuous variables. The received signals are superposed by the receiver noise, and can have any value within the dynamic range of the receiver output. In the usual mathematical description generally there is no limit assumed; the signal may be between plus and minus infinity.

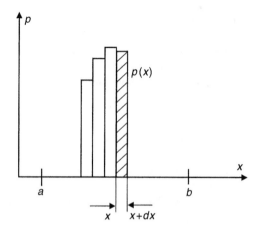

Figure 3.2   Transition to continuous variables

We assume a variable ξ to be within the interval $(x, x + \Delta x)$. Δx is assumed small compared to x (Figure 3.2). Then the probability for ξ being in the interval is:

$$P(x \leq \xi \leq x + \Delta x) = p(x)\Delta x \qquad (3.9)$$

$p(x)$ is named the probability density function (abbreviated PDF) of x if Δx becomes infinitesimally small, that is $\Delta x \rightarrow dx$. For a larger interval we then can write:

$$P(a \leq x \leq b) = \int_{a}^{b} p(x)\,dx \qquad (3.10)$$

Corresponding to equation 3.3 we have for the certain event that x has an arbitrary value:

$$\int_{-\infty}^{+\infty} p(x)\,dx = 1 \qquad (3.11)$$

The value of $p(x)$ is always positive. The probability distribution function is given by:

$$D(x) = P(\xi \leq x) = \int_{-\infty}^{x} p(\xi)\,d\xi \qquad (3.12)$$

By this distribution function we may indicate the probability of a noise voltage being below a threshold x. In reverse we have:

$$p(x) = \frac{dD(x)}{dx} = \dot{D}(x) \qquad (3.13)$$

For two variables x and y we have the joint probability density function $p(x, y)$ defined, according to equation 3.9, by:

$$P(x \leq \xi \leq x + dx, y \leq \eta \leq y + dy) = p(x, y)\, dx\, dy \qquad (3.14)$$

This may be extended to the important case of N variables. Then we have the joint probability density function $p(x_1, x_2, \ldots, x_N)$.

From equation 3.11 we have:

$$\int_{-\infty}^{+\infty} \int_{-\infty}^{+\infty} p(x, y)\, dx\, dy = 1 \qquad (3.15)$$

and integrating over all possible values y:

$$p(x) = \int_{-\infty}^{+\infty} p(x, y)\, dy \qquad (3.16)$$

The conditional probability density follows from equation 3.5:

$$p(x/y) = \frac{p(x, y)}{p(y)} \qquad (3.17)$$

If x and y are independent then:

$$p(x, y) = p(x)p(y) \qquad (3.18)$$

and:

$$p(x/y) = p(x) \qquad (3.19)$$

Example 2: probability density of the sum of two variables

As an illustration let us consider the following example. For the two variables x and y may be given $p(x, y)$. We form $z = x + y$, but what is the PDF of z? From equation 3.12 we have:

$$P(\xi \leq z) = D(z) = P(y \leq z - x) = \int_{-\infty}^{+\infty} dx \int_{-\infty}^{z-x} p(x, y)\, dy$$

With equation 3.13 follows:

$$p(z) = \dot{D}(z) = \int_{-\infty}^{+\infty} dx \frac{d}{dz} \left[\int_{-\infty}^{z-x} p(x, y)\, dy \right] = \int_{-\infty}^{+\infty} dx\, p(x, z - x)$$

For independent x and y follows $p(x, y) = p_1(x)p_2(y)$ and then:

$$p(z) = \int_{-\infty}^{+\infty} dx\, p_1(x)p_2(z - x) \qquad (3.20)$$

This integral corresponds to the convolution given by equation 2.25 in connection with the filtering of a signal.

3.2.3 Functions of random variables

We can give the PDF of a variable x and a unique function $y = y(x)$. We want to know the PDF for y. The interval $(x, x + dx)$ is mapped onto the interval $(y, y + dy)$.

The probability of the variables falling in the intervals is the same:

$$p(x)\,dx = p(y)\,dy$$

or, taking into account the fact that $p(y)$ has to be positive:

$$p(y) = p(x)\left|\frac{dx}{dy}\right| \tag{3.21}$$

Example 3: quadratic detector

As an illustration let us consider the example of noise and a quadratic demodulator (Figure 3.3). We assume:

$$y = x^2, \qquad p(x) = \frac{1}{\sqrt{2\pi}}\exp\left(-\frac{x^2}{2}\right)$$

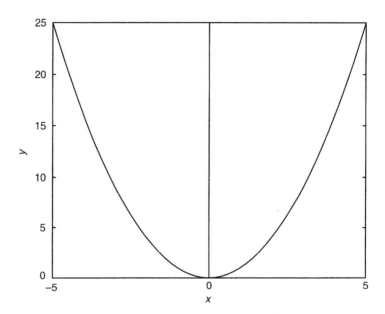

Figure 3.3   Quadratic demodulation characteristic

where $p(x)$ = Gaussian density distribution and $x = \pm\sqrt{y}$; we have to consider positive and negative x separately and add a factor of two in the result:

$$\frac{dx}{dy} = \frac{1}{2\sqrt{y}}$$

and

$$p(y) = \frac{2}{\sqrt{2\pi}} \frac{1}{2\sqrt{y}} \exp\left(-\frac{y}{2}\right)$$

Generalisation to N variables

Equation 3.21 may be generalised to N variables: $(x_1, \ldots, x_N)$ with a PDF $p(x_1 \cdots x_N)$ is transformed to $(y_1, \ldots, y_N)$ with a PDF $p(y_1 \cdots y_N)$ by functions $y_n = y_n(x_1 \cdots x_N)$ and reverse functions $x_n = x_n(y_1 \cdots y_N)$ for $n = 1, \ldots, N$. The mapping from a volume element $dx_1 \cdots dx_N$ to the transformed volume element $dy_1 \cdots dy_N$ results in equal probabilities for random vectors to fall within these elements:

$$p(y_1 \cdots y_N) \, dy_1 \cdots dy_N = p(x_1 \cdots x_N) \, dx_1 \cdots dx_N$$

According to equation 3.21 we have:

$$p(y_1 \cdots y_N) = \frac{\partial(x_1 \cdots x_N)}{\partial(y_1 \cdots y_N)} p(x_1 \cdots x_N) = \Delta p(x_1 \cdots x_N) \tag{3.22}$$

with Δ the so-called Jacobian determinant given by Reference 1:

$$\Delta = \begin{vmatrix} \dfrac{\partial x_1}{\partial y_1} & \cdots & \dfrac{\partial x_N}{\partial y_1} \\ \vdots & \ddots & \vdots \\ \dfrac{\partial x_1}{\partial y_N} & \cdots & \dfrac{\partial x_N}{\partial y_N} \end{vmatrix} \tag{3.23}$$

Example 4: Rayleigh and Rice probability density function

The orthogonal signal components x and y are assumed Gaussian distributed. We look for the PDF of the resultant amplitude and phase (Figure 3.4).

First we have:

$$\begin{aligned} x &= r \cos\varphi \\ y &= r \sin\varphi \end{aligned} \tag{3.24}$$

$$\Delta = \frac{\partial(x, y)}{\partial(r, \varphi)} = \begin{vmatrix} \dfrac{\partial x}{\partial r} & \dfrac{\partial y}{\partial r} \\ \dfrac{\partial x}{\partial \varphi} & \dfrac{\partial y}{\partial \varphi} \end{vmatrix} = \begin{vmatrix} \cos\varphi & \sin\varphi \\ -r\sin\varphi & r\cos\varphi \end{vmatrix} = r\cos^2\varphi + r\sin^2\varphi = r$$

$$\tag{3.25}$$

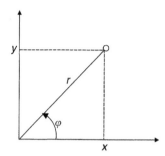

Figure 3.4   Coordinate transformation x, y to r, φ

For independent x and y we get, using the Gaussian PDF as in example 3:

$$p(x, y) = p(x)p(y) = \frac{1}{\sqrt{2\pi}} \exp\left(-\frac{x^2}{2}\right) \frac{1}{\sqrt{2\pi}} \exp\left(-\frac{y^2}{2}\right)$$

It follows:

$$p(r, \varphi) = \Delta p(x, y) = r\frac{1}{2\pi} \exp\left(-\frac{1}{2}(x^2 + y^2)\right) = r\frac{1}{2\pi} \exp\left(-\frac{1}{2}r^2\right)$$

The PDF $p(r, \varphi)$ is independent of the phase, which means that it is equally distributed. The PDF $p(r)$ follows from equation 3.16:

$$p(r) = \int_0^{2\pi} p(r, \varphi)\, d\varphi = r \exp\left(-\frac{r^2}{2}\right) \tag{3.26}$$

For the phase we have simply:

$$p(\varphi) = \frac{1}{2\pi}$$

Equation 3.26 is known as the Rayleigh probability density function. It describes the amplitude probability density of Gaussian noise signals (Figure 3.5).

We extend the above example:

A sinusoidal signal may be superposed onto the noise. It may be represented by the components $x = a$ and $y = 0$, that is its phase is assumed to be zero. Again we look for the PDF of r (Figure 3.6).

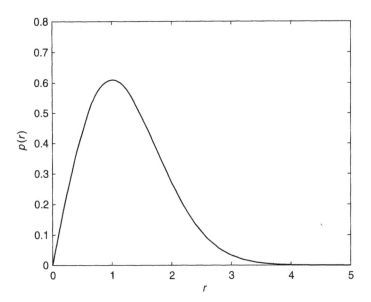

Figure 3.5   Rayleigh distribution function

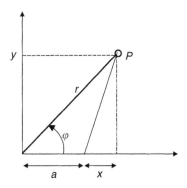

Figure 3.6   Sinusoidal signal a and Gaussian x and y

From equations 3.24 and 3.25:

$$p(x, y) = \frac{1}{2\pi} \exp\left(-\frac{1}{2}(x - a)^2 + y^2\right)$$

$$= \frac{1}{2\pi} \exp\left(-\frac{1}{2}\left(r^2 + a^2\right)\right) \exp\left(ar \cos \varphi\right)$$

$$p(r, \varphi) = \frac{r}{2\pi} \exp\left(-\frac{1}{2}\left(r^2 + a^2\right)\right) \exp\left(ar \cos \varphi\right)$$

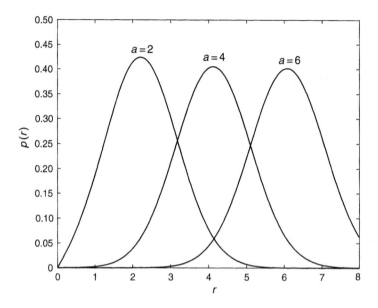

Figure 3.7   Rice distribution function for sin signal and Gaussian noise

For the PDF of r we have to integrate with respect to φ:

$$p(r) = r \exp\left(-\frac{1}{2}(r^2 + a^2)\right) \int\limits_0^{2\pi} \frac{1}{2\pi} \exp\left(ar \cos \varphi\right) d\varphi$$

The integral is known as the modified zero-order Bessel function:

$$p(r) = r \exp\left(-\frac{1}{2}(r^2 + a^2)\right) I_0(ar) \tag{3.27}$$

This is the so-called Rice distribution function for the amplitude of a sinusoidal signal and added Gaussian noise. For $a = 0$ the result equals equation 3.26. Figure 3.7 shows examples of the Rice PDF for several values of a.

Equations 3.26 and 3.27 are the statistical descriptions of very common radar signals.

3.2.4   Statistical averages

Instead of using the PDF to describe a signal we often want to characterise a signal with only a few parameters. The most frequently used parameter values are mean and variance.

We may observe one random signal for a long time or we may take samples at one instant of time from an ensemble of signals with the same properties and with as many representations as possible. If the statistical results are the same for both cases the processes are named ergodic.

We assume a function $g(x)$ of a random variable x, for x only taking discrete values from $x_1, \ldots, x_M$. By repeating N times a basic experiment and counting the occurrences as in section 3.2.1 we form the arithmetic average with:

$$g(x)_{av} = \frac{n(x_1)}{N}g(x_1) + \cdots + \frac{n(x_M)}{N}g(x_M)$$

In the limit according to equation 3.1 we get the statistical average of g or the so-called expectation:

$$\bar{g}(x) = E\{g(x)\} = \sum_{m=1}^{M} P(x_m)g(x_m) \tag{3.28}$$

For a continuous variable x we get, accordingly:

$$\bar{g}(x) = E\{g(x)\} = \int_{-\infty}^{+\infty} p(x)g(x)\,dx \tag{3.29}$$

Of special interest are the functions $g(x) = x^n$. Then equation 3.29 gives the nth moment of the variable x:

$$E\{x^n\} = \int_{-\infty}^{+\infty} x^n p(x)\,dx \tag{3.30}$$

The first moment is the mean of x:

$$m = E\{x\} = \int_{-\infty}^{+\infty} xp(x)\,dx \tag{3.31}$$

This gives us the DC component of a signal x. The signal $(x - m)$ would of course have a zero DC component. The second moment $E\{x^2\}$ gives a measure of the average power of the signal x. Of special interest is the expression:

$$E\{(x - m)^2\} = \sigma^2 \tag{3.32}$$

which gives a measure of the power of the varying component of x. It is called the variance of x. The positive square root σ is called the standard deviation and is a measure of the AC component of a signal x.

From equation 3.32 follows a useful relation:

$$\sigma_x^2 = E\{x^2\} - E\{2mx\} + m^2 = E\{x^2\} - 2mm + m^2$$

$$\sigma_x^2 = E\{x^2\} - m^2 \tag{3.33}$$

3.2.5   Correlation

If the variables x and y are somehow dependent, we say that they are correlated. A measure of the dependence is given by the so-called covariance (see section 2.5):

$$q_{xy} = E\{(x - m_x)(y - m_y)\}$$

or in more detail:

$$q_{xy} = \int\int_{-\infty}^{+\infty} (x - m_x)(y - m_y)p(x, y)\, dx\, dy \tag{3.34}$$

The mean values are subtracted and only the varying random part of the signals is included by the covariance q_{xy}.

3.2.6   Gaussian density function

A frequently occurring function is the Gaussian or normal PDF:

$$p(x) = \frac{1}{\sqrt{2\pi}\sigma} \exp\left(-\frac{(x - m)^2}{2\sigma^2}\right) \tag{3.35}$$

which has a standard deviation σ according to equation 3.32 and a mean m (Figure 3.8).

The reason for the frequent occurrence of the Gaussian probability density function in nature is given by the so-called central limit theorem [1]: in nature there

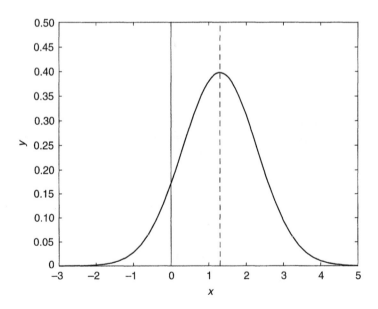

Figure 3.8   Gaussian density distribution function, $m = 1.3$, $\sigma = 1$

are many cases of the superposition of many individual contributions, for example the random movement of electrons dependent on the temperature within electronic devices and thus producing a noise signal. Another example is the complex radar echo from a ground scene or meteorological phenomena.

So we assume that there are many independent individual random processes ξ_n with equal statistical qualities which contribute by superposition to our observable variable x:

$$x = \frac{1}{\sqrt{N}} \sum_{n=1}^{N} \xi_n$$

The probability density functions of the single processes may be arbitrary, but are the same for all processes. Then by the central limit theorem it is proved that for N approaching infinity the PDF of x approaches the Gaussian density function.

If we have a signal vector with independent components $(x_1, \ldots, x_n, \ldots, x_N)$ then using equation 3.18 we get:

$$p(x_1 \cdots x_N) = p(x_1)p(x_2) \cdots p(x_N)$$

and

$$p(x_1 \cdots x_N) = \frac{1}{(2\pi)^{N/2}\sigma^N} \exp\left(-\frac{1}{2\sigma^2} \sum_{n=1}^{N}(x_n - m_n)^2\right) \qquad (3.36)$$

For a standard Gaussian PDF we set $m = 0$ and $\sigma = 1$:

$$p(x_1 \cdots x_N) = \frac{1}{(2\pi)^{N/2}} \exp\left(-\frac{1}{2} \sum_{n=1}^{N} x_n^2\right) \qquad (3.37)$$

3.2.7 Correlated Gaussian variables

We will now use the matrix notation as in chapter 2, section 2.1. Then we have:

$$\mathbf{x} = \begin{bmatrix} x_1 \\ \vdots \\ x_N \end{bmatrix}$$

and from equation 3.37 follows

$$p(\mathbf{x}) = \frac{1}{(2\pi)^{N/2}} \exp\left(-\frac{1}{2}\mathbf{x}^*\mathbf{x}\right) \qquad (3.38)$$

Up to now the signals x_n of $\mathbf{x}$ are assumed mutually independent, or in other words, they correspond to an orthogonal N-dimensional coordinate system. For the covariance matrix of $\mathbf{x}$ defined according to equation 2.10 and 3.34 by $\mathbf{Q} = E\{\mathbf{x}\mathbf{x}^*\}$ we get because of the independence in this case $\mathbf{Q} = \mathbf{I}$ (identity matrix).

We may apply to $\mathbf{x}$ now a coordinate transformation (rotation of the coordinate system) by a unitary quadratic matrix $\mathbf{U}[N, N]$:

$$\mathbf{v} = \mathbf{U}\mathbf{x}$$

and in reverse

$$\mathbf{x} = \mathbf{U}^{-1}\mathbf{v}$$

For the covariance matrix of $\mathbf{v}$ follows:

$$\mathbf{Q} = E\{\mathbf{v}\mathbf{v}^*\} = E\{\mathbf{U}\mathbf{x}\mathbf{x}^*\mathbf{U}^*\} = \mathbf{U}E\{\mathbf{x}\mathbf{x}^*\}\mathbf{U}^* = \mathbf{U}\mathbf{U}^*$$

and because $\mathbf{U}$ is unitary also $\mathbf{U}^{-1*}\mathbf{U}^{-1} = \mathbf{Q}^{-1}$, the inverse of the covariance matrix. We determine the pdf for $\mathbf{v}$ with Δ for the Jacobi-determinant from equation 3.23:

$$p(\mathbf{v}) = \Delta p(\mathbf{x}) = \begin{vmatrix} \dfrac{\partial x_1}{\partial v_1} & \cdots & \dfrac{\partial x_N}{\partial v_1} \\ \vdots & & \vdots \\ \dfrac{\partial x_1}{\partial v_N} & \cdots & \dfrac{\partial x_N}{\partial v_N} \end{vmatrix} p(\mathbf{x})$$

Because of the linear transform

$$\mathbf{x} = \mathbf{U}^{-1}\mathbf{v}$$

we have

$$\Delta = \det(\mathbf{U})^{-1} = (\det \mathbf{U})^{-1}$$

and from equation 3.38 follows

$$p(\mathbf{v}) = \frac{1}{(2\pi)^{N/2} \det \mathbf{U}} \exp\left(-\frac{1}{2}\mathbf{v}^*\mathbf{U}^{-1*}\mathbf{U}^{-1}\mathbf{v}\right)$$

or with $\mathbf{v}$ replaced by $\mathbf{x}$ and using $\mathbf{Q}$ as given above:

$$p(\mathbf{x}) = \frac{1}{(2\pi)^{N/2}(\det \mathbf{Q})^{1/2}} \exp\left(-\frac{1}{2}\mathbf{x}^*\mathbf{Q}^{-1}\mathbf{x}\right) \qquad (3.39)$$

Equation 3.39 is the more general form of equation 3.38 for correlated signal components. The correlation is described by the covariance matrix $\mathbf{Q}$.

3.2.8 *Complex Gaussian variables*

If we have complex signals:

$$\mathbf{z} = \mathbf{x} + j\mathbf{y} = \begin{bmatrix} x_1 + jy_1 \\ \vdots \\ x_N + jy_N \end{bmatrix}$$

the equation 3.39 is transformed into the important form for the Gaussian distribution [3.2], now with $\mathbf{Q} = E\{\mathbf{z}\mathbf{z}^*\}$:

$$p(\mathbf{z}) = \frac{1}{\pi^N \det \mathbf{Q}} \exp\left(-\mathbf{z}^*\mathbf{Q}^{-1}\mathbf{z}\right) \qquad (3.40)$$

Discussion for explanation and confirmation: we assume complex independent signals with zero mean. Then we have from equation 3.36, observing that we have now $2N$ variables $(x_1 \cdots x_N \, y_1 \cdots y_N)$

$$p(\mathbf{z}) = p(\mathbf{x}, \mathbf{y}) = \frac{1}{\pi^N (2\sigma^2)^N} \exp\left(-\mathbf{z}^* \frac{1}{2\sigma^2} \mathbf{z}\right)$$

In this case we get with equation 3.32

$$\mathbf{Q} = E\{\mathbf{zz}^*\} = E\{\mathbf{xx}^*\} + E\{\mathbf{yy}^*\} = \mathbf{Q}_x + \mathbf{Q}_y$$
$$= diag_N \, E\{x_n^2 + y_n^2\} = diag_N (2\sigma^2)$$

($diag_N (e_n)$ has the meaning of a diagonal matrix of dimension N with elements e_n at the main diagonal.)
   Then:

$$\det \mathbf{Q} = (2\sigma^2)^N$$

and

$$Q^{-1} = diag_n \left(\frac{1}{2\sigma^2}\right)$$

With this relation we recognise that equation 3.40 matches to equation 3.39.
   For $\sigma = 1$ and mean $\mathbf{m}$ we then have the form to be used for several applications:

$$p(\mathbf{z}) = \frac{1}{(2\pi)^N} \exp\left(-\frac{1}{2}(\mathbf{z} - \mathbf{m})^*(\mathbf{z} - \mathbf{m})\right) \qquad (3.41)$$

3.3   Likelihood-ratio test

The fundamental problem for statistical decision theory when applied to radar is the detection of a signal out of noise and interference. We measure the received signal vector $\mathbf{z}$. This may be composed of a target echo signal vector $\mathbf{s}$ and noise vector $\mathbf{n}$ or only noise alone. We define the two hypotheses:

$$H_0 : \quad \mathbf{z} = \mathbf{n}$$
$$H_1 : \quad \mathbf{z} = \mathbf{s} + \mathbf{n}$$

We want to make our decision for H_0 or H_1 based on the measured $\mathbf{z}$. In other words, we have to divide the set of possible realisations for $\mathbf{z}$ into two regions D_0 and D_1, one for H_0 and one for H_1 (see Figure 3.9). If we measure $\mathbf{z}$ in region D_0 we accept H_0 and for $\mathbf{z}$ in D_1 we accept H_1. The problem now is to find the borderline between D_0 and D_1. We now define decision rules, following e.g. Middleton [3]:

If $\mathbf{z}$ is in D_1 then we express the decision for H_1 by:

$$\delta(H_1/\mathbf{z}) = 1 \quad \text{and} \quad \delta(H_0/\mathbf{z}) = 0$$

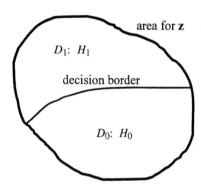

Figure 3.9   Decision areas between two hypotheses

If **z** is in D_0 then we express the decision for H_0 by:

$$\delta(H_0/\mathbf{z}) = 1 \quad \text{and} \quad \delta(H_1/\mathbf{z}) = 0$$

We always have for all **z**:

$$\delta(H_0/\mathbf{z}) + \delta(H_1/\mathbf{z}) = 1 \tag{3.42}$$

We assume as known the conditional probability densities $p(\mathbf{z}/0)$ and $p(\mathbf{z}/s)$ according to the hypotheses H_0 and H_1. Then we may compute the error probabilities for wrong decisions. The false-alarm probability α or P_F is given by:

$$\alpha = \int_D p(\mathbf{z}/0)\,\delta(H_1/\mathbf{z})\,d\mathbf{z} = \int_{D_1} p(\mathbf{z}/0)\,d\mathbf{z} \tag{3.43}$$

We integrate the probability density function valid for the case $s = 0$ over the area reserved for the decision H_1. In reverse we have the probability β for missing a target:

$$\beta = \int_D p(\mathbf{z}/s)\,\delta(H_0/\mathbf{z})\,d\mathbf{z} = \int_{D_0} p(\mathbf{z}/s)\,d\mathbf{z} \tag{3.44}$$

We now demand for a selected α a minimum of β, or *vice versa*. This is achieved by minimisation of a linear combination of α and β with an arbitrary factor η:

$$\min_\delta(\beta + \eta\alpha) = \min_\delta\left[\int_D p(\mathbf{z}/s)\delta(H_0/\mathbf{z})\,d\mathbf{z} + \eta\int_D p(\mathbf{z}/0)\delta(H_1/\mathbf{z})\,d\mathbf{z}\right]$$

With equation 3.42 follows:

$$\min_\delta(\beta + \eta\alpha) = \min_\delta\left[1 + \int_D d\mathbf{z}(-p(\mathbf{z}/s) + \eta\,p(\mathbf{z}/0))\delta(H_1/\mathbf{z})\right] \tag{3.45}$$

The expression given by equation 3.45 achieves a minimum if the integrand within the area D_1 is negative. In other words, we use all those values of **z** which contribute to a reduction of the linear combination of error probabilities. This means:

$$-p(\mathbf{z}/\mathbf{s}) + \eta \, p(\mathbf{z}/0) \leq 0$$

and we get finally the well-known decision rule [1,3,4] in the form of the ratio of probability densities. It is usually called the likelihood-ratio test (LRT):

$$\lambda(\mathbf{z}) = \frac{p(\mathbf{z}/\mathbf{s})}{p(\mathbf{z}/0)} \geq \eta \tag{3.46}$$

If the functions $p(\mathbf{z}/\mathbf{s})$ and $p(\mathbf{z}/0)$ are known, then the test function λ can be computed as the ratio. The measured signal **z** is used in the ratio and then the decision is achieved by threshold comparison. If the test value is above the threshold η a target is assumed to be present.

The choice of the threshold determines both error probabilities. A high value η results in a low α but a high β. The probability for target detection is given by:

$$P_D = 1 - \beta$$

For a selected $P_F = \alpha$ the likelihood-ratio test (LRT) achieves the maximum possible P_D.

The signal is assumed in equation 3.46 to be completely known. In many applications the signal depends on a set of parameters ϑ within the parameter space Θ. The probability density may be $p(\vartheta)$. Then equation 3.44 is extended to [3]:

$$\beta = \iint_{D\Theta} p(\mathbf{z}/\mathbf{s}(\vartheta))p(\vartheta) \, d\vartheta \, \delta(H_0/\mathbf{z}) \, d\mathbf{z} \tag{3.47}$$

or

$$\beta = \int_D \langle p(\mathbf{z}/\mathbf{s}(\vartheta)) \rangle_\Theta \delta(H_0/\mathbf{z}) \, d\mathbf{z}$$

The probability density $p(\mathbf{z}, \mathbf{s}(\vartheta))$ is averaged with respect to ϑ. Inserting equation 3.47 into equation 3.45 results in the more general decision rule:

$$\frac{\int_\Theta p(\mathbf{z}/\mathbf{s}(\vartheta))p(\vartheta) \, d\vartheta}{p(\mathbf{z}/0)} \geq \eta \tag{3.48}$$

Example 5: detection of a known signal

The target signal vector **s** may be known and the noise is assumed to be Gaussian distributed. Then we have from equation 3.40 with **s** as the mean vector:

$$p(\mathbf{z}, \mathbf{s}) = \frac{1}{\pi^N \det \mathbf{Q}} \exp\left(-(\mathbf{z} - \mathbf{s})^* \mathbf{Q}^{-1}(\mathbf{z} - \mathbf{s})\right) \tag{3.49}$$

and for the LRT test function:

$$\lambda(\mathbf{z}) = \frac{p(\mathbf{z}, \mathbf{s})}{p(\mathbf{z}, 0)}$$

$$= \exp\left(-(\mathbf{z}-\mathbf{s})^*\mathbf{Q}^{-1}(\mathbf{z}-\mathbf{s}) + \mathbf{z}^*\mathbf{Q}^{-1}\mathbf{z}\right)$$

$$= \exp\left(\mathbf{s}^*\mathbf{Q}^{-1}\mathbf{z} + \mathbf{z}^*\mathbf{Q}^{-1}\mathbf{s}\right)\exp\left(-\mathbf{s}^*\mathbf{Q}^{-1}\mathbf{s}\right)$$

The second exp factor does not depend on $\mathbf{z}$, it is only a factor for the decision threshold, and we can write for λ:

$$\lambda(\mathbf{z}) = \exp\left(2\,\mathrm{Re}\,\{\mathbf{z}^*\mathbf{Q}^{-1}\mathbf{s}\}\right) \tag{3.50}$$

$\mathrm{Re}\{z\}$ means real part of $\mathbf{z}$.

This equation teaches us to compute the expression:

$$y = \mathrm{Re}\,\{\mathbf{z}^*\mathbf{Q}^{-1}\mathbf{s}\} = \mathrm{Re}\,\{\mathbf{s}^*\mathbf{Q}^{-1}\mathbf{z}\}$$

or for uncorrelated receiver noise with $\mathbf{Q} = \mathbf{I}$:

$$y = \mathrm{Re}\,\{\mathbf{s}^*\mathbf{z}\} = \mathrm{Re}\,\left\{\sum_{n=1}^{N} s_n^* z_n\right\}$$

This has a plausible explanation. All the components s_n of $\mathbf{s}$ are vectors with a certain known phase, while the noise components have a random phase. By a back rotation of the received signal components z_n according to the phases of s_n by forming the products $z_n s_n^*$ all the signal components are aligned before the final summation and the sum results in a maximum signal value. This is shown in the Figure 3.10. The noise vectors also have a random phase after rotation according to the expected signal phase and the noise sum remains of the same order of magnitude with and without rotation. The signal voltage increases by this phase alignment by the factor N, the power increases therefore by a factor N^2. The noise power increases only by N. The ratio of the signal-to-noise power, SNR, increases, as a result, by a factor N. This increase in SNR depends on the correct phase alignment according to the expected signal, with a greater increase for increasing values of N. For the detection of a Doppler-shifted signal this rotation processing forms a filter for a certain Doppler frequency. For high N we get a sharp or narrow filter.

If the expected signal vector components differ in amplitude then an amplitude weighting is performed using the LRT according to the products $z_n s_n^*$. This means that signal components with a high signal-to-noise power ratio (SNR) are more important for the detection decision or contribute more effectively to an increase in the final SNR.

This processing by rotating the received signal vectors according to an expected signal phase and amplitude weighting is known as a matched filter. It applies to several radar procedures, for example pulse compression, Doppler filtering, beamforming and SAR focusing.

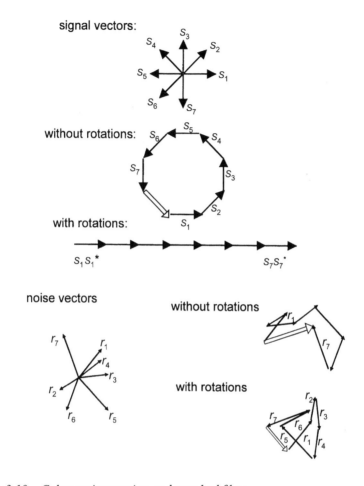

Figure 3.10   Coherent integration and matched filter

Example 6: detection of a Doppler-shifted signal series

For the detection of a Doppler signal with frequency shift f_d we have for the components s_n of $\mathbf{s}$ with $\Delta\varphi = 2\pi f_d T$ as the phase shift within one radar period T:

$$s_n = \exp(j\,\Delta\varphi\,n) \quad \text{for } n = 1, \ldots, N \tag{3.51}$$

We assume the signal amplitude to be equal to 1, because it has no influence on the structure of the test. Additionally, we usually have an unknown initial phase ψ, equally distributed within the interval $(0, 2\pi)$. We use equation 3.50 and apply equation 3.48:

$$\lambda(\mathbf{z}) = \frac{1}{2\pi} \int_0^{2\pi} d\psi \, \exp\left(2\,\mathrm{Re}\,\{\mathbf{z}^*\mathbf{Q}^{-1}\mathbf{s}e^{j\psi}\}\right) \tag{3.52}$$

This integral has been solved [6]:

$$\lambda(\mathbf{z}) = I_0\big(2|\mathbf{z}^*\mathbf{Q}^{-1}\mathbf{s}|\big) \tag{3.53}$$

with I_0 the modified zero-order Bessel function. Because I_0 is a monotonic increasing function we may simplify the test to:

$$\lambda(\mathbf{z}) = \text{Re}\,\big\{\mathbf{z}^*\mathbf{Q}^{-1}\mathbf{s}\big\}^2 + \text{Im}\,\big\{\mathbf{z}^*\mathbf{Q}^{-1}\mathbf{s}\big\}^2 \tag{3.54}$$

We have omitted constant factors for the test function, since they only change the detection threshold.

We usually also have an unknown Doppler frequency shift for the detection of the flying targets. The Doppler frequency has an unambiguous interval of $0, \ldots, 1/T$. Again, using equation 3.48, we have to average λ according to equation 3.53 over the Doppler region:

$$\lambda(\mathbf{z}) = \int\limits_0^{1/T} I_0\big(2|\mathbf{z}^*\mathbf{Q}^{-1}\mathbf{s}(f)|\big)\,df \tag{3.55}$$

In practice, we average in the frequency domain with a filter bank with N filters, for a signal series with N signal samples:

$$\lambda(\mathbf{z}) = \frac{1}{N}\sum_{n=0}^{N-1} I_0\big(2|\mathbf{z}^*\mathbf{Q}^{-1}\mathbf{s}(f_n)|\big), \quad \text{with } f_n = \frac{1}{NT}n \tag{3.56}$$

This test function may be approximated by:

$$\lambda(\mathbf{z}) = \max_n\,\big(|\mathbf{z}^*\mathbf{Q}^{-1}\mathbf{s}(f_n)|\big) \tag{3.57}$$

This version of the test function is used in practice. The maximum output from the filter bank is selected and compared with the decision threshold. If $\mathbf{Q}$ describes clutter and noise signals an optimal clutter suppression is included in this test [7]. If $\mathbf{Q} = \mathbf{I}$, that is we have as disturbance only uncorrelated receiver noise, equation 3.57 represents a conventional filter bank. This may be realised with a Fourier transform (FFT), as discussed in chapter 2, section 2.3.

3.4   Parameter estimation

After the detection of a target follows the determination of its parameters, for example the mean and variance of its amplitude, Doppler frequency shift, azimuth and elevation angle.

For the signal $\mathbf{z}$ we assume as known the probability density function, which depends on a parameter set ϑ, that is $p(\mathbf{z}) = p(\mathbf{z}; \vartheta)$.

For a measured signal $\mathbf{z}_0$, inserted in $p(\mathbf{z}; \vartheta)$, we get a function of ϑ only. The estimate ϑ^* is determined by maximising $p(\mathbf{z}_0; \vartheta)$ with respect to ϑ. This maximum is found as usual by setting the first derivative with respect to ϑ to zero:

$$\frac{\partial p(\mathbf{z}_0; \vartheta)}{\partial \vartheta}\bigg|_{\vartheta=\vartheta*} = 0$$

or equivalently:

$$\frac{\partial \ln p(\mathbf{z}_0; \vartheta)}{\partial \vartheta}\bigg|_{\vartheta=\vartheta*} = 0 \qquad (3.58)$$

This equation determines for the measured signal that estimate $\vartheta*$ with the highest probability.

For example, if ϑ is the target direction, we have to maximise $p(z_0; \vartheta)$ by varying the assumed direction. This application will be discussed in detail within chapter 11. The corresponding estimation of Doppler frequency is treated in chapter 8.

3.4.1 Variance of the estimate and Cramér-Rao limit

An estimate $\hat{\vartheta}$ will be a random variable with a certain variance, given by equation 3.33, because of the noise component of the received and measured signal $\mathbf{z}$. Under noise we may understand also measurement errors. It is possible to determine the minimal achievable variance for the estimated parameter.

We will give, for the interested reader, its derivation in the following, according to Reference 4, because the result has a fundamental importance for radar signal-processing problems.

Otherwise you may proceed to equation 3.67, which gives the minimum variance of the parameter estimate for a known pdf $p(\mathbf{z}; \vartheta)$. For a set of parameters $\vartheta = \vartheta_1 \cdots \vartheta_K$ the result is given by equations 3.70 and 3.71.

The N signal components z_n are assumed to be independent. The estimate $\hat{\vartheta}$ is a function of the measured signal vector $\mathbf{z}$, for example, but not necessarily, according to equation 3.58:

$$\hat{\vartheta} = \hat{\vartheta}(\mathbf{z}) = \hat{\vartheta}(z_1 \cdots z_N)$$

There exists a probability density function $g(\hat{\vartheta}; \vartheta)$, yet unknown, for the variable $\hat{\vartheta}$. We may transform the signal vector $(z_1 \cdots z_N)$ by a coordinate transform to a new vector $(\xi_1 \cdots \xi_{N-1}; \hat{\vartheta})$. For the probability density we may write with the conditional probability density function $h(\xi_1 \cdots \xi_{N-1}/\hat{\vartheta}; \vartheta)$ with equations 3.17 and 3.22 and using the assumption of independent signals:

$$p(z_1; \vartheta) \cdots p(z_N; \vartheta)\, dz_1 \cdots dz_N$$
$$= g(\hat{\vartheta}; \vartheta)\, h(\xi_1 \cdots \xi_{N-1} \,|\, \hat{\vartheta}; \vartheta)\, d\hat{\vartheta}\, d\xi_1 \cdots d\xi_{N-1} \qquad (3.59)$$

or with the Jacobian determinant Δ according to equation 3.23:

$$p(z_1; \vartheta) \cdots p(z_N; \vartheta)\Delta = g(\hat{\vartheta}; \vartheta)\, h(\xi_1 \cdots \xi_{N-1} \,|\, \hat{\vartheta}; \vartheta) \qquad (3.60)$$

With g known we could determine the variance of $\hat{\vartheta}$.

The mean or expectation of $\hat{\vartheta}$ is given by:

$$E\{\hat{\vartheta}\} = m(\vartheta) = \int_{-\infty}^{+\infty} \hat{\vartheta}\, g(\hat{\vartheta}, \vartheta)\, d\hat{\vartheta} = \vartheta + b(\vartheta) \qquad (3.61)$$

Equation 3.61 means that there may be a bias b as an error of our estimate compared to the correct value ϑ.

Further we have:

$$\int_{-\infty}^{+\infty} g\, d\hat{\vartheta} = 1$$

and therefore:

$$\int_{-\infty}^{+\infty} \frac{\partial g}{\partial \vartheta}\, d\hat{\vartheta} = 0$$

and therefore also:

$$\int_{-\infty}^{+\infty} \vartheta \frac{\partial g}{\partial \vartheta}\, d\hat{\vartheta} = 0 \qquad (3.62)$$

Using the last relation we get from equation 3.61:

$$\int_{-\infty}^{+\infty} (\hat{\vartheta} - \vartheta)\frac{\partial g}{\partial \vartheta}\, d\hat{\vartheta} = \int_{-\infty}^{+\infty} (\hat{\vartheta} - \vartheta)\sqrt{g}\,\frac{\partial \ln g}{\partial \vartheta}\sqrt{g}\, d\hat{\vartheta}$$

$$= \frac{\partial m(\vartheta)}{\partial \vartheta}$$

$$= 1 + \frac{\partial b(\vartheta)}{\partial \vartheta} \qquad (3.63)$$

Now we use the Schwarz inequality

$$\int f_1^2\, dx \int f_2^2\, dx \geq \left[\int f_1 f_2\, dx \right]^2$$

to give:

$$\int_{-\infty}^{+\infty} (\hat{\vartheta} - \vartheta)^2 g\, d\hat{\vartheta} \int_{-\infty}^{+\infty} \left(\frac{\partial \ln g}{\partial \vartheta} \right)^2 g\, d\hat{\vartheta} \geq \left(1 + \frac{\partial b}{\partial \vartheta} \right)^2$$

or:

$$\int_{-\infty}^{+\infty} (\hat{\vartheta} - \vartheta)^2 g\, d\hat{\vartheta} \geq \frac{(1 + \partial b/\partial \vartheta)^2}{\int_{-\infty}^{+\infty} (\partial \ln g/\partial \vartheta)^2 g\, d\hat{\vartheta}} \qquad (3.64)$$

The left-hand expression of equation 3.64 is the variance of $\hat{\vartheta}$ in relation to the true parameter value ϑ. Because g is unknown we have to express g by $p(z; \vartheta)$. This is possible by using equation 3.60. By taking the logarithm we get:

$$\sum_{k=1}^{N} \ln p(x_k; \vartheta) + \ln \Delta = \ln g + \ln h$$

Assuming Δ independent of ϑ we get after taking the derivative with respect to ϑ:

$$\sum_{k=1}^{N} \frac{\partial \ln p(z_k; \vartheta)}{\partial \vartheta} = \frac{\partial \ln g}{\partial \vartheta} + \frac{\partial \ln h}{\partial \vartheta} \qquad (3.65)$$

Equation 3.65 is squared, multiplied with equation 3.59 and integrated. We have to take into account for this operation the following:

$$\int_{-\infty}^{+\infty} p(z_k; \vartheta) \, dz_k = 1$$

and therefore:

$$\int_{-\infty}^{+\infty} \frac{\partial p}{\partial \vartheta} \, dz_k = 0$$

With:

$$\frac{1}{p} \frac{\partial p}{\partial \vartheta} p = \frac{\partial \ln p}{\partial \vartheta} p$$

we get from the second equation above:

$$\int_{-\infty}^{+\infty} \frac{\partial \ln p(z_k; \vartheta)}{\partial \vartheta} p \, dz_k = 0$$

The corresponding result applies to g and h. With this result all mixed terms from squaring the sum expression in equation 3.65 vanish and we receive from equation 3.65:

$$\sum_{k=1}^{N} \int_{-\infty}^{+\infty} dz_k \left(\frac{\partial \ln p}{\partial \vartheta} \right)^2 p = \int_{-\infty}^{+\infty} \left(\frac{\partial \ln g}{\partial \vartheta} \right)^2 g \, d\vartheta$$

$$+ \int_{-\infty}^{+\infty} d\vartheta g \int_{-\infty}^{+\infty} \cdots \int_{-\infty}^{+\infty} \left(\frac{\partial \ln h}{\partial \vartheta} \right)^2 h \, d\xi_1 \cdots d\xi_{N-1}$$

We omit the second term of the right-hand side and get the inequality:

$$\sum_{k=1}^{N} \int_{-\infty}^{+\infty} dz_k \left(\frac{\partial \ln p}{\partial \vartheta}\right)^2 p \geq \int_{-\infty}^{+\infty} \left(\frac{\partial \ln g}{\partial \vartheta}\right)^2 g \, d\vartheta \tag{3.66}$$

With this equation we accomplish our aim: we substitute the denominator in equation 3.64 by the greater expression so that the inequality holds even still more:

$$\int_{-\infty}^{+\infty} (\hat{\vartheta} - \vartheta)^2 g \, d\hat{\vartheta} \geq \frac{(1 + \partial b/\partial \vartheta)^2}{\sum_{k=1}^{N} \int_{-\infty}^{+\infty} dz_k (\partial \ln p/\partial \vartheta)^2 p(z_k; \vartheta)} \tag{3.67}$$

The equals sign is valid if the probability density function h is independent of ϑ and if in the Schwarz inequality both functions are related by $f_1 = cf_2$. That means in our case:

$$(\hat{\vartheta} - \vartheta) = c \frac{\partial \ln g}{\partial \vartheta} \tag{3.68}$$

and after integration:

$$-\tfrac{1}{2}(\hat{\vartheta} - \vartheta)^2 = c \ln g + c_1$$

where c and c_1 are arbitrary constants. They may be chosen to obtain as a solution for equation 3.68 the Gaussian probability distribution function according to equation 3.35:

$$g(\hat{\vartheta}; \vartheta) = \frac{1}{\sqrt{2\pi}\sigma} \exp\left(-\frac{(\hat{\vartheta} - \vartheta)^2}{2\sigma^2}\right) \tag{3.69}$$

For a minimal variance the probability density function of the estimate $\hat{\vartheta}$ has to be Gaussian. From equation 3.65 we conclude for h independent of ϑ:

$$\frac{\partial \ln g}{\partial \vartheta} = \sum_{k=1}^{N} \frac{\partial \ln p(z_k; \vartheta)}{\partial \vartheta}$$

$$= \frac{\partial \ln p(\mathbf{z}; \vartheta)}{\partial \vartheta}$$

and therefore from equation 3.68 we have:

$$\left.\frac{\partial \ln p(\mathbf{z}; \vartheta)}{\partial \vartheta}\right|_{\vartheta = \vartheta^*} = \frac{1}{c}(\hat{\vartheta} - \vartheta^*)$$

This expression is, according to equation 3.58, only equal to zero if the maximum-likelihood estimate ϑ^* is equal to the estimate $\hat{\vartheta}$. On the other hand, we have shown above that under the assumed conditions $\hat{\vartheta}$ is the estimate with minimum variance.

If a set of parameters $\vartheta = \vartheta_1, \ldots, \vartheta_K$ has to be estimated then the variances are determined using Fisher's information matrix [5]. This matrix $\mathbf{F}$ of dimension $[K, K]$ has elements:

$$F_{ij} = E\left\{\frac{\partial \ln p}{\partial \vartheta_i} \frac{\partial \ln p}{\partial \vartheta_j}\right\} \quad \text{with } 1 \leq i, j \leq K \qquad (3.70)$$

The matrix $\mathbf{G} = \mathbf{F}^{-1}$ contains on the main diagonal the lower bounds for all the variances, that is:

$$E\left\{(\hat{\vartheta}_i - \vartheta_i)^2\right\} = \sigma^2(\hat{\vartheta}_i) \geq G_{ii} \qquad (3.71)$$

Equation 3.70 can be given another useful form [5]. First we have:

$$F_{ij} = E\left\{\frac{\partial \ln p}{\partial \vartheta_i} \frac{\partial \ln p}{\partial \vartheta_j}\right\} = \int \frac{\partial \ln p}{\partial \vartheta_i} \frac{\partial \ln p}{\partial \vartheta_j} p \, d\mathbf{z}$$

$$= \int \frac{1}{p} \frac{\partial p}{\partial \vartheta_i} \frac{1}{p} \frac{\partial p}{\partial \vartheta_j} p \, d\mathbf{z} \qquad (3.72)$$

Now we use the identity by applying basic rules for forming derivations of functions:

$$\frac{\partial^2 \ln p}{\partial \vartheta_i \, \vartheta_j} = \frac{\partial}{\partial \vartheta_j}\left(\frac{1}{p} \frac{\partial p}{\partial \vartheta_i}\right) = -\frac{1}{p^2}\left(\frac{\partial p}{\partial \vartheta_i} \frac{\partial p}{\partial \vartheta_j}\right) + \frac{1}{p} \frac{\partial^2 p}{\partial \vartheta_i \, \partial \vartheta_j}$$

or

$$\frac{1}{p^2}\left(\frac{\partial p}{\partial \vartheta_i} \frac{\partial p}{\partial \vartheta_j}\right) = -\frac{\partial^2 \ln p}{\partial \vartheta_i \, \partial \vartheta_j} + \frac{1}{p} \frac{\partial^2 p}{\partial \vartheta_i \, \partial \vartheta_j} \qquad (3.73)$$

The second term on the right-hand side cancels p by forming the expectation. This term is then zero, because $\int p \, d\mathbf{z} = 1$ and the derivative $(\partial^2/\partial \vartheta_i \, \vartheta_j) \int p \, d\mathbf{z} = 0$. Finally we get:

$$F_{ij} = -E\left\{\frac{\partial^2 \ln p}{\partial \vartheta_i \, \partial \vartheta_j}\right\}$$

$$= -\int \frac{\partial^2 \ln p}{\partial \vartheta_i \, \partial \vartheta_j} p \, d\mathbf{z} \qquad (3.74)$$

3.5   Estimation of a signal

In some applications the signal has to be estimated in the presence of a disturbance signal. Two approaches are discussed and compared.

3.5.1   Maximum-likelihood estimation

The received signal vector $\mathbf{z}[N, 1]$ is composed of a target signal and Gaussian noise. The original signal $\mathbf{s}[M, 1]$, dependent on unknown parameters, is transformed by a known matrix $C[N, M]$ to the measurable target signal $\mathbf{Cs}$ and we have for the measured or received signal $\mathbf{z}$:

$$\mathbf{z} = \mathbf{Cs} + \mathbf{n} \tag{3.75}$$

We want an estimate for $\mathbf{s}$. For a Gaussian distributed noise $\mathbf{n}$ with covariance matrix $\mathbf{Q}$ the probability density function according to equation 3.49 is:

$$p(\mathbf{z}, \mathbf{s}) = \frac{1}{\pi^N \det \mathbf{Q}} \exp\left(- (\mathbf{z} - \mathbf{Cs})^* \mathbf{Q}^{-1} (\mathbf{z} - \mathbf{Cs})\right) \tag{3.76}$$

Following the discussion in section 3.4 we look for the maximum of p dependent on $\mathbf{s}$. That is, we have to minimise $a = (\mathbf{z} - \mathbf{Cs})^* \mathbf{Q}^{-1} (\mathbf{z} - \mathbf{Cs})$ by taking the derivative with respect to $\mathbf{s}$:

$$\left. \frac{\partial a}{\partial \mathbf{s}} \right|_{\mathbf{s} = \hat{\mathbf{s}}} = 0$$

or

$$\mathbf{C}^* \mathbf{Q}^{-1} (\mathbf{z} - \mathbf{Cs}) = 0$$

and finally resolving for $\mathbf{s}$:

$$\hat{\mathbf{s}} = (\mathbf{C}^* \mathbf{Q}^{-1} \mathbf{C})^{-1} \mathbf{C}^* \mathbf{Q}^{-1} \mathbf{z} \tag{3.77}$$

$\hat{\mathbf{s}}$ is the estimate of the signal vector $\mathbf{s}$ with maximum probability or a maximum-likelihood estimate.

3.5.2 Signal estimation with least-mean-square error

The received and measured signal is again $\mathbf{z} = \mathbf{Cs} + \mathbf{n}$. But now $\mathbf{s}$ is unknown with random components and the noise may be nonGaussian distributed. A linear estimate which minimises the mean-squared error for $\mathbf{s}$ shall be derived with a filter matrix $\mathbf{W}$ by:

$$\hat{\mathbf{s}} = \mathbf{W}^* \mathbf{z} \tag{3.78}$$

By the selection of $\mathbf{W}$ the mean-squared error F between $\mathbf{s}$ and $\hat{\mathbf{s}}$ shall be minimised:

$$F = E\{(\mathbf{s} - \hat{\mathbf{s}})^* (\mathbf{s} - \hat{\mathbf{s}})\} = trace\, E\{(\mathbf{s} - \hat{\mathbf{s}})(\mathbf{s} - \hat{\mathbf{s}})^*\}$$

With equations 3.75 and 3.78 we get with $\mathbf{I}$ as the identity matrix:

$$\mathbf{s} - \hat{\mathbf{s}} = (\mathbf{I} - \mathbf{W}^* \mathbf{C})\mathbf{s} - \mathbf{W}^* \mathbf{n}$$

With $E\{\mathbf{ss}^*\} = \mathbf{P}$, $E\{\mathbf{nn}^*\} = \mathbf{Q}$ and $E\{\mathbf{sn}^*\} = 0$ we get:

$$E\{(\mathbf{s} - \hat{\mathbf{s}})(\mathbf{s} - \hat{\mathbf{s}})^*\} = (\mathbf{I} - \mathbf{W}^* \mathbf{C})\mathbf{P}(\mathbf{I} - \mathbf{C}^* \mathbf{W}) + \mathbf{W}^* \mathbf{Q}\mathbf{W}$$

$$= \mathbf{P} + \mathbf{W}^* (\mathbf{CPC}^* + \mathbf{Q})\mathbf{W} - \mathbf{W}^* \mathbf{CP} - \mathbf{PC}^* \mathbf{W}$$

Via the identity concerning $\mathbf{W}$ [1]:

$$\mathbf{W}^*\mathbf{AW} - \mathbf{W}^*\mathbf{B} - \mathbf{B}^*\mathbf{W} = -\mathbf{B}^*\mathbf{A}^{-1}\mathbf{B} + (\mathbf{W} - \mathbf{A}^{-1}\mathbf{B})^*\mathbf{A}(\mathbf{W} - \mathbf{A}^{-1}\mathbf{B})$$

we arrive at a suitable expression for F:

$$F = trace\big(\mathbf{P} - \mathbf{PC}^*(\mathbf{CPC}^* + \mathbf{Q})^{-1}\mathbf{CP} + (\mathbf{W} - (\mathbf{CPC}^* + \mathbf{Q})^{-1}\mathbf{CP})^*$$
$$\times (\mathbf{CPC}^* + \mathbf{Q})(\mathbf{W} - (\mathbf{CPC}^* + \mathbf{Q})^{-1}\mathbf{CP})\big)$$

From this the required minimum follows if the square expression concerning $\mathbf{W}$ becomes equal to 0:

$$\mathbf{W}^* = \mathbf{PC}^*(\mathbf{CPC}^* + \mathbf{Q})^{-1} \tag{3.79}$$

$\mathbf{W}$ allows the estimation of a signal for a known or reasonable assumed signal covariance $\mathbf{P}$, a signal transformation by matrix $\mathbf{C}$ and a noise covariance matrix $\mathbf{Q}$.

For the signal estimate follows:

$$\hat{\mathbf{s}} = \mathbf{PC}^*(\mathbf{CPC}^* + \mathbf{Q})^{-1}\mathbf{z} \tag{3.80}$$

With the matrix identity:

$$\mathbf{PC}^*(\mathbf{CPC}^* + \mathbf{Q})^{-1} = (\mathbf{P}^{-1} + \mathbf{C}^*\mathbf{Q}^{-1}\mathbf{C})^{-1}\mathbf{C}^*\mathbf{Q}^{-1}$$

follows alternatively:

$$\hat{\mathbf{s}} = (\mathbf{P}^{-1} + \mathbf{C}^*\mathbf{Q}^{-1}\mathbf{C})^{-1}\mathbf{C}^*\mathbf{Q}^{-1}\mathbf{z}$$

For a higher signal power $\mathbf{P}^{-1}$ becomes insignificent and we arrive at equation 3.77.

3.5.3 Interference-suppression by the inverse covariance

The optimal detection of a target signal $\mathbf{s}$ from a measured signal $\mathbf{z}$ with the likelihood-ratio test (LRT) is given by equation 3.50. The meaning of the product $\mathbf{z}^*\mathbf{Q}^{-1}$ within this equation shall now be discussed. We will show that a correlated interference signal, for example a clutter signal, is suppressed in an optimal manner.

The interference signal including noise may be as before the vector $\mathbf{n}$ or $\mathbf{n}^* = (n_1^* \cdots n_k^* \cdots n_N^*)$ with the covariance matrix $\mathbf{Q} = E\{\mathbf{nn}^*\}$. We derive an estimate for n_k with the assumption that all values n_i are known except n_k. That is, we have a new vector $\mathbf{y}^* = (n_1^* \cdots n_{k-1}^* n_{k+1}^* \cdots n_N^*)$ with covariance matrix $\mathbf{Q}_y$. The conditional probability density for n_k is with equation 3.17:

$$p(n_k/\mathbf{y}) = \frac{p(n_k, \mathbf{y})}{p(\mathbf{y})} = \frac{p(\mathbf{n})}{p(\mathbf{y})} \tag{3.81}$$

With a Gaussian distribution for $\mathbf{n}$ we have

$$p(n_k/\mathbf{y}) = const \ \exp\left(-\mathbf{n}^*\mathbf{Q}^{-1}\mathbf{n} + \mathbf{y}^*\mathbf{Q}_y^{-1}\mathbf{y}\right)$$
$$= const \ \exp\left(-a\right)$$

The maximum value for $p(n_k)$ is found by minimising the expression a:

$$\left.\frac{\partial a}{\partial n_k}\right|_{n_k=\hat{n}_k} = (0\cdots010\cdots0)\mathbf{Q}^{-1}\mathbf{n}\Big|_{n_k=\hat{n}_k} = 0 \qquad (3.82)$$

The 1 is positioned at index k. $\mathbf{Q}^{-1}$ is first reduced (by multiplying with $(0\cdots010\cdots0)$) to all zeros except the row for index k with elements $q_{k1}\cdots q_{kk}\cdots q_{kN}$. Using equation 3.82 we get:

$$q_{k,k}\hat{n}_k = -(q_{k,1}\cdots q_{k,k-1}\,q_{k,k+1}\cdots q_{k,N})\mathbf{y} \qquad (3.83)$$

Adding and subtracting $n_k q_{k,k}$ in equation 3.83 results in:

$$q_{k,k}(n_k - \hat{n}_k) = (0\cdots010\cdots0)\mathbf{Q}^{-1}\mathbf{n} \qquad (3.84)$$

Equation 3.84 has the following interpretation. For each sample of a signal series the maximum-likelihood estimate of the interference part is determined from all other samples and then this estimate is subtracted.

For this interference estimation the correlation of the interference signal between the respective time instants or the parallel channels is effective. If there is no correlated interference we have the covariance $\mathbf{Q} = \mathbf{I}$. Then follows with $\mathbf{Q}^{-1} = I$, $q_{k,k} = 1$ and from equation 3.84:

$$n_k - \hat{n}_k = n_k \quad \text{or} \quad \hat{n}_k = 0$$

This means that there is no interference estimate to be subtracted and therefore no interference suppression, as expected.

In the case of a time series we have index $k = N$ for the newest or the actual noise signal n_N. The estimated noise or interference signal $\hat{n}_N$ is then subtracted from n_N. In other words, $\hat{n}_N$ is estimated or predicted from the passed signals $n_1, \ldots, n_{N-1}$. By subtraction for interference suppression $(n_N - \hat{n}_N)$ the prediction error results as the received signal without interference for further processing, for example Doppler filtering. The prediction error filter is given by the first column of $\mathbf{Q}^{-1}$ [8]. It offers a convenient solution for adaptive clutter suppression [9].

Equation 3.84 leads to the simple mechanism against an interference performed by $\mathbf{z}^*\mathbf{Q}^{-1}$ in the lilelihood-ratio test of equation 3.50:

Estimate and subtract the interference!

So, we have a plausible explanation for the interference suppression by applying $\mathbf{Q}^{-1}$. It applies to all problems of interference suppression where a signal vector resulting from a series of time samples or from parallel channels by spatial sampling has to be processed. Additionally, there is a normalisation by q_{kk}.

This interpretation may also help to develop suboptimal interference suppression methods.

The likelihood-ratio test given by equation 3.50 results in the case of a Gaussian distributed interference signal with covariance matrix $\mathbf{Q}$ to the simple rule: in the first step the interference must be suppressed by multiplying with $\mathbf{Q}^{-1}$ followed by matched filtering with the expected signal $\mathbf{s}$.

3.5.4 Improvement of signal-to-noise-and-interference ratio (SNIR)

For a judgement and comparison of signal-processing methods it is of interest to know the degree of improvement of the signal-to-noise-plus-interference power ratio. Starting from equation 3.50 we have to weight the received signal by a vector:

$$\mathbf{w} = \mathbf{Q}^{-1}\mathbf{s}$$

The output signal power is then given by:

$$|\mathbf{s}^*\mathbf{w}|^2 = (\mathbf{s}^*\mathbf{Q}^{-1}\mathbf{s})^2$$

The noise and interference power is given by vector $\mathbf{n}$ for the total interference signal:

$$E\{|\mathbf{w}^*\mathbf{n}|^2\} = \mathbf{s}^*\mathbf{Q}^{-1}E\{\mathbf{nn}^*\}\mathbf{Q}^{-1}\mathbf{s} = \mathbf{s}^*\mathbf{Q}^{-1}\mathbf{Q}\mathbf{Q}^{-1}\mathbf{s} = \mathbf{s}^*\mathbf{Q}^{-1}\mathbf{s}$$

Finally, we get for the *SNIR* at the output of our processing subsystem:

$$SNIR|_{output} = \frac{|\mathbf{s}^*\mathbf{w}|^2}{E\{|\mathbf{w}^*\mathbf{n}|^2\}} = \mathbf{s}^*\mathbf{Q}^{-1}\mathbf{s} \qquad (3.85)$$

At the input this ratio is:

$$SNIR|_{input} = \frac{\mathbf{s}^*\mathbf{s}}{trace(\mathbf{Q})}$$

trace($\mathbf{Q}$) means the sum of all elements on the main diagonal of matrix $\mathbf{Q}$.

The processing gain or improvement factor is the ratio:

$$G = \frac{SNIR|_{output}}{SNIR|_{input}} \qquad (3.86)$$

3.6 Summary

In this chapter we summarised the basic tools for considering problems of signal processing from an engineering viewpoint. After the introduction of basic statistics we discussed the likelihood ratio test for deriving optimal processing rules for signal or target detection. Some important examples have been treated.

The parameter estimation has to follow target detection in most applications. The basic procedure has been given, together with the determination of the achievable accuracy for the parameter estimate given by the Cramér-Rao bound.

Signal estimation is also sometimes required. The basic procedure has been described.

A general problem in radar is interference suppression. A processing rule derived from the likehood ratio test has been used for discussion of the general mechanism.

3.7 References

1 DAVENPORT, W. B., and ROOT, W. L.: 'An introduction to the theory of random signals and noise' (McGraw-Hill Book Company, Inc., New York, Toronto, London, 1958)
2 VAN DEN BOS, A.: 'The multivariate complex normal distribution – a generalisation', *IEEE Trans. Inf. Theory*, March 1995, **41** (2), pp. 537–539
3 MIDDLETON, D.: 'An introduction to statistical communication theory' (McGraw-Hill Book Company, Inc., New York, Toronto, London, 1960) chapter 19
4 CRAMÉR, H.: 'Mathematical methods of statistics' (Princeton University Press, Princeton, 1966)
5 VAN TREES, H. L.: 'Detection, estimation and modulation theory' (John Wiley and Sons, New York, 1968)
6 SELIN, I.: 'Detection of coherent radar returns of unknown Doppler shift', *IEEE Trans. Inf. Theory*, July 1965, pp. 396–400
7 WIRTH, W. D.: 'Bewegtzielerkennung bei Impulsradarsystemen', *Nachrichtentechnische Zeitschrift NTZ*, 1966, **19** (5), pp. 279–287
8 BURG, J. P.: 'A new analysis technique for time series data'. NATO Advanced Study Institute on Signal Processing with Emphasis on Underwater Acoustics, Enschede (NL), 1968
9 WIRTH, W. D.: 'Suppression of noise signals if their correlation is known', *NTZ-Communication Journal*, 1970, **23** (2), pp. 64–68 (in German)

Chapter 4
Array antennas

In this chapter we will discuss the most important and well-known relations and parameters of an array antenna which are relevant for the conception of a radar system [1–3]. Possible implementations for different applications will be summarised. We will start our discussion with a simple illustration, the basic principle of which is sketched in Figure 4.1. A set of antennas or an array of antenna elements is distributed on a metal ground plane, preferably on a regular grid. These antenna elements may be, for example, dipoles matched with the length of their arms to the operating wavelength λ of the radar system. In the case of transmission each antenna element is the source of a spherical wavefront. As a first step we assume waves of equal phase from each antenna element. Comparatively, we observe circular water waves if a group of persons standing at a linear sea wall are throwing stones at the same instant of time into a calm water. These waves superpose coherently according to the famous Huygens' Principle at each point in space. If all sources radiate in phase then at the boresight direction, orthogonal to the antenna plane, we have a linear or coherent summation of the field strength of all individual waves. That is if, at a certain distance from the antenna, the electrical field strength produced from one antenna element is E, then we would have with N antenna elements the field strength NE resulting at the boresight direction, or a power density proportional to $(NE)^2 = N^2 E^2$. Outside the boresight direction the condition of in-phase superposition is not fulfilled and there is approximately, as the spatial mean over all directions, only a superposition of the power, that is we have there a power density with an order of magnitude only proportional to NE^2. More detailed considerations for the antenna pattern will follow. This results in a factor of N for the power density between the boresight direction and all other directions. In this example our main beam is formed in the boresight direction and in the other directions only the unwanted sidelobes are produced. We recognise that only for N approaching infinity can we expect the sidelobe to mainlobe power ratio to approach zero. That would require an antenna with an infinite antenna plane. The angular width of our main beam, the region with approximately coherent superposition of the partial waves from the antenna elements, also depends on N. The angular region for an in-phase superposition of all individual waves decreases for an increasing N, resulting in a narrower beam.

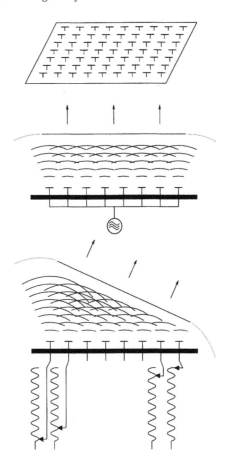

Figure 4.1   Principle of array antennas

The steering of the main beam can simply be accomplished by introducing delays to the individual waves. As shown in Figure 4.1, the main beam direction is steered clockwise into a corresponding angle by a linear increasing delay from left to right. These delays or phase shifts can be steered electronically and inertialess beam steering is the result.

After this illustrative explanation the behaviour of array antennas is described more precisely using mathematical relations as much as is necessary for system conception and signal processing. We will also learn or recall some simple and very useful rules of thumb.

4.1   Array factor

According to the discussion and illustration above, the superposition of the partial waves in the farfield shall be described in the following with mathematical relations.

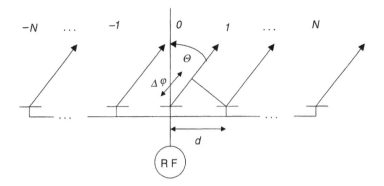

Figure 4.2   Linear regular array

The farfield range is defined as that distance where the phase differences of the waves from the individual antenna elements are similar to those for an infinite range, with a phase error of less than $\pi/8$. In Figure 4.2 the geometry for a regularly-spaced linear array is given. There are $2N + 1$ antenna elements regularly distributed with a distance d.

We assume initially that all antenna elements with index number $n = -N \cdots +N$ are fed with an RF signal with equal phase. We ask for the result of superposition in a direction Θ from the boresight. For the sake of simplicity the amplitudes are assumed to be equal to 1. The phase difference $\Delta\varphi$ of the waves from neighbouring antennas depends on Θ and is:

$$\Delta\varphi = 2\pi \frac{d}{\lambda} \sin \Theta \tag{4.1}$$

The phase from an antenna element with index number n is therefore simply $n\Delta\varphi$. The complex contributions from all antenna elements are summed and result in the so-called array factor:

$$f(\Theta) = \sum_{n=-N}^{N} e^{j\Delta\varphi n} = \sum_{n=-N}^{N} \exp\left[2\pi j \frac{d}{\lambda} \sin \Theta n\right] \tag{4.2}$$

According to equation 2.1 we recall $e^{j\varphi} = \cos \varphi + j \sin \varphi$. For $\Theta = 0$ (boresight direction) we have $\Delta\varphi = 0$ and all $e^{j\Delta\varphi n} = 1$. With equation 4.2 the array factor then becomes the maximum possible value:

$$f(0) = 2N + 1 \tag{4.3}$$

As expected, the field strength (electrical or magnetic) resulting from a single antenna is multiplied by the element number $2N + 1$.

This complicated expression given by equation 4.2 for the array factor f may be simplified using the following intermediate development:

$$S = \sum_{-N}^{N} e^{j\varphi n} \tag{4.4}$$

$$S - e^{j\varphi} S = e^{-j\varphi N} - e^{+j\varphi(N+1)}$$

$$S = \frac{e^{-j\varphi N} - e^{j\varphi(N+1)}}{1 - e^{j\varphi}} \frac{1 - e^{-j\varphi}}{1 - e^{-j\varphi}}$$

For the numerator follows:

$$num = e^{-j\varphi N} + e^{j\varphi N} - e^{+j\varphi(N+1)} - e^{-j\varphi(N+1)}$$

$$= 2(\cos N\varphi - \cos (N+1)\varphi)$$

$$= 4 \sin (2N+1)\frac{\varphi}{2} \sin \frac{\varphi}{2}$$

and for the denominator:

$$den = 1 - e^{j\varphi} - e^{-j\varphi} + 1 = 2(1 - \cos \varphi)$$

$$= 4 \sin^2 \left(\frac{\varphi}{2}\right)$$

Then the sum of equation 4.4 is:

$$S = \sum_{n=-N}^{N} e^{j\Delta\varphi n} = \frac{nom}{den}$$

$$= \frac{\sin (2N+1)(\varphi/2)}{\sin \varphi/2} \tag{4.5}$$

With equation 4.5 we get for the array factor the well-known expression:

$$f(\Theta) = (2N+1)\frac{\sin \left[(2N+1)\pi(d/\lambda) \sin \Theta\right]}{(2N+1) \sin \left[\pi(d/\lambda) \sin \Theta\right]} \tag{4.6}$$

or with the abbreviation:

$$x = (2N+1)\pi \frac{d}{\lambda} \sin \Theta$$

and the approximation $x \approx \sin x$ for small values x:

$$f(\Theta) \approx (2N+1)\frac{\sin x}{x} \tag{4.7}$$

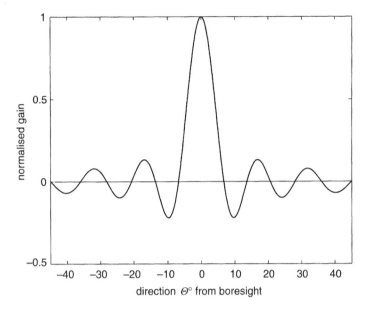

Figure 4.3   Array factor of linear array without weighting, $N = 8$

The limiting value for $(\sin x)/x$ for x approaching zero is equal to 1. So again with equation 4.6 we get $f(0) = 2N + 1$, as expected from equation 4.2. This corresponds to the centre of the main beam. The course of the array factor according to equation 4.7 is shown as an example in Figure 4.3.

For steering the main beam in direction Θ_0 we have to modify equation 4.2 or 4.5:

$$f(\Theta/\Theta_0) = \sum_{n=-N}^{N} \exp\left[2\pi j \frac{d}{\lambda}(\sin \Theta - \sin \Theta_0)n\right] \tag{4.8}$$

$$= (2N + 1)\frac{\sin\left[(2N + 1)\pi(d/\lambda)(\sin \Theta - \sin \Theta_0)\right]}{(2N + 1)\sin\left[\pi(d/\lambda)(\sin \Theta - \sin \Theta_0)\right]} \tag{4.9}$$

If $\Theta = \Theta_0$ then the main beam condition, all arguments of the exp function in equation 4.8 to be equal to zero, is fulfilled.

The steering phases for steering the beam into direction Θ_0 are from equation 4.8:

$$\varphi_n = 2\pi \frac{d}{\lambda} n \sin \Theta_0 \quad n = -N \cdots N$$

The first zero in the pattern is according to equation 4.9 given by:

$$(2N + 1)\pi \frac{d}{\lambda}(\sin \Theta - \sin \Theta_0) = \pi$$

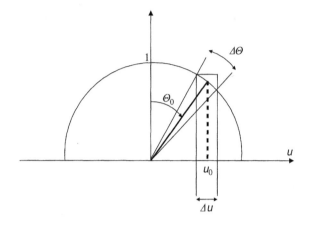

Figure 4.4   Projected direction u

To steer a beam to this first zero we have:

$$\sin \Theta_0 = \frac{\lambda}{(2N + 1)d}$$

The phase variation along the array or phase difference between the end elements is then approximately 2π.

The use of projected directions is very common. As shown in Figure 4.4:

$$u = \sin \Theta \qquad (4.10)$$

For values of Θ from $-90°$ to $90°$ the new variable u assumes values -1 to $+1$.

The array factor of equation 4.8 then becomes:

$$f(u/u_0) = \sum_{n=-N}^{N} \exp\left[2\pi j \frac{dn}{\lambda}(u - u_0)\right]$$

or with the x coordinates of the antenna elements $x_n = dn$:

$$f(u/u_0) = \sum_{n=-N}^{N} \exp\left[2\pi j \frac{x_n}{\lambda}(u - u_0)\right] \qquad (4.11)$$

The extension to two dimensions for planar arrays to steer the beam into u for azimuth and v for elevation and with antenna element coordinates (x_n, y_n) is given by:

$$f(u, v/u_0, v_0) = \sum_{n=-N}^{N} \exp\left[\frac{2\pi j}{\lambda}((u - u_0)x_n + (v - v_0)y_n)\right] \qquad (4.12)$$

The steering phase for an antenna element n is:

$$\varphi_n = \frac{2\pi}{\lambda}(u_0 x_n + v_0 y_n) \tag{4.13}$$

The vector (u_0, v_0) is the projection of the unity vector in the beam direction onto a plane parallel to the antenna plane. The steering phase is proportional to the scalar product of the directional vector and the element position vector. Or in other words: the phase is proportional to the projection of the element position vector onto the beam direction.

The third component w_0 of this beam direction vector is the projection onto the axis in the boresight direction which is normal to the antenna plane. If the antenna elements are distributed in a volume they have coordinates (x_n, y_n, z_n). The array factor then is:

$$f(u, v/u_0, v_0) = \sum_{n=1}^{N} \exp\left[\frac{2\pi j}{\lambda}((u - u_0)x_n + (v - v_0)y_n + (w - w_0)z_n)\right] \tag{4.14}$$

and the steering phase becomes:

$$\varphi_n = \frac{2\pi}{\lambda}(u_0 x_n + v_0 y_n + w_0 z_n) \tag{4.15}$$

Because $u^2 + v^2 + w^2 = 1$ we have for a selected (u, v) also given the value of w. In the case of a planar antenna, equations 4.12 and 4.13, we have $z_n = 0$ for all antenna elements and the third term in equations 4.14 and 4.15 is zero.

Usually all the single antenna elements have a certain directional pattern $e(\Theta)$. This is multiplied with the array factor resulting in the final antenna pattern $E(\Theta)$ (field strength):

$$E(\Theta) = e(\Theta)f(\Theta) \tag{4.16}$$

The element pattern $e(\Theta)$ is influenced by mutual coupling effects from the surrounding antenna elements. The complicated coupling effects are nowadays well understood and are dealt with in the literature [1–3]. In particular, the effect of the edge antenna elements is different from that of the centre elements. Therefore, equation 4.16 is exactly valid only for an infinite regular array, but it is nevertheless a good approximation.

4.2 Array parameters

In this section we derive and discuss some important array parameters by using and evaluating the formulas for the array factor from equation 4.8 or 4.9.

4.2.1 Half-power beamwidth

We first ask for the angle difference Θ_H from the steering direction Θ_0 where the power density is reduced by a factor of 2 or 3 dB. The array factor is reduced at this angle by the factor $1/\sqrt{2} = 0.707$ and from equation 4.9 as follows:

$$f(\Theta_H) = (2N + 1) \cdot 0.707$$

Because $(\sin x)/x = 0.707$, for $x = 1.39$ we have with equation 4.9:

$$\sin \Theta_H - \sin \Theta_0 = \frac{1.39}{\pi} \frac{\lambda}{L} = 0.442 \frac{\lambda}{L}$$

with $L = (2N + 1)d$. With the abbreviation $2(\Theta_H - \Theta_0) = \Theta_B$ for the half-power beamwidth and the approximation $\sin x \approx x$ for small arguments x we get:

$$\sin \Theta_H - \sin \Theta_0 = 2 \cos \left(\frac{\Theta_H + \Theta_0}{2} \right) \sin \left(\frac{\Theta_H - \Theta_0}{2} \right)$$

$$\approx 2 \cos \Theta_0 \cdot \frac{\Theta_B}{4}$$

So we have finally for Θ_B in radians:

$$\Theta_B = 0.884 \frac{\lambda}{L \cos \Theta_0} \tag{4.17}$$

or in degrees with the factor $360/2\pi = 57,296$ the simple rule of thumb:

$$\Theta_B \approx 50 \frac{\lambda}{L \cos \Theta_0} \tag{4.18}$$

The denominator in these equations is the length of the antenna which is projected into the direction Θ_0 of the beam. This shows a fundamental disadvantage of linear and planar arrays: the broadening of the main beam with scanning from the boresight.

For the projected direction u the beamwidth becomes according to Figure 4.3:

$$\Delta u = \Theta_B \cos \Theta_0 = 0.884 \frac{\lambda}{L} \tag{4.19}$$

For projected directions the beamwidth is independent of u. This is of some convenience for the development of a scanning raster for the target search functions of the radar system.

4.2.2 Bandwidth limitation with phase steering

If the beam direction is steered with time delays according to the antenna element's position, as shown in principle in Figure 4.1, there is no fundamental bandwidth limitation. But for reasons of economy usually the beam direction is steered with phase shifters operating modulo 2π. We use again equation 4.8 and look for the direction variation dependent on frequency:

$$f(\Theta, \omega) = \sum_{n=-N}^{N} \exp\left[j \left(2\pi \frac{d}{\lambda} \sin \Theta - 2\pi \frac{d}{\lambda_0} \sin \Theta_0 \right) n \right] \qquad (4.20)$$

The second term in the brackets represents the steering phase for direction Θ_0 and wavelength λ_0. This steering phase is applied usually by a phase shifter, giving a phase value independent of frequency and modulo 2π.

With c as the velocity of light first we have:

$$\omega = \omega_0 + \Delta\omega = \frac{2\pi c}{\lambda}$$

and also:

$$\frac{\lambda_0}{\lambda} = \frac{\omega_0 + \Delta\omega}{\omega_0}$$

We recall that for the main beam direction the expression in the brackets becomes equal to zero. We get $\Theta = \Theta_0 + \Delta\Theta$ and from equation 4.20:

$$\omega \sin \Theta = (\omega_0 + \Delta\omega) \sin (\Theta_0 + \Delta\Theta) = \omega_0 \sin \Theta_0$$

With:

$$\sin (\Theta_0 + \Delta\Theta) = \sin \Theta_0 \cos \Delta\Theta + \sin \Delta\Theta \cos \Theta_0$$
$$\approx \sin \Theta_0 + \Delta\Theta \cos \Theta_0$$

we get:

$$(\omega_0 + \Delta\omega)(\sin \Theta_0 + \Delta\Theta \cos \Theta_0) = \omega_0 \sin \Theta_0$$

and finally:

$$\Delta\Theta = -\frac{\Delta\omega}{\omega} \tan \Theta_0 \qquad (4.21)$$

With increasing frequency the beam shifts to the boresight direction. Let us choose an example: for $\Theta_0 = 60°$ (the maximum useful scanning angle), $\Delta\Theta$ in $°$ and frequency variation in per cent we get with $(360/2\pi) \tan 60° \approx 100$ the convenient rule of thumb:

$$\Delta\Theta[°] = -\frac{\Delta\omega}{\omega}[\%] \qquad (4.22)$$

The beam shift in degrees is equal to the frequency shift in per cent. We may tolerate a beam shift across a target within the 3 dB beamwidth Θ_B. Then we have a simple design rule: the beamwidth, expressed in degrees, is equal to the relative bandwidth, expressed in per cent.

4.2.3   Antenna element spacing without grating lobes

Again we recall the array factor:

$$f(u/u_0) = \sum_{n=-N}^{N} \exp\left[2\pi j n \frac{d}{\lambda}(u - u_0)\right]$$

The array factor assumes the maximum value if the terms in the sum are equal $\exp[2\pi jn]$, with n an integer $0, 1, 2, \ldots$. A secondary lobe, a so-called grating lobe, appears therefore in direction u_{GL} if:

$$\frac{d}{\lambda}|u_{GL} - u_0| = 1$$

An example is shown in Figure 4.5 with $d/\lambda = 0.75$ and $u_0 = 0.75$, resulting in a grating lobe at $u_{GL} = -0.58$.

Therefore we have to require $d/\lambda|u_{GL}-u_0| < 1$. We want to have no grating lobe within the visible angle area $u = -1 \cdots +1$. This means that we require $|u_{GL}| > 1$, resulting in the rule:

$$\frac{d}{\lambda} < \frac{1}{1 + u_0} \tag{4.23}$$

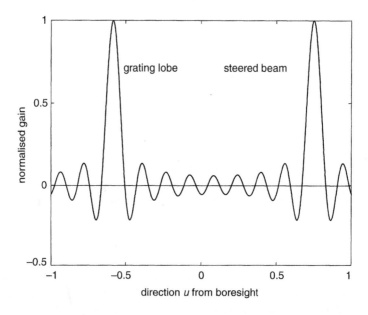

Figure 4.5   Linear array with $d/\lambda = 0.75$ with a grating lobe

This is illustrated by a numerical example: for $\Theta_0 = 60°$ or $u_0 = 0.866$ follows:

$$\frac{d}{\lambda} = \frac{1}{1.866} = 0.536$$

Our desire is to fill an antenna aperture as large as possible, to achieve a narrow beamwidth, with our precious antenna modules. But this procedure is limited by the grating-lobe effect and the rule given by equation 4.23. With a smaller required scan angle the spacing d may be increased using this rule. The standard value for antenna element spacing is $d = \lambda/2$.

4.2.4   Gain of regularly spaced planar arrays with $d = \lambda/2$

The effective antenna area A is defined [1] using the gain G by:

$$A = \frac{G\lambda^2}{4\pi} \tag{4.24}$$

N antenna elements with $\lambda/2$ spacing in both dimensions result in an occupied area of $N\lambda^2/4 = G\lambda^2/4\pi$. Therefore, in the boresight direction, the gain becomes:

$$G_0 = \pi N \tag{4.25}$$

If the beam is steered from the boresight by an angle ϑ, experience shows that, including mutual coupling effects, gain is given by:

$$G(\vartheta) \approx \pi N (\cos \vartheta)^{1.5} \tag{4.26}$$

4.2.5   Reduction of sidelobes by tapering

The antenna pattern, approximated around the main beam by the array factor, is given by equation 4.7 with the $(\sin x)/x$ function. This is plotted in Figure 4.3. It shows highly unfavourable and unwanted near sidelobes. The sidelobes may be reduced by applying amplitude tapering or weighting to the antenna elements. Then the resulting array factor is given by:

$$f(u) = \sum_{n=-N}^{N} g_n \exp\left[2\pi j \frac{d}{\lambda} un\right] \tag{4.27}$$

We recognise in equation 4.27 the Fourier transform relation of equation 2.20 in chapter 2 between the tapering function g_n and the array pattern $f(u)$.

For a regularly spaced array a so-called Taylor tapering is usually applied [3] for the sum beam. A general idea for this Taylor tapering is given in section 4.8. An example of this Taylor weighting for a linear array is shown in Figure 4.6, and the resulting array pattern (array factor) is compared in Figure 4.7 with the pattern without weighting. The sum beam is used typically for the detection of targets.

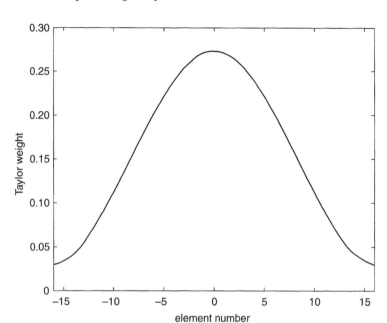

Figure 4.6   Example of Taylor weighting function for a linear array

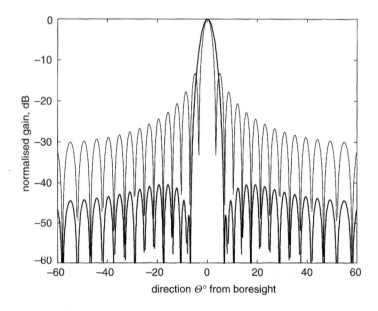

Figure 4.7   Array pattern with Taylor weighting (fat line) compared to constant weighting (thin line)

The bell-shaped Taylor weighting reduces the effective aperture length and therefore the beam is broadened a little. The gain for the boresight direction is reduced according to:

$$G_0 = \pi \frac{\left[\sum_{n=1}^{N} g_n \right]^2}{\sum_{n=1}^{N} g_n^2} \qquad (4.28)$$

For equal weights $g_n = 1$ we have again equation 4.25. This means that we have to pay a price for the reduction of sidelobes in the form of a loss in gain, given by the loss factor:

$$L_G = \frac{\left[\sum_{n=1}^{N} g_n \right]^2}{N \sum_{n=1}^{N} g_n^2} \qquad (4.29)$$

This loss is of the order of 1.5 to 2 dB.

Additionally, a difference beam is formed for the estimation of target direction. Direction estimation is discussed in chapter 11. The difference beam has at its centre an approximately linear pattern going through zero in the steering direction. For the difference beam a Bayliss tapering function (see section 4.8) is applied to achieve low sidelobes. An example of the difference pattern for a circular planar array is given in Figure 4.8.

For large regularly spaced arrays, with good angular resolution, a very high number of antenna elements would be necessary. To reduce the cost one may apply a random thinning of elements. The thinning density for the antenna elements may

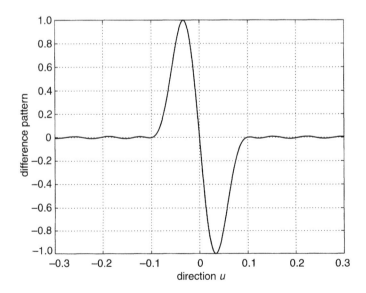

Figure 4.8   Difference pattern for circular array with Bayliss weighting

follow the Taylor weighting function. By using random element distribution, the grating-lobe effect is avoided and the grating lobes are distributed over the whole sidelobe region, increasing the sidelobe level. As a rule of thumb SLL, the sidelobe mean power level relative to the main beam, is given for an array with N antenna elements by:

$$SLL \approx \frac{1}{N} \tag{4.30}$$

4.3   Circular array

For special applications a circular array may be applied, and for the sake of completeness the array factor will also be presented. This array factor applies to cylindrical arrays in the plane orthogonal to the cylinder axis. The N antenna elements may be distributed with a regular distance d along the circumference above a metallic cylinder. The active sector for beamforming is given by the elements $i = -N_1, \ldots, N_1$, and the geometry is sketched in Figure 4.9. The z axis may be vertical. The small circles indicate the positions of the antenna elements, e.g. dipoles. With $d = \lambda/2$ the element pattern including mutual coupling effects between the dipoles is given by a cos function dependent on the azimuth angle Θ [3]. Therefore we discuss this pleasant case.

With $d = \lambda/2$ we get:

$$d = \frac{\lambda}{2} = \frac{2\pi R}{N}$$

or

$$N = \frac{4\pi R}{\lambda} \tag{4.31}$$

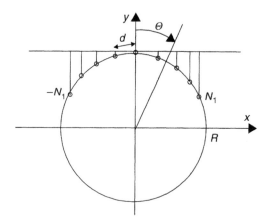

Figure 4.9   Geometry for circular array

The element pattern of antenna element i is denoted $e_i(\Theta)$. The field strength E in the azimuth direction Θ then has element coordinates (x_i, y_i) and phase ψ_i in order to achieve a linear or flat phase front:

$$E(\Theta) = \sum_{-N_1}^{N_1} e_i(\Theta) \exp\left[\frac{2\pi j}{\lambda}(x_i \sin\Theta + y_i \cos\Theta) + \psi_i \right] \qquad (4.32)$$

For the direction $\Theta = 0$ the values ψ_i, are with respect to the x axis:

$$\psi_i = -2\pi \frac{R}{\lambda} \cos\left(i\frac{2\pi}{N}\right)$$

$$= -\frac{N}{2} \cos\left(i\frac{2\pi}{N}\right)$$

From equations 4.31 and 4.32 results:

$$E(\Theta) = \sum_{-N_1}^{N_1} e_i(\Theta) \exp\left[j\frac{N}{2}\left(\sin\left(i\frac{2\pi}{N}\right) \sin\Theta \right.\right.$$

$$\left.\left. + \cos\left(i\frac{2\pi}{N}\right) \cos\Theta - \cos\left(i\frac{2\pi}{N}\right) \right) \right]$$

$$= \sum_{-N_1}^{N_1} e_i(\Theta) \exp\left[j\frac{N}{2}\left(\cos\left(i\frac{2\pi}{N} - \Theta\right) - \cos\left(i\frac{2\pi}{N}\right) \right) \right] \qquad (4.33)$$

The element pattern is given by Reference 3:

$$e_i(\Theta) = \cos\left(\Theta - i\frac{2\pi}{N}\right) \quad \text{for} \quad \left|\Theta - i\frac{2\pi}{N}\right| \leq \frac{\pi}{2}$$

$$= 0 \quad \text{otherwise}$$

The element density increases in the projection, e.g. on the x axis for $\Theta = 0$, by a factor $1/\cos(i(2\pi/N))$. On the other hand, the element pattern decreases by a factor $\cos(i(2\pi/N))$. Both effects are mutually compensating. Because with increasing index i the element contribution decreases, the active array sector should be chosen to be only about 90°, which means that $2N_1 + 1 = N/4$. An example for the resulting antenna pattern is shown in Figure 4.10.

A tapering function, e.g. a Taylor weighting, has to be applied along a line orthogonal to the beam direction. For $\Theta = 0$ this is the x axis. The weighting values along this line have to be projected onto the antenna elements. The result is given in Figure 4.11 for the same antenna configuration as for Figure 4.10.

4.4 Phase and amplitude errors

The cost of an electronically-steered array depends, among other things, on the required accuracy for phase and amplitude steering of the antenna channels. In this

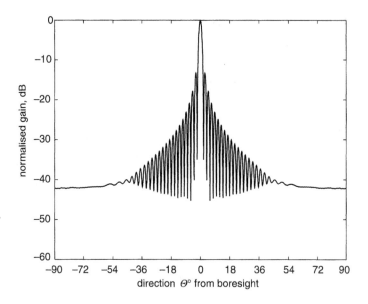

Figure 4.10   Antenna pattern of circular array, N = 256, active sector 90°

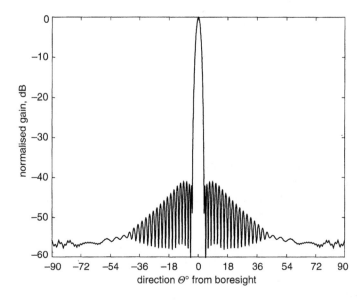

Figure 4.11   Circular array with Taylor weighting, N = 256, active sector 90°

section the effect of phase and amplitude quantisation on the gain, sidelobe level and beam direction is summarised according to the results in References 2 and 4.

Usually digital phase shifters, characterised by the number of bits P, are applied. This determines the residual phase error ε within the interval $\Delta\varphi = \pm\pi/2^P$.

The quantisation errors may be assumed as random variables, which are equally distributed within their quantisation interval. This is because the phase and amplitude values to be quantised are large compared to the quantisation interval and therefore the quantisation errors are independent from one antenna channel to the other. The sidelobes which are caused by these random errors are also a random variable and are generated independently of the designed sidelobes by tapering, and are added to these. The random sidelobes form a lower limit for the achievable sidelobes. Both sidelobe levels, designed and error sidelobes, should be matched in order to reduce cost.

For the sidelobe contribution, caused by amplitude and phase errors, we compute [4] the variance of the array factor using equations 3.33 and 4.27. This variance represents the power density transmitted into all directions u and is a measure for the average sidelobe level.

Because of the reciprocity of antennas the pattern variance also applies to the receiving case.

Without errors we have for the array factor:

$$f(u) = \sum_{n=-N}^{N} g_n \exp\left[2\pi j \frac{d}{\lambda} un\right]$$

and with errors q_n for the amplitude and ε_n for the phase:

$$\tilde{f}(u) = \sum_{n=-N}^{N} (g_n + q_n) \exp\left[2\pi j \frac{d}{\lambda} un + j\varepsilon_n\right]$$

The errors q_n and ε_n are all assumed to be random, independent and equally distributed within their intervals $\pm \Delta q$ and $\pm \Delta \varphi$, respectively. These errors may be produced by quantisation of the steering values for phase and amplitude or the weighting (tapering) factors. The variance is, according to equation 3.33 in chapter 3:

$$\sigma_{\tilde{f}}^2 = E\{|\tilde{f}|^2\} - |E\{\tilde{f}\}|^2 \tag{4.34}$$

For the computation of the expected values we use the probability densities:

$$p(q) = \frac{1}{2\Delta q}$$

and:

$$p(\varepsilon) = \frac{1}{2\Delta \varphi}$$

For the determination of $E\{\tilde{f}\}$ we observe that the average of q_n is zero and we get:

$$E\{\tilde{f}\} = f \frac{1}{2\Delta \varphi} \int_{-\Delta \varphi}^{\Delta \varphi} d\varepsilon \exp(j\varepsilon)$$

$$= f \frac{\sin \Delta \varphi}{\Delta \varphi} \tag{4.35}$$

For the first term $E\{\tilde{f}\tilde{f}^*\}$ of equation 4.34 we have a double sum:

$$\tilde{f}\tilde{f}^* = \sum_{n=-N}^{N} (g_n + q_n) \exp\left[2\pi j \frac{d}{\lambda} un + j\varepsilon_n \right]$$

$$\times \sum_{k=-N}^{N} (g_k + q_k) \exp\left[-2\pi j \frac{d}{\lambda} uk - j\varepsilon_k \right]$$

For $n \neq k$ we get, as for equation 4.35:

$$g_n g_k \exp\left(2\pi j \frac{d}{\lambda} u(n-k) \right) \left(\frac{\sin \Delta\varphi}{\Delta\varphi} \right)^2$$

For $n = k$ the phases of the exp terms cancel and the expectation for these terms becomes:

$$\int_{-\Delta q}^{\Delta q} dq \frac{1}{2\Delta q} (g_n + q)^2 = g_n^2 + 2g_n \frac{1}{2\Delta q} \int_{-\Delta q}^{\Delta q} dq\, q + \frac{1}{2\Delta q} \int_{-\Delta q}^{\Delta q} dq\, q^2$$

$$= g_n^2 + \frac{(\Delta q)^2}{3}$$

We use these intermediate results in equation 4.34: all terms with $n \neq k$ cancel, because of the subtraction of the squared contribution from equation 4.35 which are equal to the above terms of

$$E\{\tilde{f}\tilde{f}^*\} \quad \text{for } n \neq k$$

By summing the terms for $n = k$ we get finally:

$$\sigma_f^2 = (2N+1)\frac{(\Delta q)^2}{3} + \left(1 - \left(\frac{\sin \Delta\varphi}{\Delta\varphi} \right)^2 \right) \sum_{n=-N}^{N} g_n^2 \qquad (4.36)$$

We take this variance as a measure of the power in the sidelobe region. The relative sidelobe level is given with $f(0)$ for the main beam value:

$$RSL = \frac{\sigma_f^2}{|f(0)|^2} \qquad (4.37)$$

Now we recognise the first term of equation 4.36 as the sidelobe contribution by amplitude errors.

For $g_n = 1$, that is without amplitude tapering, using phase errors from equation 4.36 and by applying a Taylor-series approximation for $\sin \Delta\varphi$, we get the same formula for the sidelobes as given in Reference 1.

Table 4.1 Average sidelobe level in dB due to
quantisation errors

Number of elements	1000	5000	10000
Amplitude errors			
2 bits	−40.7	−47.7	−50.7
3 bits	−46.8	−53.8	−56.7
4 bits	−52.8	−59.8	−62.8
Phase errors			
2 bits	−33.8	−40.8	−43.8
3 bits	−39.8	−46.8	−49.8
4 bits	−45.8	−52.8	−55.8

In Table 4.1, as an illustration, the average sidelobe level due to quantisation errors for a planar array with amplitude weighting is given. This gives a feeling for the influence of the element number and the quantisation accuracy and their impact on the sidelobe level.

From equation 4.34 follows a loss in gain by phase quantisation:

$$\Delta G = \frac{G - \tilde{G}}{G}$$

$$= 1 - \left(\frac{\sin \Delta\varphi}{\Delta\varphi} \right)^2$$

With an approximation by a Taylor-series expansion $\sin x \approx x - x^3/6$ we get, as given also in References 1 and 2:

$$\Delta G \approx 1 - \left(1 - \frac{\Delta\varphi^2}{6} \right)^2 \approx \frac{1}{3} \Delta\varphi^2$$

or

$$\Delta G \approx \frac{1}{3} \frac{\pi^2}{2^{2P}} \tag{4.38}$$

For example, $P = 3$ results in $\Delta G = 0.23\,\text{dB}$. Therefore the loss in gain is insignificant.

For the special case of an array without tapering we derive for the sidelobes by phase quantisation from the second term of equation 4.36, with $g_n = 1$ and $\tilde{N} = 2N + 1$ as the total number of antenna elements, similar to above:

$$RSL \approx \frac{1}{\tilde{N}^2} \left[1 - \left(1 - \frac{\Delta\varphi^2}{6} \right)^2 \right] \tilde{N}$$

$$= \frac{\pi^2}{3 \cdot 2^{2P} \tilde{N}} = \frac{3,3}{2^{2P} \tilde{N}} \tag{4.39}$$

Equation 4.39 matches the formula given in Reference 2:

$$RSL \approx \frac{5}{2^{2P} \tilde{N}}$$

(4.40)

The beam pointing accuracy is given by [2]:

$$\frac{\Delta \Theta}{\Theta_B} = \frac{9}{2^P \tilde{N}}$$

(4.41)

For example, $P = 3$, $\tilde{N} = 100$, $\Delta \Theta / \Theta_B \approx 0.01$. Therefore, the beam pointing accuracy is also quite good enough, even for a quantisation with only three bits.

A rough estimation of the effect of element failure follows from the following consideration: for an array with tapering g_n the power density change corresponding to a single element missing is approximately equal to the power density when this one element is present alone:

$$RSL_p \approx \frac{|g_p|^2}{\left| \sum g_n \right|^2}$$

We can take the mean over all $\tilde{N}$ elements for the effect of the failure of any one element:

$$\langle RSL \rangle \approx \frac{(1/\tilde{N}) \sum_{\tilde{N}} |g_n|^2}{\left| \sum_{\tilde{N}} g_n \right|^2}$$

We introduce the efficiency η of a tapered array defined by the directivity ratio of a tapered to an untapered array:

$$\eta = \frac{\left| \sum_{\tilde{N}} g_n \right|^2}{(1/\tilde{N}) \sum_{\tilde{N}} g_n^2} \Bigg/ \frac{\tilde{N}^2}{(1/\tilde{N})\tilde{N}}$$

$$= \frac{\left| \sum_{\tilde{N}} g_n \right|^2}{\tilde{N} \sum_{\tilde{N}} g_n^2}$$

(4.42)

This efficiency has a value of about 0.5 for a tapered array and of course unity for an array without tapering. With equation 4.42 we get the simple rule of thumb for sidelobes with a missing element:

$$\langle RSL \rangle = \frac{1}{\tilde{N}^2 \eta}$$

(4.43)

If K elements are missing we have:

$$\langle RSL_K \rangle \approx \frac{K}{\tilde{N}^2 \eta}$$

(4.44)

We apply equation 4.42 to a randomly-thinned array with only N_1 out of $\tilde{N}$ elements:

$$\eta = \frac{N_1^2}{\tilde{N}N_1} = \frac{N_1}{\tilde{N}}$$

With equation 4.44 we come again to the rule of thumb, given by equation 4.30:

$$\langle RSL \rangle = \frac{\tilde{N} - N_1}{\tilde{N}^2}\frac{\tilde{N}}{N_1} = \frac{1 - (N_1/\tilde{N})}{N_1}$$

$$\approx \frac{1}{N_1} \tag{4.45}$$

4.5 Architectures of passive and active array antennas

There are different possibilities for the combination of a transmitter and a receiver with the array antenna elements and the phase shifters. Most popular has been the so-called central feed configuration. One high-power transmit tube amplifier produces and delivers the transmit pulse with its full peak power P_p for distribution to the N antenna elements. The most simple technique is a distribution from the transmitter, Tx, by radiation from a central feed antenna, for example a horn antenna, to collecting antennas assigned to the final emitting antennas, as shown in Figure 4.12.

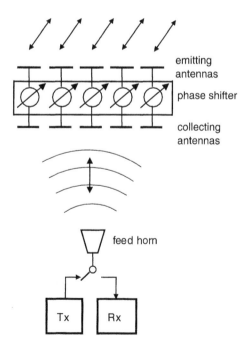

Figure 4.12   Central space-fed array configuration

The collecting antennas are behind the planar antenna structure. The phase shifters are placed between the collecting and the emitting antenna elements and are steered according to the desired direction. The whole antenna structure, with collecting and emitting antennas and the phase shifters in between, acts as an electronic switchable lens. The phase shifters have to be designed for operation with a relatively high RF power, given by P_p/N. The phase-shifter attenuation has to be as low as possible, because otherwise the transmitter efficiency would be degraded correspondingly. The preferred technique is a ferrite phase shifter, because of its power-handling capability and low attenuation. For receiving, the signals travel the reverse path and are fed into the one central receiver, Rx.

The differences in the path length from the feed horn to the individual collecting antennas have, of course, to be compensated for by considering the phase differences for the steering commands for the phase shifters. Then, for example for the broadside beam direction, all antennas must have the same phases. The tapering function is determined by the illumination of the collecting antennas from the feed horn. One can imagine that it is difficult to realise a Taylor or Bayliss tapering function exactly.

Instead of space feeding a constraint feed network may be applied; this configuration is shown in Figure 4.13. Different techniques may be selected for the implementation of the distribution and combining network. For high-power systems one has to apply waveguide techniques for this network, but for low-power systems a microstrip solution is adequate and more cost effective.

An active array has individual transmit/receive modules (TRMs) for each antenna element. Each TRM contains a transmit amplifier, a receive amplifier with variable gain and a phase shifter as the basic circuits. The configuration is given in Figure 4.14. A combining and distribution network distributes the transmit signal at a low power level to the TRMs. The phase shifters determine the beam direction. At the receiving

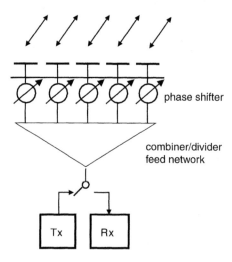

Figure 4.13   Central constraint-fed array

transmit/receive wavefront

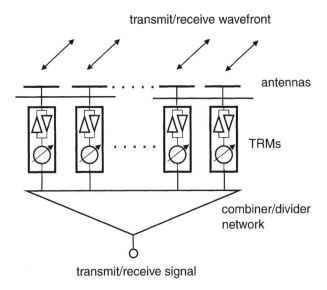

antennas

TRMs

combiner/divider
network

transmit/receive signal

Figure 4.14   Active array architecture

end the low-noise amplifier (LNA) is directly behind the antenna, so there are only
small losses resulting in a low overall noise figure.

This architecture has several advantages:

- Because the power from all TRMs is added in space, the amplifier in a single
  module has to produce only a small power, which may be achieved with solid-state
  amplifiers.
- With high element numbers the added transmit power can be very high.
- Particularly with transistor amplifiers, the duty factor may be up to about
  30 per cent resulting in a high mean power.
- The mean time between failure (MTBF) for transistor amplifiers is estimated as
  being as much as 10^6 hours. This is much longer than that for tube amplifiers.
  The high number of parallel operating amplifiers means that there is a high level
  of redundancy. Failure of a single amplifier results only in a graceful degradation
  for the whole transmitter.
- There is no high-voltage circuitry.
- The system's receiving noise figure is as low as possible because the low-noise
  amplifier is near the antenna element.
- The individual receiving channels form a basis for important array signal-
  processing procedures and beamforming. These will be discussed in detail in
  the following chapters.

The transmit/receive modules have been first developed with discrete devices for
experimental demonstrator programs for active arrays [5]. These modules have tended
to be very expensive, but they are now under industrial development in several

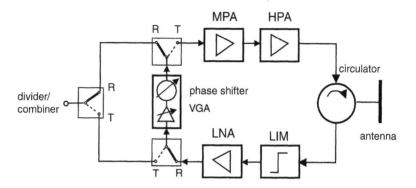

Figure 4.15   Block diagram of transmit/receive module (TRM)

places, with monolithic microwave integrated circuits (MMIC) preferable on a GaAs substrate [6].

Such a module has the typical block diagram shown in Figure 4.15. The RF reference signal to be transmitted is distributed to the modules by the dividing/combining network, usually produced using a microstrip technique. The signal path goes, for transmission, from the input, connected to the divider/combining network, *via* the first switch to the amplifier/phase shifter, then by the next switch to a medium-power driver amplifier (MPA) which is followed by the final high-power amplifier (HPA). The switches are shown in the receiving position, R. For transmission they change to position T.

The power amplifier may consist of several parallel amplifiers, depending on the power requirements. Amplifier developments are striving for a high output RF power and high efficiency, the ratio of RF output to DC prime power. Up to now field effect transistor (FET) amplifiers are the preferred choice with an efficiency of about 30 per cent. In future, hetero bipolar transistor (HBT) amplifiers may also offer an improved efficiency of more than 40 per cent [7]. The HPA output is fed to the circulator and then to the antenna element. A circulator acts for transmit/receive decoupling and reduces the mismatch which is introduced by direction-dependant mutual coupling between the antenna elements.

For receiving, the signals are fed through the circulator first to a limiter as protection for the sensitive low-noise receiver amplifier (LNA). The variable-gain amplifier in conjunction with the phase shifter may be used to calibrate all receiving channels mutually and to allow the application of a tapering function for sidelobe reduction.

All the amplifiers within a module may establish unintentional feedback loops by limited isolation of the switches (in the off position) and by radiation coupling within the module case. These feedback loops may cause dangerous spurious oscillations. These would disturb the function of the module and could even destroy it. Careful design and a proper choice for the gain of the individual amplifiers has to be applied to avoid any excitation. Extensive test procedures are necessary and have been developed, at least for the prototype modules, to assure proper function of the modules within the array [8].

The main problem for the application of TRMs in active phased arrays has been the cost for medium-scale production. The overall cost of the radar is determined by the cost of a module; this is because the number of modules is large, typically several thousands of modules. For comparison with more conventional systems the lifecycle costs, as well as the investment costs, must also be evaluated and compared.

4.5.1   Comparison of efficiency for active and passive arrays

Because of the different losses the efficiency of a centrally fed (Figure 4.13) and an active array (Figure 4.14) is different. We assume an array with $N = 5000$ antenna elements and a total mean power of 5 kW. Then the power balance is as in Table 4.2.

For mobile and airborne installations the necessary prime power is of particularly high importance. Additionally, the higher attenuation of about 4 dB in the receiving path of a centrally fed system has to be taken into account. For equal sensitivity of the centrally fed system the RF power has to be increased by a factor of 2.5. Instead of 5 kW we need 12.5 kW mean RF power.

4.5.2   Radar equation for active arrays

The well-known radar equation [9] gives the received power dependent on transmit power P_t, transmit antenna gain G_t, receiving antenna gain G_r, wavelength λ, target radar cross section σ, target range R, overall losses L:

$$P_r = \frac{P_t G_t G_r \lambda^2 \sigma}{(4\pi)^3 R^4 L} \tag{4.46}$$

The signal-to-noise ratio results for bandwidth B and noise factor F and $kT_0 = 4 \cdot 10^{-21}$ Ws:

$$\frac{S}{N} = \frac{P_r}{kT_0 F B} \tag{4.47}$$

Table 4.2   Power balance for centrally-fed and active arrays

	Centrally-fed array	Active array
Mean radiated power (kW)	5.0	5.0
Attenuation transmitter/antenna	4 dB or factor 0.4	1 dB or factor 0.79
Efficiency of transmit amplifier	0.40	0.30
DC power for transmit amplifier (kW)	$5/0.16 = 31.25$	$5/0.23 = 21$
Steering power for phase shifters (kW)	5.0	2
Total DC power (kW)	36.25	23.0

If the bandwidth B is matched to the pulse length τ, the pulse period is T, the dwell time T_d on target is nT, the loss by integrating n pulses is L_i then we have with mean power $P_m = P_t(\tau/T)$:

$$\frac{S}{N} = \frac{P_m G_t G_r \lambda^2 \sigma T_d}{(4\pi)^3 R^4 k T_0 F L L_i}$$

For an active array with the mean power p of a single module we then have $P_m = pN$ and:

$$G_t = G_r = \pi N$$

This is valid for a regularly spaced planar array in the broadside direction according to equation 4.25, for other directions and array distributions the proportionality holds. For equation 4.46 results:

$$\frac{S}{N} = \frac{N^3 p \pi^2 \lambda^2 \sigma T_d}{(4\pi)^3 R^4 k T_0 F L L_i} \tag{4.48}$$

So we have the remarkable result for active arrays that the received signal-to-noise ratio is proportional to N^3. Therefore, increasing the number of antenna elements and TRMs is the most effective means of increasing the effectiveness of a radar system with an active array antenna.

4.6   Concepts for an extended field of view

The maximum scan angle of planar arrays is limited to $60°$ from the broadside. The field of view has to be extended for different applications, such as air surveillance for air traffic control or air defence.

One direct solution to cover $360°$ in azimuth is the installation of four faces. Each array may be tilted back to extend the elevation coverage. Some systems have been built in the US for missile and air defence on land and on board ships [10,11]. This seems to be the most expensive solution.

A hybrid solution is to rotate a planar array mechanically at a high rotation rate and scan the beam electronically. This gives a certain flexibility and freedom for energy management and tracking. But the system contains again the mechanical rotation element which may cause a failure. On the other hand, this solution has the advantage of being the most cost effective (Figure 4.16).

Under investigation are so-called conformal or smart arrays with their antenna elements distributed on given surfaces of a platform, like the fuselage or wing of an aircraft or missile. The more the aperture plane is bent the more the field of view might be extended. The most popular form for the conformal array is a cylinder. But then only a part of the radiating elements would effectively contribute to beamforming and to increasing the gain. This depends on the individual element orientation and its pattern with respect to the desired beam direction. Because transmit/receive modules are very expensive for the time being they have to be switched to the selected effective aperture as discussed in connection with the circular or cylindrical array in section 4.3.

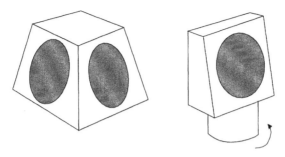

Figure 4.16   Concepts for 360° coverage: four planar arrays or one rotating planar array

4.6.1   Volume array for complete azimuth coverage

At FGAN the concept of a volume array has been developed with contribution by the author. The fundamental ideas are the following:

- all antenna elements shall contribute to beamforming for all directions without any shadowing
- the pattern of the main beam shall be independent of the scan direction
- the beam shall be scanned for complete azimuth coverage

These requirements can be fulfilled by an array with antenna elements distributed within the volume of a sphere of air without any metallic grounded surface. Because there exists no regular spatial grid with an equally spaced projected grid for all directions we selected a randomly-thinned distribution of the antenna elements within the volume space. The projection of the element positions onto a plane which is perpendicular to the beam direction determines the array factor and the beam pattern. For all directions we have the same projected density distribution of the antenna elements in a statistical sense. The beam pattern is according to equations 4.14 and 4.27 essentially given by the Fourier transform of the element density. Therefore, the main beam pattern is independent of the beam direction.

The antenna elements need, of course, individual feeding lines. We selected vertical coaxial feeding lines. For the antenna elements we used vertical magnetic dipoles, achieved with a horizontal microstrip loop antenna. These produce a horizontal electrical field vector which is orthogonal to the feeding lines. By this choice the antenna elements are decoupled from the feeding lines. The mechanical fixing of the antenna elements may be made with hard foam, which behaves with a relative dielectric constant of about 1.05, electrically similar to air. The antenna feeding lines may be connected to TRMs to build an active array or they are connected to a divider/combiner network to create a constraint feed array, as in Figure 4.13.

There are some further advantages of this array concept:

- The array is randomly thinned in space, therefore the sidelobes are determined by the element number according to equation 4.30. The pattern is robust with respect

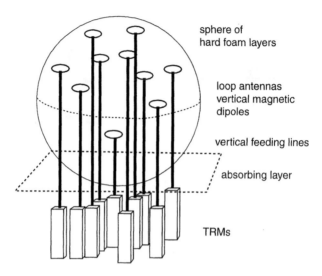

sphere of
hard foam layers

loop antennas
vertical magnetic
dipoles

vertical feeding lines

absorbing layer

TRMs

Figure 4.17   Volume array (crow's nest) with transmit/receive modules

to failure of some elements and the accuracy requirements on the transmit/receive modules are low.

- If the antenna elements are equally distributed within the sphere, the projected element distribution decreases naturally from the centre to the edge of the aperture and creates a bell-shaped tapering for lower sidelobes. The element distribution may be further chosen to achieve a Taylor density distribution in projection [12].
- A monopulse pattern may be produced which is also independent of the steering directions [12].
- A frequency change will cause no direction change in contrast to linear or planar arrays. In contrast to our discussion in section 4.2.2 there results no shift to the boresight direction with increasing frequency because in this case there exists no boresight direction. But there will be a loss in gain dependent on the frequency shift.

There are also some disadvantages:

- there is a certain moderate loss in the coaxial feeding lines
- the polarisation is fixed to horizontal
- the sidelobes cannot be reduced to extremely low levels by amplitude tapering

An experimental volume array antenna system is being developed at FGAN [12] with 512 antenna elements at X band (10 GHz). 128 TRMs have been developed in the laboratory and are connected to the feeding lines. First experiments for pattern measurements have been successful. Investigations with respect to some parameters and losses are given in Reference 13. An industrial prototype demonstrator is described in Reference 14.

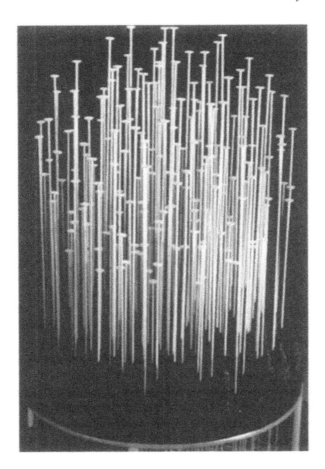

Figure 4.18   Model of volume array crow's nest

Figure 4.18 shows a demonstration model of the volume array crow's nest: all the vertical feeding lines result in only a small attenuation for the horizontal electrical field which is radiated by the circular antennas. In Figure 4.19 a diagram of the radiating antenna element is given. In the centre is the matching device between the coaxial feeding line and the microstrip loop for the circular RF current producing a vertical magnetic dipole. In Figure 4.20 the experimental antenna is shown [12]. The elements are fixed by adhesive styropor granulate (small foam spheres).

4.7   Monitoring of phased-array antennas

Failure of steering elements for the array antenna, such as phase shifters or complete transmit/receive modules, would degrade the performance of the array antenna. Because there are thousands of antenna channels for operational systems, the monitoring of all these steering elements is not trivial. The original test and calibration of

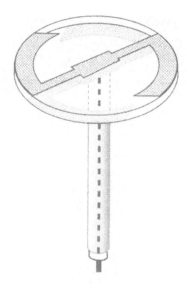

Figure 4.19   Loop antenna for crow's nest

the steering elements and the whole antenna in the factory cannot be repeated with the same methods in the field.

Test monitoring during operation may concentrate on the individual TRMs and on the antenna pattern. Any degradation of the gain, radiated power, beamwidth and sidelobe level must be noted and an adequate repair organised if the antenna's performance decreases below the requirements.

4.7.1   Antenna measurement

For conventional reflector antennas the beam pattern is given by the mechanical properties and dimensions and will be the same for all mechanically-steered directions. For phased arrays the beam is steered into every direction with a new phase shifter setting and the beam shape, especially the sidelobes, may change for different directions. Otherwise the beam shape can be measured only by moving around a probe or source in the farfield. The probe or source has to receive or transmit, with an auxiliary antenna, a test signal for measuring the array's transmit or receive beam, respectively. This pattern measurement must be repeated in principle for all steered beam directions. This procedure is therefore too complicated and expensive for an application in the field.

The other more adequate possibility is to have a probe or source installed in the farfield and to scan the beam across its position. The result is then a scan pattern. This gives equal information about the beamwidth of the main beam and the sidelobe level. It is nevertheless inconvenient, especially for mobile radars, to install an auxiliary probe antenna in the farfield.

Figure 4.20   Experimental crow's nest at X band

Another possibility would be to install a source or probe by a small auxiliary antenna in the nearfield, that is about the aperture dimension in front of the array, at a fixed known position. Then the array may be focused by an additional phase into the nearfield measurement point. This additional phase has to follow approximately a quadratic function across the array. It may be measured once in the factory and stored to include the phase characteristic of the auxiliary antenna. The additional phase would, of course, be applied by the digital phase shifters for beam steering by adding the corresponding values to the usual steering commands if the test procedure has to be performed. The measuring equipment could be combined with the array structure. It is therefore recommended to measure the scan pattern in the nearfield to test the complete array.

4.7.2   Transmit/receive module (TRM) monitoring

The antenna scan-pattern measurement is not sufficient to test the performance of the individual TRMs. There are many ways of measuring the performance of modules or subsystems indirectly: current of driver circuits for phase shifters, power consumption, parity errors in digital control words, state of fuses. By inspection of these items severe faults may be detected quickly, but many types of fault will remain undetected and additional means are necessary.

For an active array antenna we need a test for transmitting and receiving. This may be accomplished by a test signal coupled to or from the individual antenna channel. A feed network may be coupled to the lines between the TRMs' outputs and the antenna elements, as indicated in Figure 4.21. The selection of the TRM to be tested must be made by switching the module on or off.

A feeding network of power dividers and directional couplers is expensive and may itself be a source of failure. Moreover, the antenna element and its connection to the module is excluded from the test.

An alternative solution is the distribution of test signals by radiation from the auxiliary antenna in the nearfield, discussed above for antenna scan-pattern measurement. The receiving path may be tested as follows: an RF test signal is radiated from the auxiliary antenna to all modules. The phase shifter of the module to be tested is switched sequentially to all its states, thus modulating the received test signal while all

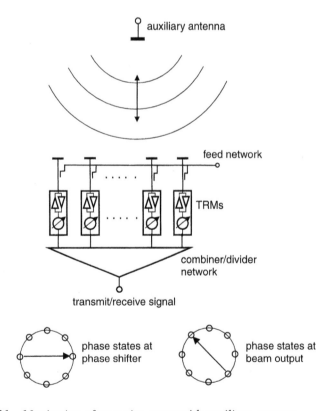

Figure 4.21   Monitoring of an active array with auxiliary antenna

the other phase shifters are kept in their position. The sum beam output is sampled and its modulation is evaluated by a special algorithm to derive the relative gain, the transmission phase, the orthogonality of the I and Q signal component and the accuracy of the phase shifters. A description of the algorithm for obtaining the transmission phase, the incremental phase steps and the channel gain is given in Reference 15. For the transmitting channels the procedure is similar: the module to be tested is phase switched alone and the superposed signal from all channels is measured at the auxiliary antenna and its modulation is evaluated.

One important parameter is the transition phase of the individual modules. This may be simply measured by the following configuration: the phase shifter is first steered for zero phase and then to 180° as shown in Figure 4.21 on the lower left diagram. The corresponding output difference vector then gives the transition phase. These phase values are stored and are used as correction phases for the beam-steering operation. If there is any phase change from temperature effects or change of cable length or exchange of modules these correction phases are easily measured and the proper array focusing will result again. The only precondition for this method is the possibility of steering the phase shifters individually.

The following points have to be taken into account for this test method. The dynamic range of the measuring unit must be sufficiently high to evaluate the modulation of the module under test among all other module contributions. To achieve a higher accuracy, integration of a series of test pulses may be considered. If the output of subarrays is available the ratio of the modulated test signal to the unmodulated signal produced by all other elements is improved. In the case of jamming noise the measuring performance may be maintained by steering a zero of the main beam into the jammer direction. This test procedure could be interleaved within multifunction operation with the other radar functions.

Finally, we show a special main beam with sidelobes, which has developed by chance at FGAN to illustrate the spatial pattern of an array antenna (Figure 4.22).

4.8 Appendix: Taylor and Bayliss weighting

Only a general idea of Taylor and Bayliss tapering can be given here. Details of Taylor weighting can be found in Reference 3, page 128 ff. From equation 4.6 the array factor without tapering is, for a linear array with $d = \lambda/2$, $\tilde{N} = 2N + 1$ and $\tilde{u} = \tilde{N} u/2$:

$$f(\tilde{u}) = \tilde{N}\frac{\sin \pi \tilde{u}}{\pi \tilde{u}} = \tilde{N} \operatorname{sinc} \pi \tilde{u}$$

The pattern envelope is thus given by $1/\pi \tilde{u}$, resulting in high near-in and low far-out sidelobes. Pattern zeros are at integer values of $\tilde{u}$. To reduce the near-in sidelobes the zeros must be shifted together in a manner to achieve $\bar{n}$ near-in sidelobes at the same and reduced level. The desired pattern may be synthesised by:

$$f_T(\tilde{u}) = \operatorname{sinc} \pi \tilde{u} \prod_{n=1}^{\bar{n}-1} \frac{1 - \tilde{u}^2/z_n^2}{1 - \tilde{u}^2/n^2}$$

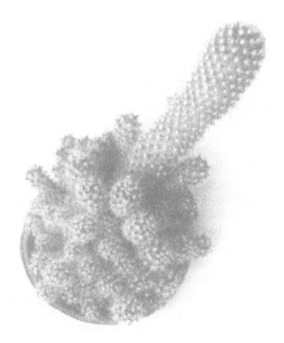

Figure 4.22   'Main beam and sidelobes'

The zeros are positioned with $\sigma = \bar{n}/\sqrt{A^2 + (\bar{n} - 1/2)^2}$ at:

$$z_n = \pm\sigma\sqrt{A^2 + (n - 1/2)^2} \quad \text{for } 1 \le n \le \bar{n}$$

and:

$$z_n = \pm n \quad \text{for } n \ge \bar{n}$$

The sidelobe ratio b determines A:

$$\cosh(\pi A) = b$$

By a Fourier series, according to the Fourier transform relation between the array pattern and tapering function equation 4.27, the required tapering function is then derived from f_T, with x as the normalised coordinate along the array $(-1 \le x \le 1)$:

$$g(x) = 1 + 2 \sum_{m=1}^{\bar{n}-1} F(m, A, \bar{n}) \cos(m\pi x)$$

with:

$$F(m, A, \bar{n}) = \frac{[(\bar{n} - 1)!]^2}{(\bar{n} - 1 + \tilde{N})!(\bar{n} - 1 - \tilde{N})!} \prod_{i=1}^{\bar{n}-1} \left(1 - \frac{m^2}{\tilde{u}_i^2}\right)$$

The Bayliss pattern is synthesised for a linear array in a similar way (see Reference 3, page 136 ff.):

$$f_B(\tilde{u}) = \pi \tilde{u} \cos(\pi \tilde{u}) \frac{\prod_{n=1}^{\bar{n}-1} \left(1 - (\tilde{u}/\sigma z_n)^2\right)}{\prod_{n=1}^{\bar{n}-1} \left(1 - (\tilde{u}/(n+1/2))^2\right)}$$

with:

$$\sigma = \frac{\bar{n} + 1/2}{\left(A^2 + \bar{n}^2\right)^{0.5}}$$

The tapering function is then again derived by a Fourier series.

For planar circular arrays corresponding techniques have been developed (Reference 3, page 157 ff.).

4.9   References

1 SKOLNIK, M. (Ed.): 'Radar handbook' (McGraw-Hill, New York, USA, 1990, 2nd edn) chapter 7
2 'Phased arrays – mutual coupling effects' in GALATI, G. (Ed.): 'Advanced radar techniques and systems' (Peter Peregrinus Ltd., London, UK, 1993) chapter 10
3 MAILLOUX, R.: 'Phased array antennas' (Artech House, Boston-London, 1994)
4 GRÖGER, I.: Internal FGAN-FFM report 392
5 'Active aperture arrays' in GALATI, G. (Ed.): 'Advanced radar techniques and systems' (Peter Peregrinus Ltd., London, UK, 1993) chapter 10
6 MCQUIDDY, D. N., GASSNER, R. L., HULL, P., MASON, J. S., and BEDINGER, J. M.: 'Transmit/receive module technology for X-band active array radar', *Proc. IEEE*, March 1991, **79** (3), pp. 308–341
7 HIGGINS, J. A.: 'GaAs heterojunction bipolar transistors: a second generation microwave power amplifier transistor', *Microw. J.*, May 1991, pp. 176–194
8 WILDEN, H.: 'Microwave tests on prototype T/R modules'. Proceedings of IEE international conference on *Radar*, Edinburgh, UK, October 1997, pp. 517–521
9 SKOLNIK, M. (Ed.): 'Radar handbook' (McGraw-Hill, New York, USA, 1990, 2nd edn) chapter 1
10 ROBINSON, C. A.: 'Missile defence radar system tests set', *Aviation Week & Space Technology*, September 20, 1976
11 SCUDDER, R. M., and SHEPPARD, W. H.: 'AN/SPY-1 phased-array antenna', *Microw. J.*, May 1974, pp. 51–55
12 ENDER, J., and WILDEN, H.: 'The crow's nest antenna – experimental results'. Proceedings of IEEE international conference on *Radar*, Arlington VA, USA, May 1990, pp. 280–285
13 ENDER, J., and WILDEN, H.: 'Die Krähennest-Antenne: Aspekte zu Antennen-verlusten und Monopulsschätzung'. 9. Radarsymposium der DGON, Stuttgart, Germany, April 1997, pp. 155–162

14 MASCHEN, R.: 'Realisierung einer Räumlichen Phased-Array-Antenne (Krähennestantenne)'. ITG-Fachberichte 111, Antennentagung, Wiesbaden, Germany, March 1990, pp. 255–259

15 SANDER, W.: 'Monitoring and calibration of active phased arrays'. Proceedings of IEEE international conference on *Radar*, Arlington, Va, USA, May 1985, pp. 45–51

Chapter 5

Beamforming

Beamforming with all antenna elements of an array for transmitting or receiving means reproducing a desired beam pattern as closely as possible. Generally, a narrow main beam with high gain at the centre in the desired direction combined with low sidelobes for all other directions is required. The fundamental procedure for beamforming is, as described in chapter 4 section 4.2.5, weighting the individual signals of all antenna elements and then summing all these weighted signals.

By phase steering or phase rotation the complex target echoes from all antenna elements are aligned in phase for the desired direction to produce a maximum sum signal, as shown in Figure 5.1. Then, according to example 5 in chapter 3 (detection

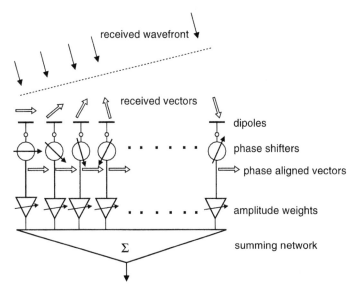

Figure 5.1   Basic principle of beamforming: phase alignment, amplitude weighting (tapering) and summing

of a known signal) and illustrated in Figure 3.10, the maximum signal-to-noise ratio is achieved for target echoes from the desired directions. This procedure therefore optimises target detection out of noise.

The phase rotations are usually provided by microwave phase shifters. In principle, phase shifting may also be achieved by multiplying the received signal with a corresponding complex number. The signal of each antenna element then has first to be converted down to the baseband and analogue-to-digital converted. The signal is then represented by its orthogonal components as given in chapter 2 equation 2.1. For all N antenna elements we get the signals:

$$s_n = s_{x_n} + j s_{y_n} \quad \text{for } n = 1, \ldots, N$$

With back rotation of the phase $n\varphi$ for element n and summing all signals we get for the beamforming output (without amplitude tapering):

$$y = \sum_{n=1}^{N} s_n \exp\left(-j\varphi n\right) = \sum_{n=1}^{N} (s_{x_n} + j s_{y_n})(\cos\left(\varphi n\right) - j \sin\left(\varphi n\right))$$

or:

$$y = \sum_{n=1}^{N} s_{x_n} \cos \varphi n + s_{y_n} \sin \varphi n + j(s_{y_n} \cos \varphi n - s_{x_n} \sin \varphi n) \tag{5.1}$$

For the purpose of direction estimation by monopulse processing with a planar array the difference beams in azimuth and elevation are additionally formed, as mentioned already in chapter 4 section 4.2.5. Direction estimation will be discussed in more detail in chapter 11.

A compromise between high gain and low sidelobes is achieved by application of an amplitude tapering or weighting function. For the sum beam the Taylor function is usually applied and for the difference beams the Bayliss function. A loss in gain and a widening of the beam has to be accepted for the achievement of lower sidelobes as described in chapter 4 section 4.2.5 and Figure 4.7.

Likewise for the transmitting beam, the signals from all antenna elements are aligned with the phases of their field vectors for the desired direction by phase shifters, as described in chapter 4. These field vectors are then superposed in space. A maximum field strength and therefore power density is achieved in the centre of the steered transmit beam to maximise the pulse power radiated to the target and thereby also maximising the echo signal.

In the case of transmission, beamforming is applied at microwave or radio frequency (RF) by the distribution network and phase shifters, as shown in Figures 4.13 and 4.14. Amplitude tapering may be achieved for centrally fed arrays by a specially designed distribution network or for active arrays by variable gain settings of the individual power amplifiers. The technology used in the distribution network has to be matched to the respective power level. Therefore for high-power systems waveguide distribution is necessary.

In the case of space-fed systems, such as that in Figure 4.12, the amplitude tapering is determined by the radiation pattern of the feed horn. It is obviously difficult to create a distinct amplitude tapering function in this case.

For the receiving case there are several possibilities for beamforming that will be discussed in the following.

5.1   Single receiving beam

The usual radar system designs use only one beam at a time to perform the multi-function operation by time multiplex. In the following sections possible techniques applicable to an active receiving array, as illustrated in Figure 4.14, are discussed.

5.1.1   RF beamforming

A constraint network for weighting and combining the received signals may be applied at RF. Because the receiving power level is low, microstrip techniques will be used as a light weight and cost effective solution. The transmit/receive modules (TRMs) directly follow the antenna elements and therefore determine the receiving noise figure. The steered phase shifters within the TRMs determine the beam direction. A certain loss in the following combining network will be tolerable if the gain of the low-noise amplifier (LNA) within the TRMs is high compared to this loss. Complete beamforming is performed in this case at RF without mixing the received signals down to intermediate frequency (IF) or baseband (BB). The amplitude tapering function may be realised by variable gain receiving amplifiers (VGA) within the TRMs. By using the passive beamforming microstrip network for summing all signals from the TRM outputs any problem with dynamic range in connection with the summing operation is avoided. After beamforming the output signals are within only one receiving channel and then converted to IF, amplified and filtered, converted to baseband with I and Q components and finally analogue-to-digital converted (ADC).

5.1.2   Subarrays and partial digital beamforming

An important concept for beamforming which is also useful for special radar tasks by active array processing is based on partitioning the receiving array into a suitable number of subarrays. Within each subarray are summed the outputs of neighbouring antenna elements after phase steering. This concept makes possible the following functions:

- adaptive suppression of multiple jammers to minimise the degradation of the radar coverage (discussed in chapter 10)
- adaptive suppression of main beam jammers (discussed in chapters 10 and 11)
- correction of monopulse direction estimation under main beam jamming (discussed in chapter 11)
- angular superresolution (discussed in chapter 12)
- antenna pattern shaping

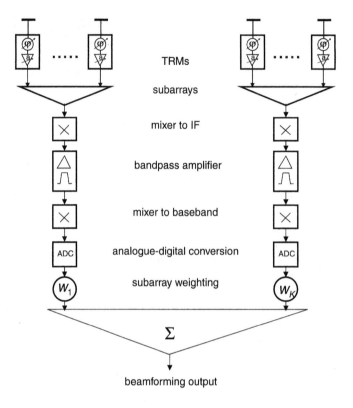

Figure 5.2   Concept for partial digital beamforming

- forming a beam cluster for the acceleration of the target search
- one set of weights for antenna element tapering for forming one sum and two difference beams
- correction of the antenna pattern in case of failure or degradation of antenna parts
- redundancy in case of failure of complete receiving channels
- modular construction

This most promising concept is illustrated in Figure 5.2. Within the transmit/receive modules here are indicated only those elements which are relevant for receiving beamforming: the phase shifter and the variable gain amplifier for amplitude weighting. All phase shifters are steered according to the desired beam direction. The summing of the subarray antenna channels is performed at microwave frequency (RF) by a microstrip combining network. Each subarray output has its own receiving channel with a mixer to intermediate frequency (IF) and a bandpass IF amplifier. This is followed by a mixer to the baseband to obtain the two orthogonal components I and Q. Finally, the signal is converted with both components from analogue to digital format. All further processing then is digital. The signals from the subarrays may be weighted by a set of subarray weight factors $w_1, \ldots, w_K$ and finally summed. If all weight factors are set to one, then this concept corresponds in effect to that of section 5.1.1

for forming the usual sum beam. But special computed weights could also be applied to form both the difference beams and adapted beams against jammers.

The digital outputs of the subarrays could be used in parallel for several sets of weights to form a cluster of beams or they may be used for special further processing such as superresolution.

5.1.3 Dynamic range requirements

For this important subarray concept we consider in the following the approximate dynamic range requirements (ratio of maximum signal without limiting by saturation to RMS noise value) at the different amplifier stages.

The most demanding task of beamforming with respect to dynamic range is the suppression of noise signals received from hostile jammers. Generally, these jamming signals are received by the sidelobes of our antenna. To apply additional jammer suppression methods, discussed in chapter 10, the jamming signals should not be limited by saturation as a kind of severe nonlinear distortion. Therefore the dynamic range of the receiving channels has to be matched to the expected or tractable jammer noise level for suppression by adaptive processing.

The dynamic range of a single receiving channel may be matched to the jammer-to-noise power ratio $d_1 = J_1/N_1$ which we want to address. With K receiving channels the dynamic range at the beamformer output for a main beam jammer is then given by:

$$d_{mb} = \frac{K^2 J_1}{K N_1} = K d_1 \qquad (5.2)$$

because of the coherent integration of the jammer signal from all receiving channels within the main beam.

We assume for our antenna a main beam to sidelobe power ratio a. For a jammer in the sidelobe region the dynamic range at the beam output then is:

$$d_{sl} = \frac{d_{mb}}{a} = \frac{K d_1}{a} \qquad (5.3)$$

If this jamming signal is reduced to the noise level then we have $d_{sl} = 1$. From this follows the required dynamic range at the input of the receiving channels:

$$d_1 = \frac{a}{K} \qquad (5.4)$$

Remark: for a randomly-thinned array we have $a = K$ according to chapter 4 equation 4.44 and it follows that the required $d_1 = 1$.

With additional required jammer attenuation by tapering or pattern nulling as discussed in chapter 10 by a factor b the required dynamic range at the receiving channel input is increased to:

$$d_1 = \frac{ab}{K} \qquad (5.5)$$

Example

number of antenna elements: $K = 2000$
achievable sidelobe ratio by tapering: $a = 45\,$dB
achievable additional sidelobe suppression in jammer directions: $b = 25\,$dB

It follows that:

$$d_1 = 45 + 25 - 33 = 37\,\text{dB}$$

Because peak noise signals should also be processed without limiting by saturation we have to add about 15 dB and we get for the dynamic range requirement for the individual receiving channel:

$$d_1 = 52\,\text{dB}$$

If the subarrays have K_1 receiving elements then the channels at the output of the subarrays must have a dynamic range according to equation 5.2 of:

$$d_{sa} = d_1 K_1 \tag{5.6}$$

because the sidelobe jammer with respect to the final narrow main beam may be within the broader main beam of the subarrays.

For $K_1 = 50$ (corresponding to 17 dB) results for the dynamic range of the receiving channels following the subarrays:

$$d_{sa} = 69\,\text{dB}$$

This corresponds to an analogue-to-digital conversion with 12 bits and sign.

5.1.4   Subarray configuration for digital sum and difference beamforming

Subarray configurations and their properties with respect to different radar functions have been studied extensively by Nickel [1]. The main considerations and results are given in the following.

The subarray configuration has to fulfil a series of different requirements:

- The sum beam should have low sidelobes achieved by amplitude tapering, e.g. with a Taylor function.
- For adaptive jammer suppression the number of subarrays must be large enough to provide the necessary number of signals according to the required degrees of freedom. As a rule of thumb the number of subarrays should be twice the number of jammers to be suppressed.
- For the adapted pattern the designed low sidelobe level by tapering should be preserved as far as possible.
- The phase centres of the subarrays should be randomly distributed on the antenna plane to avoid grating-lobe or grating-notch effects.

- For difference beams in azimuth and elevation formed from the subarray outputs the sidelobe level should be low enough.
- Angular superresolution techniques should be possible.
- The subarray patterns should be as broad as possible in favour of adaptive jammer suppression and the formation of beam clusters.
- The antenna hardware should be as uncomplex as possible.

The first and the last requirements lead to the decision to apply Taylor amplitude tapering at the element level, that is within the TRMs, in favour of low sidelobes for the sum beam. All other beams, especially the difference beams, shall then be formed by a set of weighting coefficients applied at the subarray output level.

There are different ways of forming subarrays. A straightforward example would be a chessboard partition. But by this selection the grating-lobe problem would be created. For circular arrays only the outer subarrays are cut off by the circular boundary and would have unregular phase centres. Also, the centres of gravity and therefore also the phase centres are shifted away from the regular grid by the bell-shaped Taylor amplitude weighting.

To obtain other proposals a quantised version of the amplitude tapering function may be selected. The Taylor weighting function $w(x, y)$ for antenna element positions (x, y) is approximated by a staircase with a set of q elements $\{v_1, v_2, \ldots, v_i, \ldots, v_q\}$. The rotational bell-shaped weighting is transformed into rings $(1, \ldots, i, \ldots, q)$ with constant weights. All antenna elements with positions $(x, y)_i$ belonging to a ring i may be first added and afterwards weighted with v_i. Each ring may be considered as a subarray. With only six rings the sidelobe level of a 40 dB Taylor pattern may be kept below -40 dB. We see by this result that a low sidelobe pattern may be obtained with a relatively small number of subarrays of a suitable form.

This idea of a quantised weighting function may also be applied to form the difference beams. The Bayliss weighting function $d_a(x, y)$ for the difference pattern in the azimuth direction may be achieved by a weighting function at the TRM outputs (with their internal weighting w):

$$r_a(x, y) = \frac{d_a(x, y)}{w(x, y)} \tag{5.7}$$

Now, the function r_a may be quantised as above resulting in a certain set of subarrays, described by their boundaries, suited to form the difference beam in azimuth. The same procedure may be applied for the difference beam in elevation, resulting in another pattern of subarray boundaries. Both sets of boundaries are combined and give the final set of boundaries or the subarray configuration.

These configurations have been studied for different numbers of subarrays with respect to the resulting sidelobe levels. Grating-lobe effects were produced by using regular structures in the central part, resulting on one hand in a relatively high sidelobe. On the other hand these structures would turn into grating-notch effects in connection with jammer suppression: a suppressed jammer in a certain direction would cause a reduced gain or notch in another grating direction, which could be a target direction. This would result in a target loss and has to be avoided.

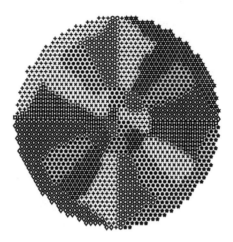

Figure 5.3   Optimised subarray configuration with 32 subarrays (courtesy U. Nickel)

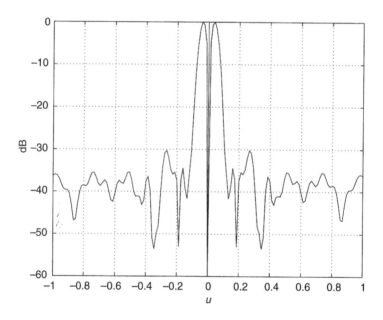

Figure 5.4   Difference pattern with 32 subarrays for configuration in Figure 5.3

By making changes in the subarray boundaries, using trial and error, these effects can be reduced. A final improved proposal for a subarray configuration with 32 subarrays is shown in Figure 5.3. It is based on a 40 dB Taylor and 40 dB Bayliss taper. In Figure 5.4 we see the azimuth cut of the resulting difference pattern. Most sidelobes are below $-35\,\mathrm{dB}$, the two maximum sidelobes are at $-30\,\mathrm{dB}$. All other cuts look similar and have their maximum sidelobes also below $-30\,\mathrm{dB}$.

It is not possible to achieve the same low sidelobe level as for the sum beam by this subarray architecture for the difference beams. Otherwise the target detection is performed with the sum beam. Its sidelobes would cause false alarms by clutter echoes or other targets within the sidelobe area. The difference beam is activated after detection only for target acquisition and tracking for a short range gate. Therefore the probability of false decisions by the sidelobes of the difference beam is low. Jamming signals received by sidelobes are suppressed adaptively by a dedicated procedure, as will be described in chapters 10 and 11.

It remains for us to discuss the computation of the weighting factors after the subarrays for the difference beams. According to Reference 1 the transformation for the Taylor weighting of the antenna element signals and forming the subarray outputs may be described by a transformation matrix T. We assume L subarrays with element numbers K_l ($l = 1, \ldots, L$):

$$T = \begin{pmatrix} \mathbf{w}_1 & 0 & \cdots & 0 \\ 0 & \mathbf{w}_2 & & \vdots \\ \vdots & & \ddots & 0 \\ 0 & \cdots & 0 & \mathbf{w}_L \end{pmatrix} \tag{5.8}$$

The column subvectors $\mathbf{w}_1, \mathbf{w}_2, \ldots, \mathbf{w}_L$ contain the Taylor weights for the antenna elements. A measured signal at the elements, $\mathbf{z}$, results at the subarray output as $\mathbf{T}^*\mathbf{z}$. We are now looking for a vector $\mathbf{r}$ which best approximates, together with the subarray transformation $\mathbf{T}$, the difference weighting $\mathbf{d}$ for azimuth or elevation, respectively. In other words, the weighting vector $\mathbf{r}$ according to equation 5.7 should minimise the quadratic error:

$$Q = \|\mathbf{Tr} - \mathbf{d}\|^2$$

According to chapter 3 section 3.5.1 and equation 3.77 a solution is given by:

$$\tilde{\mathbf{r}} = (\mathbf{T}^*\mathbf{T})^{-1}\mathbf{T}^*\mathbf{d} \tag{5.9}$$

Because we want to form a difference beam this should have a null in the look direction. A signal at the antenna elements from the look direction may be given by the signal vector $\mathbf{a}_0$. We express this additional demand or constraint by:

$$\mathbf{a}_0^*\mathbf{T}\hat{\mathbf{r}} = 0 \tag{5.10}$$

This may be fulfilled by inserting in equation 5.9 a term for an orthogonal projection with respect to the vector $\mathbf{x}^* = \mathbf{a}_0^*\mathbf{T}(\mathbf{T}^*\mathbf{T})^{-1}\mathbf{T}^*$:

$$\hat{\mathbf{r}} = (\mathbf{T}^*\mathbf{T})^{-1}\mathbf{T}^*\left(\mathbf{I} - \frac{\mathbf{x}\mathbf{x}^*}{\mathbf{x}^*\mathbf{x}}\right)\mathbf{d} \tag{5.11}$$

Explanation: orthogonal projection means that a signal with a component $\mathbf{x}^*$ is projected by a projection matrix $\mathbf{P}$ in such a way as to cancel $\mathbf{x}^*$:

$$\mathbf{x}^*\mathbf{P} = \mathbf{x}^*\left(\mathbf{I} - \frac{\mathbf{xx}^*}{\mathbf{x}^*\mathbf{x}}\right) = \mathbf{x}^* - \frac{(\mathbf{x}^*\mathbf{x})\mathbf{x}^*}{\mathbf{x}^*\mathbf{x}} = \mathbf{x}^* - \mathbf{x}^* = 0$$

Then equation 5.11 fulfils, according to equation 5.10, the added requirement. For a suitable array normalisation it turns out that the simpler equation 5.9 is sufficient.

The subarray output signals will be used for adaptation against jammer signals, as described in chapter 10. Using adaptation the algorithms equalise the noise power at the subarray outputs. This is in contrast to bell-shaped Taylor weighting for the antenna elements, with small weighting factors at the edge of the array. This weighting is disturbed by increasing the effective weight of elements at the array edge. This effect may be avoided by first normalising the subarray outputs to an equal noise power, performing the adaptation and then undoing the normalisation by applying the inverse of the normalisation factors [1,8]. In Figure 5.5 the sum beam pattern after adaptation without noise normalisation before adaptation is shown. In Figure 5.6 noise normalisation at the subarray outputs has been performed, followed by rescaling after adaptation.

5.1.5   Correction of antenna failures

If single transmit/receive modules change their amplitude gain or even subarrays show a complete failure there is a possibility of recovering the shape of the antenna pattern,

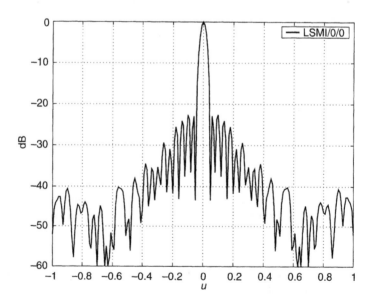

Figure 5.5   Adapted sum beam pattern with 32 subarrays without noise normalisation (courtesy U. Nickel)

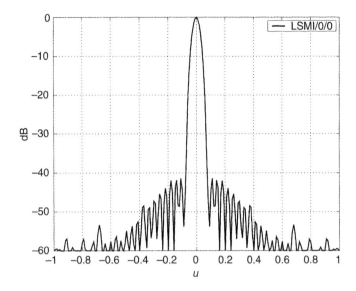

Figure 5.6   Adapted sum beam pattern with noise normalisation (courtesy U. Nickel)

especially the main beam part of the difference pattern in favour of the monopulse direction estimation. The weighting factors at the subarray outputs have to be changed for this correction. This technique is described in detail in chapter 11 section 11.2.

5.1.6   Digital beamforming at element level

As an extreme version complete digital beamforming using all individual array elements would in principle result in maximum flexibility for shaping the beam pattern, both main beam and sidelobes, by weighting all received signals with complex coefficients using digital multipliers. This concept requires a complete receiving and processing chain (mixers to IF and baseband, amplifiers, filters, ADCs for I and Q channel and a digital complex multiplier) for each array element. The statements made on word length in section 5.1.3 also apply. With available technology and for arrays with several thousand elements this concept would be not a cost-effective solution for the foreseeable future.

5.2   Broadband beamforming

For a multifunction radar with an active receiving array, a high signal bandwidth is of interest to achieve high range resolution within a wide observation area. A high range resolution is required especially for classification and imaging of targets, for example for synthetic aperture radars (SAR). SAR will be discussed in chapter 14.

With the usual direction steering by modulo 2π phase shifters the bandwidth is limited according to chapter 4 section 4.2.2 and equation 4.22 by the rule: bandwidth in per cent is equal to beamwidth in degree.

The array pattern for a linear array is given according to chapter 4 by equations 4.8 and 4.20 for phase steering by:

$$f(u/u_0) = \sum_{n=-N}^{N} g_n \exp\left[j\left(\frac{2\pi d}{\lambda}u - \frac{2\pi d}{\lambda_0}u_0\right)n\right]$$ (5.12)

The bandwidth limitation is caused by a beam shift towards the broadside direction if the frequency is increased. The pattern, main and sidelobes, is not changed but only rescaled with respect to u/λ.

By applying a true time-delay steering to each antenna element, bandwidth limitation by direction steering is avoided; the beam direction is then independent of frequency. The array pattern is now given by:

$$f(u/u_0) = \sum_{n=-N}^{N} g_n \exp\left[j\left(\frac{2\pi d}{\lambda}(u - u_0)\right)n\right]$$ (5.13)

Time-delay steering of all array elements is an expensive solution. A compromise solution is time-delay steering after subarrays, because the subarrays themselves have a broader main beam pattern.

We now assume a linear array divided into K equal subarrays. Each subarray has an aperture length reduced by a factor K compared to the complete array. Therefore the main beam of the subarrays is broader by the factor K compared to the main beam formed by the complete array. The bandwidth of the subarrays therefore is, according to the above mentioned rule, also increased by the factor K. By suitably combining the subarrays we can also expect to achieve the increased bandwidth for the complete array. This is accomplished by adding to the subarrays a phase matched to the subarray phase centres. This additional phase has to be produced by a time delay.

The necessary steering phase along the array is given from equation 5.13 and with $\lambda = c/f$ (c = velocity of light) by:

$$\varphi_n = 2\pi \frac{d}{\lambda} u_0 n = 2\pi \frac{df}{c} u_0 n$$ (5.14)

This phase is proportional to frequency f and n. The phase shifters at the antenna elements have to be of the switched delay line type, which is the usual solution. Within the subarrays no 2π steps should appear because this would be transformed at the frequencies at the high and low end of the signal bandwidth (edge frequencies) into corresponding phase differences compared to the 2π steps. These phase differences disturb the intended weighting function and would produce sidelobes increasing towards the edge frequencies. This will be illustrated by the following example.

The received frequency f is assumed to be composed of the carrier at f_0 and a signal at f_s: $f = f_0 + f_s$. The normalised signal frequency is $\alpha = f_s/f_0$. Further, we assume $d = \lambda_0/2$. The correct element steering phase is then:

$$\varphi_n = \pi u_0 n (1 + \alpha) \tag{5.15}$$

The element phase shifter with modulo 2π produces the phase shift:

$$\hat{\varphi}_n = ((\pi u_0 n) \bmod 2\pi)(1 + \alpha) \tag{5.16}$$

An additional phase for subarray i, produced by switched delays, should be matched to its phase centre at element n_i:

$$\varphi_{s_i} = (\pi u_0 n_i - (\pi u_0 n_i \bmod 2\pi))(1 + \alpha) \tag{5.17}$$

The phase error at the single element is then:

$$\Delta \varphi_n = \varphi_n - (\hat{\varphi}_n + \varphi_{s_i}) \tag{5.18}$$

A computed example is given for $N = 32$, $K = 13$, $u_0 = 0.707$, $\alpha = 0.05$ (bandwidth = 10 per cent) in Figure 5.7. It is demonstrated that with the additional subarray delays the direction of the main beam is independent of frequency, but that the sidelobes increase at the edge frequency if the element phase shifters are

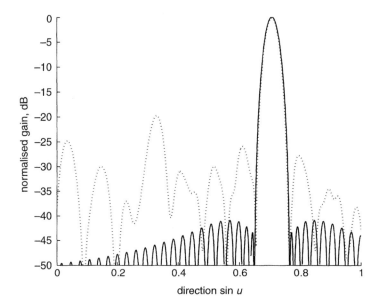

Figure 5.7   *Beamforming with additional delays at subarrays, element phase shifter modulo 2π, $\alpha = 0.05$ (10 per cent bandwidth), pattern at centre frequency (solid line), pattern at edge frequencies (dotted line)*

modulo 2π. The element phase shifters have to be extended by additional 2π or 4π steps (corresponding to one or two wavelength λ), depending on the subarray size.

The additional switched delays after the subarrays are still expensive. For future systems a more convenient solution is offered. The signals at the subarray outputs are in any case sampled and AD converted according to the signal bandwidth B. Therefore it is easy to delay the digitised signals by multiples of the sampling period $\Delta t = 1/B$. The necessary delay follows from equation 5.17:

$$\tau_i = \frac{\varphi_{s_i}}{2\pi f_0}$$

The digital delay is then:

$$\hat{\tau}_i = \Delta t \cdot \text{int} \left(\frac{\tau_i}{\Delta t} \right)$$

with int (x) giving the integer part of x. MATLAB uses the function fix (x) for the integer value of x:

$$\hat{\tau}_i = \Delta t \, \text{fix} \left(\frac{\tau_i}{\Delta t} \right)$$

The additional φ_{s_i} phase is now quantised in multiples of $2\pi f_0 \Delta t = 2\pi (f_0/B)$ into the phase $\hat{\varphi}_{s_i}$. Therefore an additional time-delay phase shift $\tilde{\varphi}_i$ with up to $(f_0/B)2\pi$ in steps of 2π is necessary to close this gap. The final concept for the implementation is given in Figure 5.8.

After each receiving channel R for single antenna elements the time-delay phase shifters are steering the phase within each subarray without any 2π steps. After

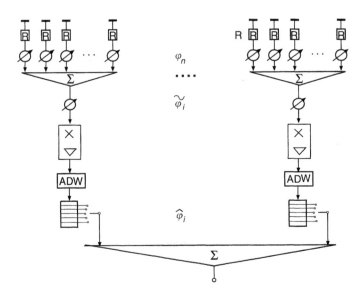

Figure 5.8   Concept for broadband beamforming: subarrays and digital delays

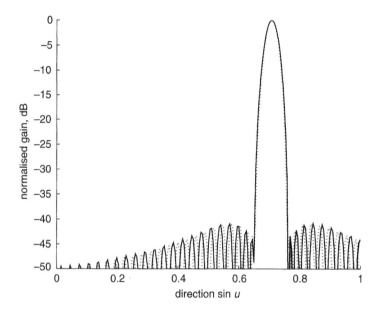

Figure 5.9 *Beamforming with additional delays at subarrays, concept as in*
Figure 5.8, α = 0.05 (10% bandwidth), pattern at centre frequency
(solid line), pattern at edge frequencies (dotted line)

forming the subarrays with an RF combining network, a further phase shifter follows
with some 2π steps to bridge the gap to the digital delays. The bulk time delay then
is achieved using digital delays. The digital delay is by a factor:

$$V = u_{0_{max}} N(B/f_0)$$

higher than the phase shift $\tilde{\varphi}_i$. For example, for $u_{0_{max}} = 0.866 \, (60°)$, $N = 32$,
$B/f_0 = 0.25$ we get $V \approx 7$. With this beamforming concept the antenna pattern
is kept as required over the whole frequency band and a signal spectrum would be
preserved. Figure 5.9 shows as an example the antenna pattern at the centre and at
the edge frequency, now in agreement.

For transmission, the signals fed to the subarrays have to be synthesised with the
necessary relative delays corresponding to the beam direction. Then the signals from
the subarrays will be aligned in time for the correct superposition in space.

5.3 Multiple beams

The demand for multiple receiving beams arises with modern multifunction radar
systems and may result from the following requirements:

- high angular resolution by a cluster of narrow pencil beams to cover the
  observation space illuminated by a broader transmit beam

- high number of pulses N to improve Doppler resolution between different targets and between targets and clutter
- energy management by varying the dwell time by selecting the number of pulses with a mean value $N \gg 1$
- high unambiguous range R and therefore long pulse repetion interval T_P
- the frame time T_f for surveillance is constrained

With M beams and BP beam positions to cover the observation space the frame time T_f is given by:

$$T_f = \frac{BP \cdot N \cdot T_P}{M}$$

All the above listed requirements may be fulfilled by using a cluster of multiple beams which are fixed in their relative angular separation.

Multiple beams steerable independently within a narrow angular sector may be applied for superresolution, especially for implementing the PTMF algorithm, discussed in chapter 12.

Multiple independent beams steerable within the complete field of view may be applied for interference suppression, see chapter 10.

A cluster of multiple receiving beams covering the whole field of view is required for special system concepts:

- multistatic radar
- passive surveillance and reconnaissance
- floodlight transmitting with an omnidirectional antenna and silent operation as discussed with the OLPI system in chapter 18
- interference suppression in beamspace

Multiple beamforming may be implemented at radio frequency (RF), at intermediate frequency (IF) or at baseband (BB).

For transmission only one beam is necessary. Short contiguous pulses may be directed into different beam positions serially, or the transmit beam may be broadened compared to the receive beam. Broadening of the transmit beam may be achieved by:

- Phase steering with a parabolic phase function for defocusing the beam. This results in a loss of gain and is most useful for short and medium-range search tasks.
- Applying a smaller transmit array by switching off a part of the transmit elements. This has a similar effect to defocusing and may be easier to achieve.

We may combine a regularly spaced transmit array with an enlarged and randomly thinned receiving array, which would produce narrower beams. A cluster of receive beams then should be matched to the broader transmit beam to make use of the illuminated angular space.

5.3.1   RF multiple beamforming

Active receiving modules containing low-noise amplifiers (LNAs) and phase shifters may feed several RF combining networks in parallel. These networks produce several mutually squinted beams by introducing a progressive phase shift by a certain added line length from element to element. For example, a phase shift of 2π across the aperture results in a beam shift of about one beamwidth. A beam cluster produced by this scheme is then scanned by the phase shifters within the modules.

Special RF networks (Rotman lens, Blass or Butler matrix) are used for the generation of multiple beams without phase shifters. They are applicable for linear arrays with equidistant antenna elements. For planar arrays with regularly distributed antenna elements two sets of such networks have to be cascaded. These networks produce mutually orthogonal beams which cover the field of view dependent on the antenna element spacing [2].

The Rotman lenses are, in principle, wideband systems since their design is based upon electrical path length and corresponds to time-delay steering of the array. The operating frequency band may cover one octave. For low-power receiving applications, constructions with solid dielectric are applicable. The main disadvantage is the relatively low number of achievable beams and difficulty of achieving low sidelobes and low losses. Descriptions of successful Rotman lenses are given in References 3–5.

The Butler matrix can be treated as a microwave realisation of the fast Fourier transform. It consists of a network of couplers and fixed phase shifters (added line length) with N input ports, connected to the antenna elements, and an equal number of beam outputs. The main advantage of this matrix is the minimum number of couplers. The number of ports is restricted to powers of two. Individual beam shaping is only possible by weighting and combining neighboured output ports. At FGAN-FFM an experimental microstrip Butler matrix has been made [6], serving as a multiple beamformer for a regularly spaced linear array with 64 radiators in the S band. The mean loss for each beam is about 3.2 dB. The relative bandwidth, defined here as the frequency range where the gain reduction of the main beam does not exceed 1 dB, is about 20 per cent.

There are some problems with these RF matrices:

- the sidelobes are relatively high because there is no amplitude tapering
- there is a certain loss of the order of 3–5 dB
- there is no individual beam shaping

By providing an individual low-noise amplifier for each antenna channel before beamforming the signal-to-noise ratio (SNR) could be preserved. These amplifiers must have a stable gain and transmission phase. Within these amplifiers tapering by a steerable gain could be applied to the output of the antenna elements. Additional processing after beamforming (beam space processing) could improve the individual beam pattern.

5.3.2 RF multiple beamforming using subarrays

To form a beam cluster subarrays could be used; their outputs (RF) being fed to a Rotman lens or Butler matrix. A unique beam cluster is formed within a space according to the grid of the phase centres of the subarrays. Periodic repetitions are attenuated by the subarray pattern down to the sidelobe level.

5.3.3   IF multiple beamforming by a resistive network

The signal-to-noise ratio may already be preserved by the receiver channels associated with each antenna element and losses in the beamforming network can therefore
be tolerated. The receiving channels must have stable gain and phase. The main
function of the beamforming network (BFN) is to sum all received signals after complex weighting. For this weighting the signals from each receiving channel must be
split into two components which are in phase and quadrature relative to an arbitrary
phase reference. Both signals are then combined by a resistor network which determines the attenuation and phase of the weighting coefficients. The weights can be
chosen arbitrarily, and as many simultaneous beams as desired may be formed with
no restrictions on the crossover level. N receiving channels and M beams require a
BFN with $2N$ inputs and M outputs. For practical reasons, the number of signals to
be combined in a single summing stage is limited. Thus, if N is high, partial sums
have to be formed which must be combined in a second summing stage. An example
of a working IF beamforming system is the AN/FPS 85 radar at Eglin AFB which
employs 4660 receivers and generates a cluster of nine closely-spaced beams which
can be commonly steered by the phase shifters assigned to the antenna elements [7].

5.3.4   Baseband multiple beamforming

Beamforming with a BFN can also be performed at baseband. Compared to the IF
solution some problems such as mutual interference, coupling and proper shielding
are reduced. On the other hand baseband beamforming is more sensitive to DC offset
effects.

5.3.5   Time-multiplex beamforming for arbitrary directions

The method can be considered as a special case of baseband beamforming and has
been achieved within the ELRA system (chapter 17) [8]. A single beam is generated
by summing the outputs of all the receiver modules at baseband. This beam is then
time multiplexed by fast switching the phase shifters operating at IF. If M directions
are to be observed, the switching rate has to be M times the sampling rate given
by the signal bandwidth. Therefore time multiplexing is only applicable with small
signal bandwidths, since the output bandwidth is M times the signal bandwidth. To
preserve the SNR the multiplexed phase shifters have to follow the bandwidth-defining
filters. For further processing the signal samples from different directions have to be
demultiplexed. It should be mentioned that these multiplexed beams are not parallel
in time, resulting in a partial decorrelation of interference signals at the beam outputs
depending on the interference bandwidth.

5.3.6   Digital multiple beamforming using subarrays

Using the subarray concept of Figure 5.2 several sets of complex weights could be
applied in parallel to the subarray outputs. The subarrays may be seen as elements of

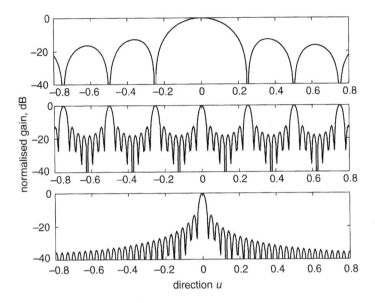

Figure 5.10   *Combined pattern after subarray weighting without direction shift,*
(N = 64, K = 8)
above: pattern of subarrays
middle: pattern of superarray
below: final pattern

a superarray, which are used to generate the final beam by applying the corresponding
phase shifts by the complex weighting factors. The beam pattern of this superarray
multiplies with the pattern of the subarrays. This pattern of the superarray has of
course grating lobes because the distance between the phase centres of the subarrays
is greater than $\lambda/2$, as discussed in chapter 4 section 4.2.3. If all the weights for the
subarray outputs are equal to unity the result is the conventional sum beam. This is
illustrated for a linear array in Figure 5.10. The grating lobes of the superarray are
cancelled by notches in the subarray pattern.

If the weights for the subarray outputs introduce a phase shift then a squinted beam
is produced. Now the grating lobes produced by the superarray are not cancelled and
higher sidelobes result as demonstrated in Figure 5.11. Therefore disadvantages result
for beams formed outside the centre of the subarray beam (direction determined by
phase shifter setting), because grating lobes due to the array factor for the subarrays
are emerging as relatively high sidelobes.

In the case of unequal subarrays their phase centres are distributed randomly. The
grating lobes of the superarray merge into sidelobes randomly distributed in space.
Then this concept shows advantages if multiple beams forming a cluster within the
subarray beamwidth are formed.

In principle, the subarrays could also be formed from overlapping weighting
functions for the antenna elements. Then some of the antenna elements would be

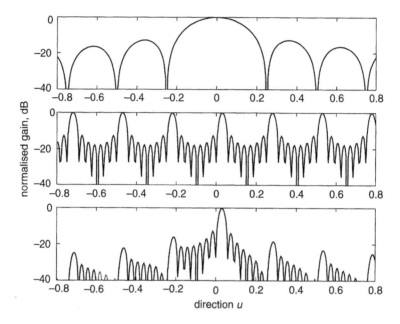

*Figure 5.11   Combined pattern after subarray weighting with a direction shift,
(N = 64, K = 8)*
above: pattern of subarrays
middle: pattern of superarray
below: final pattern

used for neighbouring subarrays. By applying a tapering function such as $(\sin x)/x$
or (squared cos) the subarray main beam could be broader and the sidelobes could
be reduced [8]. Thus the sidelobes of the final beams produced by the grating-lobe
effect of the superarray would also be reduced.

An obvious limitation of using subarrays for multiple beams is that it is only
possible to generate multiple beams within the beamwidth of the subarray beams.
Small subarrays with a broader subarray beam pattern allow the formation of a higher
number of beams.

5.3.7   Multiple beam cluster for target search

The formation of a receiving beam cluster is an interesting application for target
search. This results in an accelerated search for short-range application, applicable
especially for higher elevation angles. This is illustrated by Figure 5.12. For transmitting we assume, as an example, a fan beam and for receiving a cluster of b pencil
beams stacked one above the other. The advantage given by the multiple receiving
beams in comparison to only one fan beam for receiving may be estimated by the
signal-to-noise ratios for the two cases.

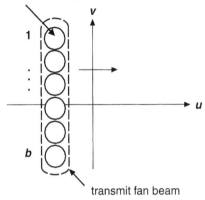

stacked receiving pencil beams

transmit fan beam

Figure 5.12   Search with stacked receiving beams

Fan beam for transmitting and receiving

This corresponds to a conventional search radar, producing the fan beam by a reflector antenna. P_m is the mean transmit power, G the antenna gain, T_d the dwell time and σ the target's radar cross section. In the *const* are summarised all the other parameters for the radar equation (see chapter 4 equation 4.46):

$$\text{SNR} = const\, P_m G^2 T_d \sigma$$

Fan beam for transmitting and a cluster of pencil beams for receiving

Now the gain of the receiving beam is higher by the factor b. The corresponding aperture for a narrow beam in both dimensions is, by a factor b, greater than for producing a fan beam. We simply get:

$$\text{SNR} = const\, P_m G(bG) T_d \sigma$$

In comparison to a conventional fan beam radar we obtain the advantage given by factor b, which may be used for reduced transmit power, radar cross section or dwell time. The price is the parallel processing of b receiving channels.

In the case of a transmit pencil beam we would also gain another factor b for the SNR. But then we would have to spend a search time which is longer by a factor b.

5.4   Deterministic spatial filtering

Using the Taylor tapering function the sidelobes are reduced everywhere outside the main beam, as discussed in chapter 4 section 4.2.5. Sometimes it could be useful to have a low sidelobe region only for a certain angular sector from which permanently disturbing signals from clutter or jammers are expected. Typically, this would be in the horizontal direction. Also for the transmit beam it could be necessary to reduce the sidelobe radiation in horizontal or otherwise specified directions.

We will give a short discussion on how to derive the array weight vector for this application [9]. First we have to assume an interference situation described by an angular interference power density $p(\Theta)$. The complex gain of the ith array element in the direction Θ may be given by $a_i(\Theta)$. For interference suppression we apply from chapter 3, section 3.3.1, equation 3.50. Interference suppression was also further discussed in section 3.5.3. Therefore, we need the covariance matrix $\mathbf{Q}$ describing the correlation between antenna elements i and k. For the element q_{ik} of $\mathbf{Q}$ we may write with Ω for the visible angular space:

$$q_{ik} = \int_{\Omega} p(\Theta) a_i(\Theta) a_k^*(\Theta) \, d\Theta \qquad (5.19)$$

For those regions Ω_1 where we expect interference we assume for example $p = 1$, otherwise we select a small value $p = \varepsilon \ll 1$. For a circular area containing the main beam we must also select $p = \varepsilon$. The components of $\mathbf{a}(\Theta_0)$ are the complex gains for a target in the direction Θ_0. The array weight vector $\mathbf{w}$ for a main beam in the direction Θ_0 then follows with equation 3.50:

$$\mathbf{w} = \mathbf{Q}^{-1} \mathbf{a}(\Theta_0)$$

This solution also corresponds to the optimum solution of adaptive methods (see chapter 10).

If we assume for a general reduction of sidelobes $p(\Theta) = 1$ for all directions in Ω, the corresponding covariance matrix $\mathbf{Q}$ becomes the identity matrix. This assumes a linear antenna at $\lambda/2$ spacing and if the complex gains are described by pure phase shifts corresponding to the time delay between the elements. The optimum weighting is then equal to the weighting for a plane wave phase front and there results no sidelobe reduction.

The sidelobe reduction is achieved by a redistribution of the sidelobes: a reduction within Ω_1 leads to an increase of the sidelobes within the region $\Omega_2 = \Omega - \Omega_1$. The larger the extent of Ω_1 and the required degree $1/\varepsilon$ of sidelobe suppression in this region, the greater will be the increase in sidelobe levels in the region Ω_2.

For this application an accurate measurement of the antenna in the real environment is necessary including mutual coupling effects. The complex gains $a_i(\Theta)$ as a function of the direction have to be known. For a regular array, these measurements can be reduced to one single element, as all elements behave equally in this case. With such measured complex gains we can calculate the element weighting for a deterministic sidelobe reduction in given solid angles. The required accuracy and stability for the transmitting or receiving channels is in accordance with the values of chapter 4, section 4.4. For example, we need a quantisation of four bits for phase and amplitude for a sidelobe reduction below $-50\,\text{dB}$ with a 10 000 element array.

The large number of antenna elements of an operational array corresponds to a large number of degrees of freedom. For cost reasons such a realisation is not acceptable. The expense can be reduced by forming subarrays. The number of degrees of freedom given by the number of subarrays has to be sufficiently larger by a factor of two to four than the number of overlapping main beams covering the sector Ω_1 with reduced sidelobes.

The complex weighting at subarray level $\mathbf{v}$ follows with equation 5.8. The subvectors $\mathbf{w}_l$ describe in this case the complex gains of subarray l, and $\mathbf{s} = (1, 1, 1, \ldots, 1)^*$ is the steering vector at the subarray output:

$$\mathbf{v} = (\mathbf{TQT}^*)^{-1}\mathbf{s} \tag{5.20}$$

The covariance matrix depends on the look direction of the antenna Θ_0. Therefore, each look direction requires a matrix inversion. It could be reasonable to store the weightings for different look directions and different shapes of angular regions with reduced sidelobes (e.g. horizon, part of horizon, sectors).

The most probable region with interference is the horizon or part of it, because jammers are mostly ground based or at long ranges. For a stationary radar in a fixed environment the storage of the different weightings should be no problem. The complex gains $a_i(\Theta)$ can be measured once. The situation is different for mobile radar systems. The nearfield environment is changing and thus, within certain limits, also the transmission factors. For each mechanical position of the antenna we need a set of weighting vectors for all look directions for a given region of required low sidelobes.

Deterministic sidelobe reduction is susceptible to all the errors of the excitation and the construction of the antenna including nearfield effects. For successful reduction of the sidelobes field measurements of the antenna are necessary. Long-term stability of these characteristic values of the array, the array manifold, is also necessary.

Spatial filtering has also been applied to improve low-level tracking [10]. The outputs of a vertical linear array are processed with the objective to reduce the antenna gain in the direction of the image of the target reflected from sea surface, relative to the gain in the direction of the target itself. Let the vertical array have N elements, which could be obtained by dividing a planar phased array into N horizontal rows which are stacked vertically. The required spatial filtering is obtained by in effect forming another array of m elements ($m \leq N$) in which each element is a linear combination of the outputs of the elements in the first array. These m spatial filters have a high attenuation below the horizon. Experimental trials have demonstrated that this technique can significantly reduce elevation errors in low-level tracking over the sea. A more effective solution is superresolution discussed in chapter 12.

5.5 References

1 NICKEL, U.: 'Subarray configurations for digital beamforming with low sidelobes and adaptive interference suppression'. Proceedings of IEEE international conference on *Radar*, Alexandria, USA, 1995, pp. 714–719

2 'Phased array antennas' in SKOLNIK, M. I. (Ed.): 'Radar handbook' (McGraw-Hill, New York, 1990, 2nd edn) chapter 7

3 NIAZI, A. Y., SMITH, M. S., and DAVIES, D. E. N.: 'Microstrip and triplate rotman lenses'. Proceedings of conference on *Military microwaves*, London, UK, October 1980, pp. 3–12

4 MAYBELL, M. J.: 'Printed Rotman lens-fed array having wide bandwidth, low sidelobes, constant beamwidth and synthesised radiation pattern'. IEEE international symposium Digest on *Antennas and propagation*, 1983, pp. 373–376

5 SOLE, G. C., and SMITH, M. S.: '3-D multiple beamforming lenses for planar arrays'. Proceedings of 14th European conference on *Microwaves*, Liège, Belgium, September 1984, pp. 686–690

6 WIRTH, W. D.: 'Omnidirectional low probability of intercept radar'. International conference on *Radar*, Paris, France, April 1984, pp. 27–30

7 REED, J. E.: 'The AN/FPS-85 Radar System', *Proc. IEEE*, March 1969, **57** (3), pp. 324–335

8 SANDER, W.: 'Beam forming with phased array antennas'. Proceedings of IEE international conference on *Radar*, London, UK, 1982, pp. 403–407

9 GRÖGER, I.: 'Antenna pattern shaping'. International conference on *Radar*, Paris, France, 1989, pp. 283–287

10 PEARSON, A., BARTON, P., WADDOUP, W. D., and SHERWELL, R. J.: 'An X-band array signal processing radar for tracking targets at low elevation angles'. Proceedings of IEE international conference on *Radar*, London, UK, 1982, pp. 439–443

Chapter 6

Sampling and digitisation of signals

At the output of beamforming and the receiver channel, with a bandpass filter at the intermediate frequency IF, the received signals are available for further processing. They are at this stage continuous analogue signals. For digital processing we need sampled values of the orthogonal components I and Q as described in chapter 2 with equation 2.1. In this chapter we will discuss the necessary sampling rate and methods to derive the orthogonal components I and Q.

For the interested readers and for completeness we start here with an explanation of the term analytical signal. This material is discussed in more detail within several text books, e.g. Reference 1.

6.1 Analytical signal

A real signal at the frequency ω_0 may be represented by:

$$f(t) = a(t) \cos(\omega_0 t + \alpha) \tag{6.1}$$

In the case of radar systems we have a narrow bandwidth. That means $a(t)$ is slowly varying compared to the carrier frequency ω_0.

To this real signal corresponds a spectrum $F(j\omega)$ given by the Fourier transform:

$$f(t) = \frac{1}{2\pi} \int_{-\infty}^{\infty} F(j\omega) \exp(j\omega t) \, d\omega \tag{6.2}$$

The integral may be partitioned with respect to the integration limits into the form:

$$f(t) = \frac{1}{2\pi} \left[\int_{0}^{\infty} d\omega + \int_{-\infty}^{0} d\omega \right]$$

and by using $F^*(j\omega) = F(-j\omega)$ to achieve a real signal $f(t)$ with equation 6.1 we get:

$$f(t) = \frac{1}{2\pi} \int_0^\infty [F(j\omega)\exp(j\omega t) + F^*(j\omega)\exp(-j\omega t)]\,d\omega$$

$$= \mathrm{Re}\,\frac{1}{\pi} \int_0^\infty F(j\omega)\exp(j\omega t)\,d\omega \qquad (6.3)$$

This may be written:

$$f(t) = \mathrm{Re}\, f_+(t)$$

The complex signal $f_+(t)$ is named analytical. It is composed of:

$$f_+(t) = f(t) + jg(t) \qquad (6.4)$$

We have to add to the original real signal $f(t)$ an orthogonal or imaginary part $g(t)$. For the signal $f_+(t)$ there is a corresponding spectrum in the range $\omega = -\infty \cdots +\infty$:

$$F_+(j\omega) = F(j\omega) + jG(j\omega) \qquad (6.5)$$

We now want to determine $G(\omega)$.

From equation 6.3 follows, because $f_+(t)$ is defined only by positive frequencies:

$$F_+(j\omega) = 0 \quad \text{for } \omega < 0 \qquad (6.6)$$

Because $f(t)$ and $g(t)$ are real we have $F^*(j\omega) = F(-j\omega)$ and $G^*(j\omega) = G(-j\omega)$, and it follows:

$$F^*(j\omega) + jG^*(j\omega) = 0 \quad \text{for } \omega > 0 \qquad (6.7)$$

The complex conjugate form of equation 6.7 results in:

$$F(j\omega) - jG(j\omega) = 0 \quad \text{for } \omega > 0 \qquad (6.8)$$

The result in equation 6.8 is used in equation 6.5:

$$F_+(j\omega) = 2F(j\omega) \quad \text{for } \omega > 0 \qquad (6.9)$$

We already had above:

$$F_+(j\omega) = 0 \quad \text{for } \omega < 0 \qquad (6.10)$$

From equation 6.8 follows directly, after multiplication with j, our result:

$$G(j\omega) = -jF(j\omega) \quad \text{for } \omega > 0 \qquad (6.11)$$

and with equations 6.5 and 6.10:

$$G(j\omega) = jF(j\omega) \quad \text{for } \omega < 0 \qquad (6.12)$$

With these equations the imaginary part of the spectrum $F_+(j\omega)$ is determined. Both equations may be combined with the signum function sign ω, which is equal to 1 for $\omega > 0$ and equal to -1 for $\omega < 0$:

$$G(j\omega) = F(j\omega)(-j \, sign \, \omega) \quad \text{for all } \omega \qquad (6.13)$$

This relation may be transformed into the time domain. According to chapter 2 section 2.4 equations 2.23 and 2.25 the product of the frequency spectra of two functions corresponds to the convolution in the time domain. The time function $s(t)$ for the second term in equation 6.13 may be determined by a Fourier transform. For the sign function we use as an intermediate help function [1]:

$$h(\omega) = \text{sgn}(\omega)\exp[-\alpha \, sign(\omega)\omega] \quad \text{with } \alpha > 0$$

Then we get according to equation 6.13:

$$
\begin{aligned}
s(t) &= \frac{1}{2\pi} \int\limits_{-\infty}^{\infty} (-jh(\omega))\exp(j\omega t)\,d\omega \\[2mm]
&= \frac{j}{2\pi}\left\{ \frac{\exp(\alpha + jt)\omega|_{-\infty}^{0}}{\alpha + jt} - \frac{\exp(-\alpha + jt)\omega|_{0}^{\infty}}{-\alpha + jt} \right\} \\[2mm]
&= \frac{j}{2\pi}\left\{ \frac{1 - (-1)}{\alpha + jt} \right\}
\end{aligned}
$$

For $\alpha \to 0$ follows:

$$s(t) = \frac{1}{\pi t} \qquad (6.14)$$

The time function of the imaginary part $g(t)$ results from the convolution of $f(t)$ and $s(t)$:

$$g(t) = \frac{1}{\pi} \int\limits_{-\infty}^{\infty} \frac{f(\tau)}{t - \tau}\,d\tau \qquad (6.15)$$

This equation is well known and called the Hilbert transform. For example, the Hilbert transform of $\cos(\omega t)$ is $\sin(\omega t)$.

Equation 6.13 gives us $G(j\omega)$ for $\omega = -\infty \cdots \infty$ in the frequency domain and it follows also:

$$g(t) = \frac{1}{2\pi} \int\limits_{-\infty}^{\infty} G(j\omega)\exp(j\omega t)\,d\omega \qquad (6.16)$$

Summarising we have the following result. To a signal $f(t)$, given by equation 6.1, corresponds a spectrum $F(j\omega)$. We want to know the corresponding analytical signal $g(t)$ as the imaginary addition to $f(t)$ to give the signal $f_+(t)$ according to equation 6.4, which has a spectrum only for $\omega > 0$. The real part of $f_+(t)$ is the original signal.

6.2   Sampling and interpolation

A signal $f(t)$ may have a spectrum $F(j\omega)$ with $F = 0$ for $|\omega| > \Omega/2$ according to Figure 6.1 [1,2]. It may appear at the output of a low-pass filter.

$f(t)$ is sampled at a period T resulting in a series of values f_n. For a time segment with $n = -N \cdots N$ the Fourier transform may be formed under the assumption of a periodic repetition of the signal series given by the chosen segment $-N \cdots N$:

$$F_1(j\omega) = \sum_{-N}^{N} f_n \exp(-j\omega nT) \tag{6.17}$$

The spectrum F_1 after sampling is repeating periodically with $\omega = 2\pi/T$ as sketched in Figure 6.2.

To avoid overlapping of spectral parts (aliasing) it is necessary to choose:

$$\frac{2\pi}{T} \geq \Omega \tag{6.18}$$

For $N \to \infty$ the spectrum F_1 converges for $|\omega| < \pi/T$ to $F(j\omega)$, the spectrum of the signal $f(t)$ without limits in time and periodic repetition. Aliasing, overlapping of spectral parts, would make it impossible to separate the original spectrum to achieve an unambiguous reconstruction.

To recover the original spectrum F from F_1 back again one must filter out using a low-pass filter only the part of F_1 in the frequency range $|\omega| < \pi/T$. This may be written with the function rect (x) defined by: rect $(x) = 1$ for $|x| \leq 1$ and rect $(x) = 0$ otherwise:

$$F(j\omega) = T \text{ rect} \left[\frac{\omega T}{\pi} \right] F_1(j\omega) \tag{6.19}$$

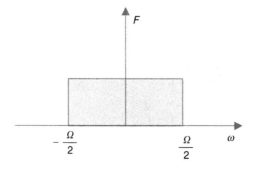

Figure 6.1   Signal spectrum after low pass

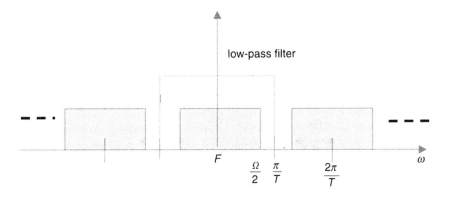

Figure 6.2   Spectrum after sampling with period T

We transform equation 6.19 into time domain:
  The Fourier transform of the first factor of equation 6.19 is given by

$$h(t) = \frac{1}{2\pi} \int\limits_{-\pi/T}^{\pi/T} T \exp(j\omega t)d\omega = \frac{T}{2\pi} \left. \frac{\exp(j\omega t)}{jt} \right|_{-\pi/T}^{\pi/T} = \frac{\sin\left(\dfrac{\pi t}{T}\right)}{\left(\dfrac{\pi t}{T}\right)} = \mathrm{si}\left(\frac{\pi t}{T}\right)$$

So we have to perform a convolution of the series f_n with the filter function h given by the si-function (using the abbreviation $\mathrm{si}(x) = \sin x/x$):

$$f(t) = \sum_{n=-\infty}^{\infty} f_n \, \mathrm{si}\left(\pi \frac{t - nT}{T}\right) \tag{6.20}$$

The si function has zeros for all points $t = mT$ with $m \neq n$. Therefore, the sampled signal values f_n are not changed by the interpolation. For all other values of t the original signal is reconstructed and we may have the continuous signal back again. But we have to consider according to equation 6.20 the infinite series for this reconstruction. Any truncation of this series for practical reasons will limit the accuracy of reconstruction.
  From equation 6.18 the Shannon sampling theorem for the sampling rate follows, as is well known.

Frequency for sampling $\geq$ bandwidth of the complex spectrum   (6.21)

Also, for signals at the intermediate frequency after a bandpass filter, as shown in Figure 6.3, the necessary sampling rate may be determined correspondingly. The interpolation for reconstruction is more complicated [1]. Then, for the necessary sampling rate, the same rule follows as above. We must also take into account the band for negative frequencies, so that the sampling rate is $2B$. If we sample the two orthogonal components I and Q the sampling frequency has to be B for each component.

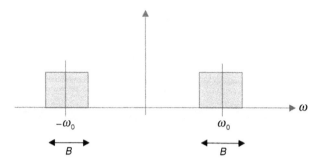

Figure 6.3   Spectrum of bandpass IF signal

Remark: an IF signal with a spectrum as shown in Figure 6.3 must be sampled according to the above rule at a rate $2B$. If this signal is mixed down to baseband the frequency extension is only $-B/2 \cdots +B/2$. So according to the above rule at a first glance the sampling rate must be B. But then we would be losing the assignment of positive and negative frequencies with respect to the IF carrier. By mixing down with two orthogonal references we get a complex signal with two components which both have to be sampled at a rate B.

6.3   Extraction of the components *I* and *Q* in digital format

For signal processing the samples of the orthogonal components are needed in digital form. They are formed usually corresponding to the block diagram shown in Figure 6.4. The IF signal is converted with the two orthogonal reference signals $\cos(\omega_0 t)$ and $\sin(\omega_0 t)$ down to the baseband (or an IF equal to zero) to result in the I and Q component. The signals' original spectrum with bandwidth B around f_0 and $-f_0$ is converted to the frequency range $f = -B/2 \cdots + B/2$. The unwanted products at the sum frequency $2\omega_0$ are suppressed by the following low-pass filter. Then follows sampling and finally analogue-to-digital conversion. With this usual concept arise several disadvantages, especially if there are high accuracy requirements for special signal processing methods, for example superresolution, adaptive jammer suppression or clutter suppression:

- two analogue-to-digital-converters are necessary
- it is difficult to meet the accurate orthogonality requirement
- the amplitude matching of the I and Q channel may be poor
- there may be DC offsets which cannot be compensated for

Nowadays ADCs are available for higher conversion rates and it is possible to sample and convert the IF signal directly and then to derive the I and Q components by digital processing. By this concept the above mentioned disadvantages may be avoided.

A band-limited signal may be represented by:

$$s(t) = a(t) \cos(\omega_0 t + \phi(t)) = I(t) \cos(\omega_0 t) - Q(t) \sin(\omega_0 t) \tag{6.22}$$

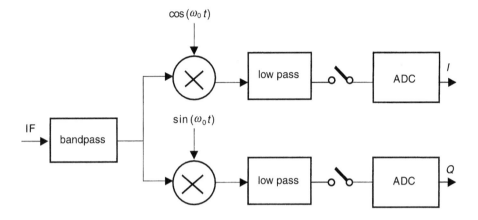

Figure 6.4   Conventional derivation of I and Q signals

In equation 6.22 $a(t)$ is the time-varying amplitude, ω_0 is the carrier frequency $2\pi f_0$, $\phi(t)$ the signal's phase and I, Q the resulting orthogonal components:

$$I(t) = a(t)\cos\phi(t)$$
$$Q(t) = a(t)\sin\phi(t)$$

(6.23)

One needs for further processing samples of I and Q at the same instants of time and at a rate which is equal to or greater than the signal bandwidth B.

First the necessary rate for sampling the IF signal has to be determined [3]. The signal spectrum $S(f)$ is sketched in Figure 6.5a. Its extension is between the low and high-band stop frequency f_l and f_h, respectively. Of course, the difference $f_h - f_l$ is equal to the bandwidth B. The real signal also needs a mirror image at the corresponding negative frequencies. In the first processing step the signal is sampled with frequency f_s. For f_s a value lower than the IF frequency f_0 may be assumed, but it has to meet the bandwidth requirement of equation 6.21. The sampling means multiplying the time signal with the series of sampling pulses or convolution with the corresponding frequency spectra. The sampling pulse series has a spectrum $A(f)$ which consists of periodically repeated lines at frequencies with an interval equal to the sampling frequency f_s, that is at frequencies mf_s (with m any integer).

We may imagine shifting $A(f)$ along $S(f)$ and multiplying both to give the convolution result. The sampling frequency must be chosen to avoid any overlapping of convolution results from $S(f)$. For this operation we also have to consider spectral parts at negative frequencies. We imagine, starting from Figure 6.5a/b, a shift of the A lines to the right. The line at mf_s meets after a shift Δ the low-band stop frequency f_l and then starts imaging the signal spectrum. The result is shown in Figure 6.5c. During this shift movement no other line of $A(f)$ should meet $S(f)$. After that the left most line of A at a negative frequency starts to image the left part of S at a frequency $0.5 f_s + \Delta$. From this consideration follows for the sampling frequency f_s, fulfilling

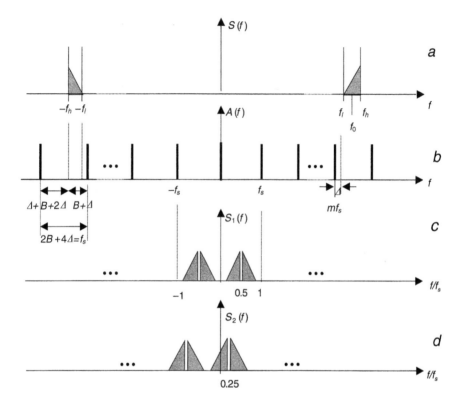

Figure 6.5   Signal spectra before and after sampling: (a) signal spectrum at IF frequency; (b) spectrum of sampling pulses; (c) spectrum of sampled signal; (d) signal spectrum around baseband

also the condition of equation 6.21:

$$f_s = 2B + 4\Delta \tag{6.24}$$

On the other hand, we have according to Figure 6.5*a*/*b*:

$$f_s = \frac{f_l - \Delta}{m} \tag{6.25}$$

From equations 6.24 and 6.25 follows:

$$\Delta = \frac{f_l - 2Bm}{4m + 1} \tag{6.26}$$

$$f_s = \frac{4f_l + 2B}{4m + 1} \tag{6.27}$$

The integer *m* has to meet the condition:

$$m \leq f_l / 2B$$

because Δ from equation 6.26 has to be positive. The sampled signal then has the spectrum S_1, presented in Figure 6.5c.

For the ratio f_0/f_s then follows from Figure 6.5a/b and equation 6.24:

$$\frac{f_0}{f_s} = \frac{mf_s + \Delta + B/2}{f_s} = \frac{mf_s + f_s/4 - B/2 + B/2}{f_s}$$

$$= m + \tfrac{1}{4} \quad \text{with } m \le f_l/2B \qquad (6.28)$$

Using equation 6.28 we get for a given IF frequency f_0 the possible values for f_s with an integer m.

In the following step the sampled signal series according to equation 6.22 will be considered. The instants of time for sampling are $t = nT_s = n/f_s$. From equations 6.22 and 6.28 follows with integer m:

$$s(n) = s(nT_s) = I(n) \cos\left(2\pi\left(mf_s + \frac{f_s}{4}\right)nT_s\right)$$

$$- Q(n) \sin\left(2\pi\left(mf_s + \frac{f_s}{4}\right)nT_s\right)$$

$$= I(n) \cos\left(2\pi mn + \frac{\pi}{2}n\right) - Q(n) \sin\left(2\pi mn + \frac{\pi}{2}n\right)$$

and it follows because m and n are integers:

$$s(n) = I(n) \cos\left(n\frac{\pi}{2}\right) - Q(n) \sin\left(n\frac{\pi}{2}\right) \qquad (6.29)$$

From this series we have to derive I and Q. Because $n = 0, 1, 2, 3, \ldots$ we have at once:

$$s(n) = (-1)^{n/2} I(n) \qquad \text{for } n \text{ even}$$

$$s(n) = (-1)^{(n+1)/2} Q(n) \quad \text{for } n \text{ odd}$$

We get the desired component I and Q for alternating instants of time n with an alternating sign. There is necessarily additional processing to derive both components for equal instants of time. According to Reference 3 $s(n)$ is multiplied with the function $-2 \sin(n\pi/2)$. This means a frequency shift by $f_s/4$. The resultant signal $s_2(n)$ is:

$$s_2(n) = -2 \sin\left(n\frac{\pi}{2}\right)\left[I_n \cos\left(n\frac{\pi}{2}\right) - Q_n \sin\left(n\frac{\pi}{2}\right)\right]$$

$$= -(I_n \sin(\pi n) + Q_n \cos(\pi n)) + Q_n$$

The resulting spectrum is sketched in Figure 6.5d. The first term is positioned around $0.5 f_s$, the desired second around zero. Q_n may be separated from the first term by a low-pass filter with a stop band at $0.25 f_s$.

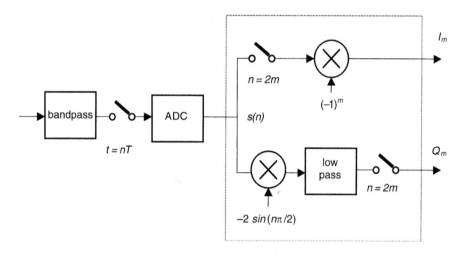

Figure 6.6   Processor structure for the derivation of I and Q

The bandwidth of the real signals I and Q is $B/2$, because the spectrum is between $-B/2$ and $+B/2$. To distinguish between positive and negative frequencies of the signal the two components are necessary, as discussed above. We can skip each second sample; this is called decimation and may be achieved using a switch. The whole configuration is given in Figure 6.6. The low-pass filter also serves to provide the interpolation for obtaining Q for the same instants of time as for I.

The digital mixing, low-pass filtering and decimation may be performed with an integrated processor, for example HSP 43216 of Harris Semiconductors [7]. This has a low-pass filter in the time domain (half-band filter) with 67 taps. The stop band attenuation is 91 dB and the maximum sampling frequency is 52 MHz.

Example for the choice of frequencies:

$f_0 = 10\,\text{MHz}$
$B = 1.6\,\text{MHz}$
$f_l = 9.2\,\text{MHz}$
$f_h = 10.8\,\text{MHz}$

For m follows first: $m \le 9.2/3.2 = 2.8$.

For $m = 2$ results, using equation 6.28, a sampling frequency $f_s = 4.44\,\text{MHz}$. Then the safety distance is $\Delta = 0.311\,\text{MHz}$.

For low values of f_0 we may also apply a sampling with $f_s > f_0$. The considerations for this case corresponding to the above discussion are presented in Figure 6.7. The spectrum after sampling repeats periodically with f_s. For the necessary f_s we derive:

$$f_s = \Delta + B + 2\Delta + B + \Delta = 4\Delta + 2B$$

By this sampling frequency the signal spectrum for the ADC and the derivation of the I and Q components is the same as in Figure 6.5.

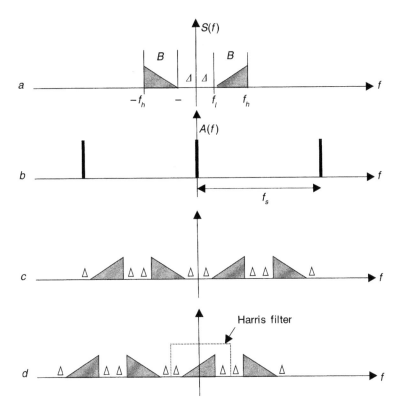

Figure 6.7   *Spectra of the signal for sampling with $f_s > f_0$: (a) signal spectrum at IF frequency; (b) spectrum of sampling pulses; (c) spectrum of sampled signal; (d) signal spectrum around baseband*

For the above example but an IF frequency $f_0 = 2\,\text{MHz}$ follows again a sampling frequency $f_s = 4.44\,\text{MHz}$.

6.4   Third-order intercept point (TOI) and dynamic range

In chapter 5 section 5.1.3 we have already given an estimation for the required dynamic range (defined as the ratio of the maximum signal without clipping or limitation and the RMS noise voltage) of the receiver channel for the purpose of jammer cancellation. For other demanding signal processing methods high linearity standards are also required. Nonlinearities may be described by an intermodulation causing spurious signals at harmonic frequencies. Because the term third-order intercept (TOI) is now commonly used to define the dynamic range it shall be discussed in the following.

We assume the output signal y for an input signal x of our amplifier. Then we have a characteristic in the general form:

$$y = c_0 + c_1 x + c_2 x^2 + c_3 x^3 + \cdots \tag{6.30}$$

The nonlinearities are determined by the coefficients c_2 and c_3.

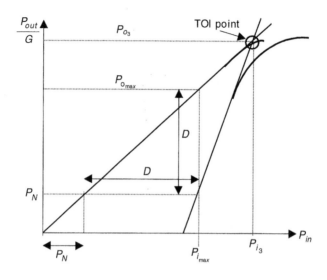

Figure 6.8   Amplifier characteristic: input and output powers in dB

In a band-pass amplifier, for example the IF amplifier, there may be two signals with similar frequencies f_1 and f_2. Nonlinearities produce spurious signals at frequencies $2f_1 - f_2$ and $2f_2 - f_1$ which are then also within the pass band of the IF amplifier.

The input x may be composed out of two signals:

$$x = x_1 + x_2 = a_1 \cos(\omega_1 t) + a_2 \cos(\omega_2 t) \tag{6.31}$$

The quadratic term in equation 6.30 results, for a signal of the form of equation 6.31, in signals with frequencies $f_1 + f_2$ and $f_1 - f_2$ which are outside the pass band and may be neglected. Therefore the term with $c_3 x^3$ results in the important contribution. The name third-order intercept results from this fact. By this term we have for two signals:

$$c_3 x^3 = c_3(x_1 + x_2)^3 = c_3\left(x_1^3 + 3x_1^2 x_2 + 3x_1 x_2^2 + x_2^3\right) \tag{6.32}$$

Owing to the effect of the band pass only those intermodulation products remain which result from both centre terms. With equation 6.31 and basic trigonometric addition theorems we get the following terms:

$$\begin{aligned}
&\tfrac{3}{4}c_3 a_1 a_2^2[\cos((2\omega_2 + \omega_1)t) + \cos((2\omega_2 - \omega_1)t)] \\
&\tfrac{3}{4}c_3 a_1^2 a_2[\cos((2\omega_1 + \omega_2)t) + \cos((2\omega_1 - \omega_2)t)]
\end{aligned} \tag{6.33}$$

These terms increase for $a_1 = a_2$ with the third power of the signal amplitude or in the logarithmic dB scale with a threefold gradient compared to a linear function. The ratio of the powers shall be discussed with Figure 6.8 according to Reference 4.

The power output to input characteristic of the amplifiers has a gradient of 1 because of the normalisation on the gain G. The asymptote for the third harmonic has a triple gradient. It crosses the amplifier characteristic at the TOI point (small circle). At this point the intermodulation signal would have the same power as that of the signal. The horizontal line at the level P_N indicates the noise power level at the input and output. If we assume the intermodulation power to be equal to the noise power then we have with D the available dynamic range. The further the TOI point lies to the right, the higher is the resulting dynamic. Therefore, a high value for P_{i_3} is desired. On the other hand, the required DC power increases with P_{i_3}.

The relevant parameters of an amplifier are:

- gain G (dB)
- bandwidth B (Hz)
- noise figure NF (dB)
- loss L (dB)
- dynamic range D (dB)

There result the following relations:

The noise power is given in dB with $kT = 4.4\ 10^{-21}$ for $45°C$ by:

$$P_N = 10 \log_{10}(kTB) + NF + L \tag{6.34}$$

The maximum input power (in dB) with acceptable distortion is:

$$P_{i_{\max}} = P_N + D \tag{6.35}$$

The accompanying output power is $P_{o_{\max}} = P_{i_{\max}} + G$. With the little triangle at $P_{i_{\max}}$ we find the relation:

$$P_{i_{\max}} = \frac{2}{3} P_{i_3} + \frac{P_N}{3} \tag{6.36}$$

Combining equations 6.35 and 6.36 gives finally the desired relationship between the TOI parameter and dynamic range:

$$D = \tfrac{2}{3}(P_{i_3} - P_N) \tag{6.37}$$

6.5   References

1 FETTWEIS, A.: 'Elemente nachrichtentechnischer Systeme' (Teubner, Stuttgart, 1996)
2 'Basic concepts on detection, estimation and optimum filtering' in GALATI, G. (Ed.): 'Advanced radar techniques and systems' (Peter Peregrinus Ltd., London, UK, 1993) chapter 1
3 HO, K. C., CHAN, Y. T., and INKOL, R.: 'A digital qudrature demodulator system', *IEEE Trans. Aerosp. Electron. Syst.*, 1996, **32** (4), pp. 1218–1227
4 COLLIER, D.: 'Avoid overkill when specifying LNAs for phased arrays', *Microw. RF*, September 1992, pp. 76–85

Chapter 7

Pulse compression with polyphase codes

7.1 Introduction

Future electronic steerable array radars will offer multifunction operation capabilities. They will be achieved with active array antennas applying individual transmit/receive modules for each antenna element. The transmit power amplifiers will be made with solid-state devices. The future cost-effective solution is the application of monolithic microwave integrated circuits (MMIC). The advantages of distributed transmit power generation are:

- highest efficiency for generation of RF power from DC
- high radiated power by superposition of the wave field from all elemental contributions in space
- extreme long systems' MTBF by very long individual MTBF for solid-state amplifiers combined with graceful degradation of the total array by high numbers of parallel channels
- avoidance of high-voltage power supplies
- variable pulse length for different radar tasks

The peak power of transistor amplifiers is at maximum twice the mean power. In the case of GaAs FET amplifiers the possible mean power can even equal the peak power. Therefore, to utilise the available mean power the transmit pulses have to be as long as possible, especially for long-range operations. Furthermore, a multiplicity of pulse waveforms are to be applied, dependent on the specific radar task, e.g. search for new targets and acquisition or tracking of already detected targets at various ranges [1]. With long pulses arises the problem of blind ranges which afford special waveform combinations. Despite long pulses the range resolution has to be achieved using suitable modulation of the pulses in combination with pulse compression on the receiving side.

For special radar concepts, e.g. omnidirectional low probability of intercept (OLPI) radar for an ARM resistant operation (see chapter 18), even a CW signal is transmitted [2].

In combination with digital processing a modulation with a discrete phase code for single pulses or with periodic repetition in CW mode has to be applied. The length of the subpulse determines the range resolution and the codelength the total pulse length.

Within the experimental phased-array radar system ELRA developed at FGAN, described in chapter 17 in more detail, an active solid-state array antenna is operated. The length of the subpulse may be chosen as 2 or 10 µs. The total pulse length is given by the subpulse length multiplied by the code length. With a selectable codelength of 1, 16, 64, 128 or 256 a high variability for signal modulation results. These parameters will be typical also for future operational systems with a solid-state active array.

7.2   Requirements and basic structure for pulse compression

Because the amplitude of the transmit signal should be constant to achieve high efficiency of the power amplifier, which is operated preferably in saturated class B or C mode, only phase modulation has to be applied.

The main radar task is to detect aircraft targets. We have therefore to take into account the unknown Doppler frequency shift of the echoes which is given, with v the radial velocity, by:

$$f_d = \frac{2v}{\lambda} \qquad (7.1)$$

The phase shift within a pulse of duration T is given by $2\pi f_d T$. For a radial velocity of 200 m/s and a wavelength $\lambda = 0.11$ m (S band, chosen for the ELRA system) results a Doppler frequency shift of $f_d = 3.64$ kHz.

The relevant parameter is given by:

$$d = f_d T \quad \text{(Doppler shift } \times \text{ pulse length)}$$

For $d = 1$ the phase rotates just one circle or 2π within the pulse. For a pulse length of, for example, 500 µs we get $d = 1.82$. Therefore the value of d will usually be >1 for longer pulses corresponding to the parameters given above.

Requirements for pulse compression are:

- low SNR loss in the whole Doppler region
- preservation of range resolution corresponding to the subpulse length

Minimal range sidelobes should be achieved, especially for the following cases and conditions:

- if possible for all range cells and possible Doppler shifts, especially for search tasks
- for the search mode against ground clutter with low sidelobes at $f_d = 0$
- for all range cells and at least for only one specified Doppler shift, e.g. for tracking of targets, where the Doppler frequency is already known
- for a certain range interval around the target and for a known Doppler shift for tracking single or several targets

Low sidelobes should be achieved in the mean-square sense or by minimising the maximum values.

The compression of a received pulse which contains a series of the phase-modulated subpulses of length τ is performed with a transversal filter according to chapter 2 section 2.4. The subpulses determine the range resolution and correspond to range cells, range elements or range bins. The samples of the subpulses are the signals $s_0, s_1, \ldots, s_{N-1}$ which are travelling through a shift register. The signals are weighted at the shift register taps by the coefficients $h_0, h_1, \ldots, h_{N-1}$ of the filter function $\mathbf{h} = \mathbf{s}^*$ according to equation 2.26. The sum of the weighted signals is the output signal. This operation is named the convolution of $\mathbf{s}$ and $\mathbf{h}$.

This operation may also be performed in the frequency domain: $\mathbf{s}$ and $\mathbf{h}$ are transformed by a Fourier transform into the frequency domain, the spectra $\mathbf{S}$ and $\mathbf{H}$ are multiplied and the result transformed into the time domain with an inverse Fourier transform, as discussed in chapter 2 section 2.4.

At that instant of time when the signal series is just completely within the shift register and $\mathbf{h}$ and $\mathbf{s}$ are matched we have as the output signal the compression peak:

$$y = \sum_{n=0}^{N-1} s_n s_n^* \tag{7.2}$$

Each subpulse s_n is rotated back to zero phase by multiplying with s_n^* according to our example in chapter 3 and Figure 3.10, where we explained the effect of a matched filter by forming the maximum possible sum of signal vectors after back rotation and phase alignment. All other shift states result in sidelobes.

7.3 Binary phase codes

The simplest phase modulation would be a binary phase code ($0°$, $180°$ or $+1$, -1). The binary code may be produced as a Barker code or as a pseudorandom series by a shift register code [3]. If there is an unknown Doppler shift of the echo signal and a corresponding phase rotation of the signal within the pulse and we try to compress the pulse using a filter function without the Doppler phase shift, then the products in equation 7.2 are not phase aligned, but show phase rotation according to the Doppler phase. Then by the final summation the vector sum may be small or even zero. Therefore, with any binary codes cancellation results within the compression filter for an increasing Doppler frequency. This is demonstrated in Figure 7.1 showing a waterfall representation of the compressed code for a series of increasing Doppler values d. In the front we see the curve along the range dimension r, measured in range cells, for $d = 0$ and in the rear for $d = 2$. The increment for d is 0.05. This kind of presentation will also be used for the following figures.

A binary phase code is obviously applicable only for $d < 0.5$. This would limit the pulse length for the above given Doppler frequency shift to the value:

$$T < \frac{0.5}{3.64\,\text{kHz}} = 137\,\mu\text{s}$$

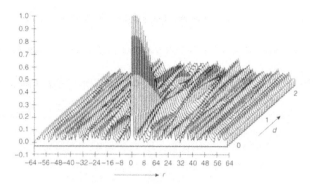

Figure 7.1   Ambiguity function of binary phase code, N = 64

For longer pulses one possibility is the parallel application of compression filter functions which are matched to a set of Doppler frequencies covering the Doppler range of interest. Then one has to select the maximum output from this filter bank. This is a kind of nonlinear operation which has to be taken into account for the following processing steps.

7.4   Polyphase code as an approximation of linear frequency modulation

A Doppler tolerance can be achieved with linear frequency modulation (LFM), or to a certain extent with nonlinear frequency modulation (NLFM). From these frequency modulations the corresponding phase may be derived and is then named the polyphase code, because multiple different phase values are applied in contrast to the binary phase codes.

For a linear frequency modulation the frequency is linearly increasing within a pulse of length T up to the frequency deviation Δf, as shown in Figure 7.2 (named up chirp):

$$f(t) = \frac{\Delta f}{T}t \quad 0 < t < T \tag{7.3}$$

or

$$f(t) = kt$$

with $k = \Delta f/T$. The compressed pulse will have the length τ. This will also be the length of a subpulse of a digital discrete modulation. The series of contiguous subpulses with its phases forms the complete pulse.

The Doppler tolerance can now be explained: a Doppler-shifted echo signal may be shifted by f_d. By a time shift Δt this signal is nearly matched to the transmitted signal, apart from a part at the beginning and end of the signal. Because this part is

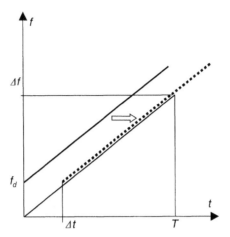

Figure 7.2   Linear frequency modulation

relatively small according to the frequency ratio $f_d/\Delta f$ (3.64 kHz/1 MHz = 0.0036) there will be almost no loss, there will be only a shift in time Δt as an error for the range measurement. This error will be expressed in time by:

$$\Delta t = \frac{T f_d}{\Delta f} = \tau f_d T = \tau d \tag{7.4}$$

With a down chirp the error would be negative and thus both results could be used for a correct range measurement. The phase function follows from equation 7.3 by:

$$\varphi(t) = 2\pi \int_0^t f(t') \, dt' = \pi \frac{\Delta f}{T} t^2 \tag{7.5}$$

The compression is performed by a convolution with the matched filter already explained in section 7.2. For a continuous or analogue compression filter the output may be computed analytically. We will apply the convolution to this problem. The signal is with equation 7.3:

$$x(t) = \exp\left(j\pi k t^2\right) \quad \text{for } t = 0, \ldots, T \tag{7.6}$$

The filter function for compression is given by $h(t) = x^*(t)$ and the output of compression follows by the convolution:

$$y(t) = \int_0^t d\tau \, x(\tau) h(T - t + \tau) \quad \text{for } 0 < t < T$$

$$y(t) = \int_{t-T}^T d\tau \, x(\tau) h(T - t + \tau) \quad \text{for } T < t < 2T \tag{7.7}$$

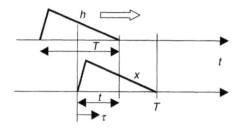

Figure 7.3   Convolution of x and h for 0 < t < T

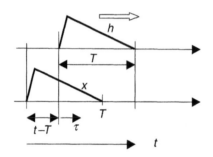

Figure 7.4   Convolution of x and h for T < t < 2T

This is illustrated by Figures 7.3 and 7.4 for both areas of time t: for a certain relative time shift t between filter h and signal x, represented with the triangles as an example for a signal, the products are integrated for the variable τ. For increasing t we observe the output $y(t)$.

For $0 < t < T$ we get:

$$y(t) = \int_0^t d\tau \, \exp\left(j\pi k\tau^2\right) \exp\left[-j\pi k(T - t + \tau)^2\right]$$

$$= \int_0^t d\tau \, \exp\left[j\pi k\left(\tau^2 - (T - t)^2 - 2(T - t)\tau - \tau^2\right)\right]$$

$$= \exp\left[-j\pi k(T - t)^2\right] \int_0^t d\tau \, \exp\left[-j2\pi k(T - t)\tau\right]$$

$$= \exp\left[-j\pi k(T - t)^2\right] \left[\frac{\exp\left[-2\pi jk(T - t)t\right] - 1}{-2\pi jk(T - t)}\right]_0^t$$

$$= t \exp\left[-j\pi k((T - t)^2 + (T - t)t)\right]$$

$$\times \frac{\exp\left[-j\pi k(T - t)t\right] - \exp\left[j\pi k(T - t)t\right]}{-2j\pi k(T - t)t}$$

and finally we get:

$$y(t) = t \exp\left[-j\pi k(T-t)T\right] \frac{\sin\left[\pi k(T-t)t\right]}{\left[\pi k(T-t)t\right]} \qquad (7.8)$$

For $T < t < 2T$ we get correspondingly:

$$y(t) = (2T - t) \exp\left[-j\pi k(T-t)T\right] \frac{\sin\left[\pi k(T-t)(2T-t)\right]}{\left[\pi k(T-t)(2T-t)\right]} \qquad (7.9)$$

The result for the compressed signal is a $(\sin x)/x$ function of time or range. The compression peak is at $t = T$. The zeros have a distance Δt from the peak given approximately by $\Delta f \Delta t = 1$. The pulse width between the zeros is then 2τ. The half-power pulse width then is approximately equal to $\tau = 1/\Delta f$.

The Doppler tolerance results from the shift similarity by the modulation with a continuously increasing or decreasing frequency. This is fulfilled by the linear frequency modulation (LFM) [4] and also by a nonlinear frequency modulation (NLFM), discussed in section 7.6.

With the discrete modulated subpulses, by the polyphase code, we have a sampled version of the continuous case and we get after compression only the sampled values of the $(\sin x)/x$ function.

The continuous phase law may be sampled at the subpulse period τ resulting in a discrete phase code of length $N = T/\tau$:

$$\varphi_n = \frac{\pi}{N} n^2 \bmod 2\pi \quad \text{for } n = 0, \ldots, N-1 \qquad (7.10)$$

The phase is taken mod 2π, which means that the phase values are taken within the first interval $0, \ldots, 2\pi$. Integer multiples of 2π are subtracted from φ_n. The code length N may be chosen arbitrarily. Because n^2 is an integer value a quantisation of the phase has to be performed with $2N$ steps. The resultant digital phase code is known under the name P3 code [5,6].

Variations are derived by a vertical shift of the function $f(t)$, for example:

$$f(t) = \frac{\Delta f}{T} t - \frac{\Delta f}{2} \qquad (7.11)$$

results in the discrete form into the so-called P4 code [5].

The well-known Frank code is a classical choice for the polyphase code [3,7], which needs minimal phase quantisation bits. It corresponds to a stepped frequency approximation of a linear chirp. For a phase quantisation with binary numbers, the codelength $N = T/\tau$ is restricted to values $N = n^2$ with $n = 2^m$ (m integer). The phase is then quantised into m bits. The phase values are given by:

$$\varphi_i = c_i \frac{2\pi}{n}$$

with

$$c_i = (i \operatorname{div} n)(i \bmod n) \quad \text{for } i = 0, \ldots, N-1 \qquad (7.12)$$

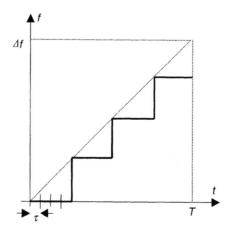

Figure 7.5   Frank polyphase code, $N = 16$, $n = 4$

The term (i div n) means the integer part of the result after performing the division i/n. (i mod n) means the residue after subtracting as many integer multiples of n as possible.

The factor (i div n) $= k$ determines the step frequency according to equation 7.12:

$$f_k = \frac{1}{2\pi} \frac{\Delta \varphi_k}{\Delta t} = \frac{1}{2\pi} \frac{2\pi}{n} \frac{k}{\tau} = \frac{1}{2\pi} \frac{2\pi}{n} \frac{kn^2}{T} = \frac{nk}{T} \tag{7.13}$$

The frequency steps above time are sketched in Figure 7.5. With the choice $n = 2^m$ there is an equivalence to the FFT algorithm. The Fourier coefficients of the n signal segments of length n have to be computed and then added.

The compression result or the ambiguity function of a polyphase-coded pulse dependent on range and Doppler shift is shown in Figure 7.6, again by a waterfall presentation. The curves over the range are computed for increasing values of d in steps of 0.05 from 0 to 2. The curve in front applies to fixed targets (ground clutter). One recognises sidelobes with an extension in range according to the codelength before and behind the signal peak. These sidelobes appear for all Doppler values.

For $d = 1$ the range error from equation 7.4 is one range increment corresponding to τ. For $d = 0.5 + p$ (p any integer) the maximum compressed signal is between two adjacent range samples of the $(\sin x)/x$ function discussed above for continuous LFM. There are only two signals on the flanks of the $(\sin x)/x$ peak, which are decreased by about 4.4 dB.

To avoid this effect one may apply only half the frequency shift Δf or double the sampling rate of the quadratic phase law. For a given sampling rate, this results in a reduced range resolution by a factor 2. Then for the P3 code the phase would be corresponding to equation 7.10:

$$\varphi_n = \frac{\pi}{2N} n^2 \bmod 2\pi \quad n = 0, \ldots, N - 1 \tag{7.14}$$

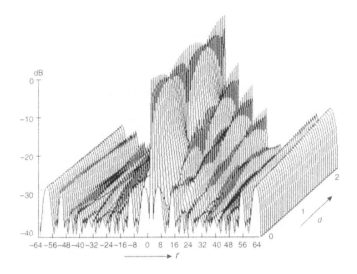

Figure 7.6   Ambiguity function (vertical axis in dB scale) for Frank polyphase-coded pulse, $N = 64, d = 0, \ldots, 2$

The compression result or ambiguity function is shown in Figure 7.7. The depressions at $d = 0.5 + p$ are now reduced substantially, but the compression peak is now broader.

7.5   Reduction of range sidelobes

7.5.1   Application of a weighting function

With a polyphase code derived from LFM with a code length N the maximum sidelobe power is by a factor $\pi^2 N$ lower than the maximum output signal peak. In the search mode with a phased-array radar it is advisable, in consideration of signal processing complexity and also to achieve a high detection efficiency, to choose only a moderate range resolution. The final high range resolution will then be applied in the acquisition and tracking mode. The code length N for the search mode therefore may be relatively short, e.g. with a subpulse length 10 μs the value for N may be between 16 and 64.

To avoid unnecessary acquisition orders by false alarms from sidelobe echoes the sidelobes should be reduced as far as possible. Sidelobe reduction by applying a weighting function is the simplest means. The compression filter function **w** is multiplied by the Hamming function [3]:

$$g(n) = 0.54 - 0.46 \cos\left(2\pi n/N\right)$$

The achievement is shown in Figure 7.8: the sidelobes are reduced by nearly 10 dB, but there is a loss in the peak signal/noise power ratio (SNR) of about 1.5 dB and there is also a loss in range resolution.

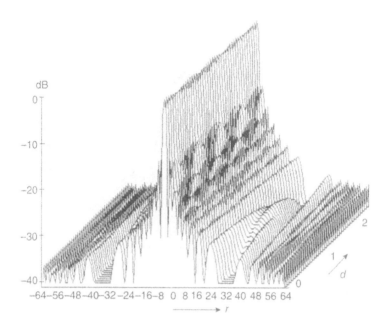

Figure 7.7   Ambiguity function of P3 code (dB scale), N = 64, with doubled sampling rate

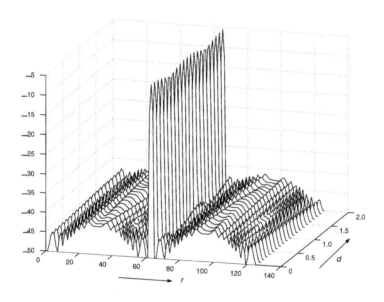

Figure 7.8   Ambiguity function of P3 code (dB scale), N = 64, compression with Hamming weighting

7.5.2 Application of a mismatched LS filter

Sidelobe reduction with expansion of the compression filter vector in time implies increasing the number of taps and weighting coefficients and leads to a mismatched filter. The sidelobes are produced by all possible relative shift positions between the signal vector and the compression filter vector. Only for the signal peak is there complete overlapping. By expanding the length of the filter vector there appear additional degrees of freedom which can be used to compute the mismatched compression filter **w** for an improved sidelobe attenuation.

We first define a code matrix **C** which contains in its rows the code series shifts step by step from left to right [8]. We assume as an example a threefold length $3N$ of the filter vector compared to the length N of the code series. The rows in matrix **C** describe the shift states of a received code series through the shift register of a transversal filter discussed in chapter 2 section 2.4:

$$\mathbf{C} = \begin{bmatrix} c_{N-1} \cdots 0 & 0 \cdots 0 & 0 \cdots 0 \\ & \ddots & & & \\ c_0 \cdots c_{N-1} & 0 \cdots 0 & 0 \cdots 0 \\ & \ddots & \ddots & & \\ 0 \cdots 0 & c_0 \cdots c_{N-1} & 0 \cdots 0 \\ & & \ddots & \ddots & \\ 0 \cdots 0 & 0 \cdots 0 & c_0 \cdots c_{N-1} \\ & & & \ddots & \\ 0 \cdots 0 & 0 \cdots 0 & 0 \cdots c_0 \end{bmatrix} \tag{7.15}$$

The dimension of matrix **C** is $[4N - 1, 3N]$. The compression result with filter vector **w** $[3N, 1]$ is given by:

$$\mathbf{Cw} = \mathbf{s} \tag{7.16}$$

We want vector **s** to approximate an ideal solution given by the design vector $\mathbf{s}_d$, with dimension $[4N, 1]$, containing as elements only zeros and a one in the centre. Vector **s** shall approximate $\mathbf{s}_d$ in the least-square sense. We have also to consider a noise vector **n**. The complete error voltage e is then given by:

$$e = |\mathbf{Cw} - \mathbf{s}_d| + \mathbf{n}^*\mathbf{w}$$

and the mean error power by:

$$F = E\{e^2\} = (\mathbf{Cw} - \mathbf{s}_d)^*(\mathbf{Cw} - \mathbf{s}_d) + \mathbf{w}^*\mathbf{E}\{\mathbf{nn}^*\}\mathbf{w}$$

F is a measure of the sidelobe and noise power at the compression filter output. F shall be minimised dependent on **w**. Therefore, the derivative of F with respect to **w** shall be zero. With $E\{\mathbf{nn}^*\} = \mathbf{Q}$ we get for the derivative of F to $\mathbf{w}$:

$$\Delta_w F = \mathbf{C}^*(\mathbf{Cw} - \mathbf{s}_d) + (\mathbf{Cw} - \mathbf{s}_d)^*\mathbf{C} + \mathbf{Qw} + \mathbf{w}^*\mathbf{Q}$$

$$\text{Re}\,\{\mathbf{C}^*(\mathbf{Cw} - \mathbf{s}_d) + \mathbf{Qw}\} = 0$$

This is fulfilled for:

$$\mathbf{w} = (\mathbf{C}^*\mathbf{C} + \mathbf{Q})^{-1}\mathbf{C}^*\mathbf{s}_d \qquad (7.17)$$

The noise power is represented by $\mathbf{Q}$. For receiver noise we have $\mathbf{Q} = q\mathbf{I}$ ($\mathbf{I}$ = identity matrix). The factor q represents the ratio of noise to signal power. For strong receiver noise $\mathbf{Q}$ outweighs $\mathbf{C}^*\mathbf{C}$ and the compression filter vector turns into the noise optimal matched filter $\mathbf{w} = \mathbf{c}^*$ (the centre column of $\mathbf{C}^*$ is selected by multiplication with $\mathbf{s}_d$). A result is shown in Figure 7.9 for an extension factor for a filter length of 3. The codelength is $N = 64$. Regrettably, the sidelobes are low only for $d = 0$ and the close vicinity. Nevertheless this quality would be effective against sidelobes from ground clutter and could be interesting for the radar search mode.

By matching the compression filter vector to a selected Doppler frequency by multiplying the code elements c_n with the Doppler phase shifters:

$$c_n^D = c_n \exp\,(2\pi j dn/N) \quad \text{for } n = 0, \ldots, N - 1$$

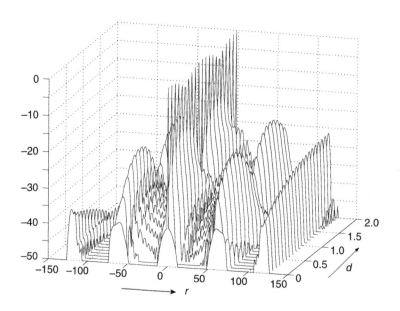

Figure 7.9   *Compression function of P3 (dB scale) with LS-mismatched filter, filter extended by factor 3, $N = 64$*

the low sidelobes could be shifted to Doppler factor d. This possibility may be interesting for target tracking where the Doppler frequency shift is already estimated.

As an alternative to the least-square solution the computation has also been performed by Chebycheff approximation, to keep the sidelobes down under a certain level. The results have been very similar when compared to the least-square method presented above [7].

Further, the approach given in equation 7.17 can be extended to minimise the sidelobes for a set of Doppler shift values ($d = d_0, d_1, d_2$) assumed in the signals under consideration in an expanded matrix $\mathbf{C}$ [5]. Equation 7.16 changes to equation 7.18. The solution is again computed by equation 7.17:

$$
\begin{bmatrix}
\mathbf{C}(d_0) \\
-- -- -- \\
\mathbf{C}(d_1) \\
-- -- -- \\
\mathbf{C}(d_2)
\end{bmatrix}
\mathbf{w} =
\begin{bmatrix}
\mathbf{s}_d(d_0) \\
-- -- \\
\mathbf{s}_d(d_1) \\
-- -- \\
\mathbf{s}_d(d_2)
\end{bmatrix}
\tag{7.18}
$$

Because the sidelobes have been maximum in the previous approach for $d = \pm 0.5$, these values have been chosen for d_1, d_2 together with $d_0 = 0$. The typical result is given by Figure 7.10.

Indeed, the sidelobes are decreased for $d = 0.5$ in comparison to Figure 7.9. The same is true for $d = -0.5$ because of symmetry but omitted in Figure 7.10. But unfortunately the sidelobe level is increased for $d = 0$ and there is no obvious advantage compared to the simple matched filter (see Figure 7.7).

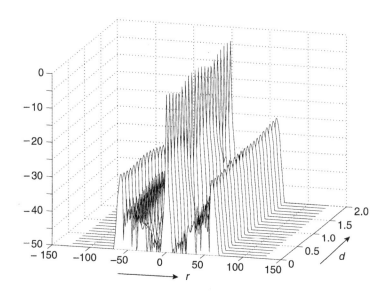

Figure 7.10   Compression of P3 code (vertical axis in dB scale), N = 64, LS filter extended by factor 3 and d = −0.5, 0, 0.5

It is concluded that a global attenuation of sidelobes for a broader Doppler region is not possible using mismatched filtering, even with filter vectors which are long compared to the signal vector representing the phase code.

7.5.3   Application of an estimation filter

The measured signal vector $\mathbf{z}$ is again given by:

$$\mathbf{z} = \mathbf{C}^*\mathbf{s} + \mathbf{n}$$

as discussed in chapter 3 section 3.5.2. Signal vector $\mathbf{s}$ has a length $[2N + 1]$ corresponding to a range interval and may contain a one at the target position. The target position shall be determined by estimating $\mathbf{s}$. For matrix $\mathbf{C}$, now only with dimension $[2N + 1, 3N]$, we have similar to the preceding section:

$$\mathbf{C} = \begin{bmatrix} c_0 \cdots c_{N-1} & 0\cdots 0 & 0\cdots 0 \\ & \ddots & \\ 0\cdots 0 & c_0 \cdots c_{N-1} & 0\cdots 0 \\ & \ddots & \ddots \\ 0\cdots 0 & 0\cdots 0 & c_0 \cdots c_{N-1} \end{bmatrix}$$

The target may be at any of the $2N + 1$ positions with the same probability. For the matrix $\mathbf{P} = E\{\mathbf{ss}^*\}$ then follows the identity matrix $\mathbf{I}[2N+1, 2N+1]$. With equation 3.80 then follows:

$$\hat{\mathbf{s}} = \mathbf{C}^*(\mathbf{CC}^* + \mathbf{Q})^{-1}\mathbf{z} = \mathbf{W}^*\mathbf{z}$$

A numerical inspection shows that $\mathbf{W}^*$ has very similar rows, shifted similarly to the codes in matrix $\mathbf{C}$. With such a row $\mathbf{w}^*$ we perform a convolution with $\mathbf{z}$. We use the centre row of $\mathbf{w}^*$ as the compression filter function, given with $\mathbf{e}^*[1, 2N + 1] = (0, \ldots, 010, \ldots, 0)$ by:

$$\mathbf{w}^* = \mathbf{C}^*(\mathbf{CC}^* + \mathbf{Q})^{-1}\mathbf{e} \tag{7.19}$$

The matrix $\mathbf{Q}$ represents as above the receiver noise and measurement errors and is selected $\mathbf{Q} = q\mathbf{I}$. The following Figures 7.11 and 7.12 illustrate the compression function for a P3 code for $N = 64$ in the linear and dB scales. The sidelobes are zero for $d = 0$ for the interval $\pm N$. For the other shift positions or ranges and other Doppler values d the sidelobes appear again.

7.6   Reduction of sidelobes by a phase code from nonlinear frequency modulation

The application of nonlinear frequency modulation (NLFM) to achieve lower sidelobes has been proposed already in Reference 3. Because of difficulties in practical

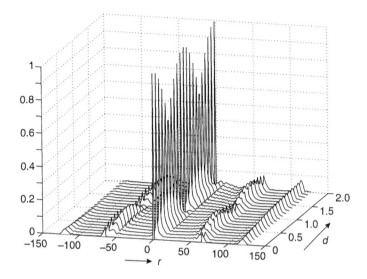

Figure 7.11   Compression function of P3 code by LS estimation, zero sidelobes around peak for d = 0, N = 64, linear scale

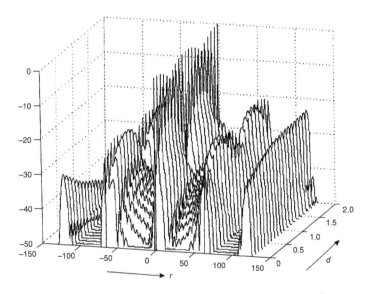

Figure 7.12   Compression function of P3 code by LS estimation, zero sidelobes around peak for d = 0, N = 64, dB scale

realisation this possibility has so far not found much interest. With digital-processing techniques there is no longer a problem for application. The corresponding polyphase code may be derived from a suitable form of NLFM as above for the LFM [8,9]. As a further step PNL phase codes derived from nonlinear frequency modulation were investigated. In this case matched filtering for compression is appropriate and any loss in SNR is avoided.

Because there may be future interest in this type of phase modulation the basic development is given here following the fundamental treatment in Reference 3.

The complex modulation to be determined is $u(t) = a(t) \exp(j\Theta(t))$. The accompanying spectrum is:

$$U(\omega) = \int a(t) \exp[j(-\omega t + \Theta(t))]\, dt \tag{7.20}$$

The phase term in equation 7.20 is oscillating with frequency $d/dt(-\omega t + \Theta(t))$ as a function of time t. The contribution to the integral above is maximum if there is no mutual cancellation by rapid oscillations, that is if $d/dt(-\omega t + \Theta(t)) = 0$ or:

$$\omega - \Theta'(t) = 0 \quad \text{or} \quad d\omega = \Theta''(t)\, dt \tag{7.21}$$

Under the condition of equation 7.21 we have a stationary phase. By the function $\Theta(t)$ the contribution of the integrand to $U(\omega)$ at frequency ω is determined. After matched filtering the spectrum will be $|U(\omega)|^2$. In favour of low sidelobes this spectrum should be bell shaped.

The phase term in equation 7.20 is now expanded with a Taylor power series around (t_0, ω_0):

$$\omega t - \Theta(t) = \omega_0 t_0 - \Theta(t_0) - \Theta''(t_0) 1/2(t - t_0)^2 + \cdots$$

The linear term vanishes with equation 7.21. For a time interval δ around t_0 we write for the integral equation 7.20:

$$U(\omega, \omega_0) = a(t_0) \int_{t_0-\delta}^{t_0+\delta} \exp\left[-j\left(\omega_0 t_0 - \Theta(t_0) - \Theta''(t_0) 1/2(t - t_0)^2\right)\right] dt$$

Following Reference 3 this integral is solved by a transformation of variables after placing terms independent of t before the integral. The Fresnel integral can be solved approximately:

$$|U(\omega)|^2 = 2\pi \frac{a^2(t)}{|\Theta''(t)|} \tag{7.22}$$

With equation 7.21 follows:

$$a^2(t)\, dt = \frac{1}{2\pi} |U(\omega)|^2\, d\omega$$

and further assuming the existence of a monotonically increasing (ωt):

$$\int_{-\infty}^{t} a^2(\tau)\, d\tau = \frac{1}{2\pi} \int_{-\infty}^{\omega(t)} |U(\Omega)|^2\, d\Omega \tag{7.23}$$

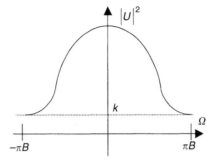

Figure 7.13   Bell-shaped function of $|U(\Omega)|^2$

This equation gives a function between t and ω describing the required frequency modulation $\omega(t)$. We assume for $a(t)$:

$$a(t) = \frac{1}{\sqrt{T}} \quad \text{for } 0 \le t \le T$$

$$= 0 \quad \text{elsewhere}$$

For $|U(\Omega)|^2$ we assume a bell-shaped function (Figure 7.13):

$$|U(\Omega)|^2 = \frac{1}{B}\left[k + (1-k)\cos^n\left(\frac{\Omega}{2B}\right)\right]$$

With parameters k and n we may influence the shape of the bell. With $x = \Omega/2B$ the integral of equation 7.23 becomes:

$$\frac{t}{T} = \frac{1}{\pi} \int_{-\pi/2}^{\pi f/B} [k + (1-k)\cos^n x]\,dx \tag{7.24}$$

As an example the integral is solved for $n = 2$ (Hamming window). With $\cos^2 x = (1 + \cos 2x)/2$ we get:

$$\frac{t}{T} = \frac{1}{\pi} \int_{-\pi/2}^{\pi f/B} \left[1 + \frac{k}{2} + \frac{1-k}{2}\cos 2x\right] dx$$

and after integration and inserting the integration limits:

$$\frac{t}{T} = \frac{1}{\pi}\left[\frac{1+k}{2}\left(\frac{\pi f}{B} + \frac{\pi}{2}\right) + \frac{1-k}{4}(\sin(2\pi f/B) - \sin(-\pi))\right]$$

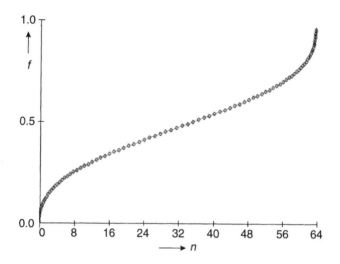

Figure 7.14   NLFM: frequency function above time

Omitting a constant term and factor we arrive at:

$$\frac{t}{T} = \frac{f}{B} + a \sin\left(2\pi \frac{f}{B}\right) \quad \text{with } a = \frac{1-k}{2(1+k)} \quad \text{for } -\frac{B}{2} \leq f \leq \frac{B}{2}$$

or as is more usual:

$$\frac{t}{T} = \frac{f}{B} - a \sin\left(2\pi \frac{f}{B}\right) \quad \text{for } 0 \leq f \leq B \tag{7.25}$$

An example of frequency function dependence on time is given by Figure 7.14. The steeper function at the beginning and the end results in higher $\Theta''(t)$. These frequency parts then exist only for a short time and therefore have less power within the whole spectrum. This leads, for $|U(\omega)|^2$, to the desired bell shape of the spectrum. A sufficient Doppler tolerance can be expected by the shift similarity, similar to the LFM as discussed above.

A program computed to a given frequency sets the corresponding time values. In order to derive finally a phase code for a series of discrete time values, according the sampling rate, we need the accompanying frequency values. These frequency values are interpolated out of the nearest-neighbour values to the selected time values; by integration of the frequency function the phase values follow. The integration is approximated by a summation as follows with τ the subpulse length:

$$\varphi_n = 2\pi \sum_{i=1}^{n} \frac{1}{2}(f_{i-1} + f_i)\tau \bmod 2\pi$$

$$\varphi_0 = 0 \tag{7.26}$$

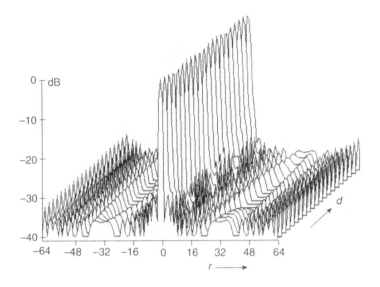

Figure 7.15   *Compression function for NLFM, N = 64, a = 1, vertical axis in dB scale*

We name this polyphase code the PNL code (phase from nonlinear frequency modulation). Simulation studies and results are given in Reference 9.

In Figure 7.15 the sidelobes are now low for all Doppler frequencies. The price for the PNL code is a reduced range resolution. There is also a slow decrease of the compressed signal for increasing d.

Depending on the parameter a in equation 7.25 which determines the degree of nonlinearity, relatively low sidelobes are achieved for all values of Doppler shift. A favourable choice turns out to be $a = 0.965$ for $N = 256$. For $a = 0$ we arrive at linear frequency modulation.

For phase quantisation resolution of the phase code, a choice of $2N$ steps has been found by computer simulation to be sufficient. But even $N/2$ steps result in an improvement compared to the Frank or P3 codes. Also of importance for a hardware realisation of the pulse compression is the necessary quantisation of the complex components of the filter weights. The computer simulation resulted in 3 bit + sign to be sufficient.

7.7   Complementary codes

Complementary codes are sometimes discussed [10] with respect to achieving zero sidelobes and obtaining responses from targets only at unambiguous ranges. A set of orthogonal subcodes is transmitted in a series. The individual compression results are coherently added afterwards with the effect of cancelling all sidelobe and ambiguous range contributions. Because of possible Doppler tolerance, our interest would be

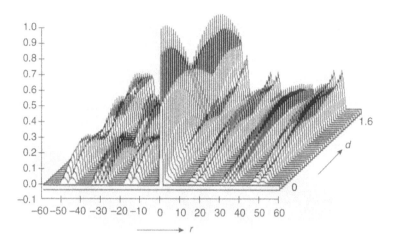

Figure 7.16   *Ambiguity function of complementary codes (4 × 4 Frank), vertical axis in linear scale [8]*

in a realisation with a polyphase code. The set of individual codes which have to be processed and combined may be derived from the Frank code or the P3 code. The behaviour of this code has been evaluated for Doppler-shifted echoes by the compression or ambiguity function as in the cases above [7]. An example is given in Figure 7.16 where four partial codes each of length 4, derived from the Frank code of length $N = 16$, are used. The length of the subpulse was assumed 1 μs and the period for the subcode is $T_c = 66.7$ μs according to an unambiguous range interval of 10 km. The Doppler shift was assumed from zero up to $f_{d_{max}} = 6$ kHz. Then the maximum Doppler factor for the complete code follows to:

$$d_{max} = 4 \cdot T_c \cdot f_{d_{max}} = 4 \cdot 66.7 \cdot 10^{-6} \cdot 6 \cdot 10^3 = 1.6$$

In Figure 7.16 the parameter d varies from 0 to 1.6. The Doppler-dependent phase shift between the subcodes has to be considered. Only for $d = 0$ is the cancellation of unwanted signals achieved, but with increasing d there are also increasing sidelobes and a decreasing target pulse. Additionally, there appear responses at unambiguous ranges. For the vertical axis the linear scale is chosen, otherwise the figure would be too confusing because of the high sidelobes.

7.8   Polyphase code with periodic repetition

For applications with continuous wave (CW) operation, e.g. the omnidirectional low probability of intercept radar (OLPI radar), which is discussed in chapter 18 as an ARM-resistant radar concept, periodic repetition of the P3 or Frank codes results in the advantage of zero sidelobes for echoes with a zero Doppler shift [3]. Therefore,

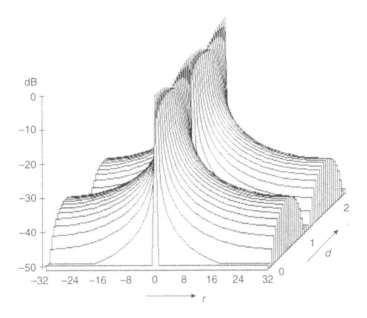

Figure 7.17   Ambiguity or compression function of periodic P3 code, $N = 64$

echoes from ground clutter or fixed targets would have, in theory, no range sidelobes after compression. The code is transmitted in a series of repetitions without any interval. The compression function is shown for the P3 code in Figure 7.17 for a code length $N = 64$. But for Doppler-shifted echoes the sidelobes increase in a similar manner as for the aperiodic polyphase-coded pulses discussed so far.

For Doppler-shifted echoes the sidelobes may be substantially reduced by applying a Hamming weighting function to the compression filter vector. This is demonstrated in Figure 7.18.

7.9   Pulse eclipsing

When transmitting long pulses there follows a corresponding blind range. Target echoes within this blind range are received as an incomplete pulse, because the receiver is shut off during the time of transmission. The following figure illustrates the situation (Figure 7.19). These mutilated pulses may nevertheless be used for target detection [11]. If, for example, only half of the coded pulse is available for pulse compression then the peak compression signal has a value of only half of the maximum amplitude. The range resolution is also reduced. For targets at shorter ranges there would be no detection problem, because by the radar equation the echo power is sufficient. Figure 7.20 shows an example of the compressed code signal. The sidelobes from the far out half of the outgoing phase of the signal series are the same as those without eclipsing. The compression peak is reduced by 6 dB.

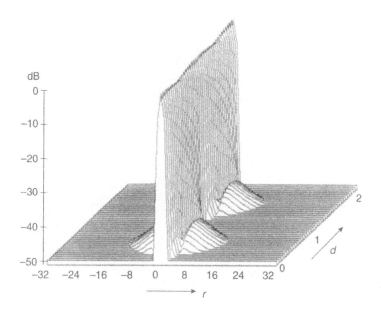

Figure 7.18   Compression function of periodic P3 code, with Hamming weighting,
N = 64 [8]

7.10   High range resolution by oversampling and
LS pulse compression

For some radar applications a high range resolution is required, for example for
the separation of closely-spaced targets. Then usually a higher signal bandwidth is
required. This has the consequence that the antenna system with its transmit/receive
modules has to be designed for the higher bandwidth. There have been suggestions for improving the range resolution by inverse filtering, that is by a filter with
a frequency transfer function inverse to the signal spectrum, to extend the signal's bandwidth subsequently after receiving [12] or by applying the methods of
superresolution [13].

   In the following we examine whether and how without an increase of the design
bandwidth of the antenna and receiving system, and only with an increased sampling rate of the analogue-to-digital converter, can an improved range resolution be
achieved. This is implemented by a transversal compression filter, designed by the
least-mean-square method. Accurate knowledge of the received signal function of a
point target is assumed [14].

   With practical radar systems the intermediate frequency and video signal bandpass
filters are matched to the length of the subpulse and are not exactly band limiting. If
the signal is sampled, as usually only with the Nyquist or bandwidth rate, some information will be lost. This additional information in the frequency sidebands outside
the design bandwidth may be exploited to a certain extent by oversampling.

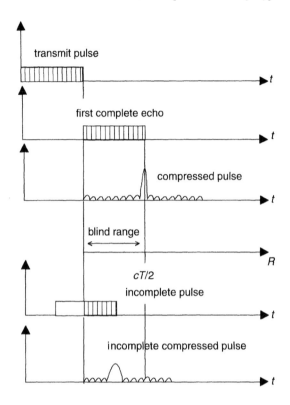

Figure 7.19   The effects of a blind range on target echoes

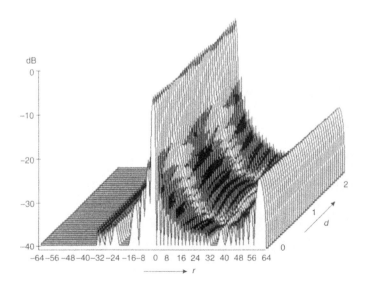

Figure 7.20   Compressed pulse with 50 % eclipsing (dB scale), P3 code, N = 64

Within the experimental phased-array radar system ELRA, developed at FFM and described in chapter 17, the matched filters for the rectangular subpulses for the receiving channels are, for example, made with surface acoustic wave filters (SAW). These filters have in the time domain a rectangular filter function and therefore a frequency transfer function of the well-known $(\sin x)/x$ form with high sidelobes in the frequency domain. The 3 dB bandwidth of this type of filter is given by $B = 1.2/\tau$. This is nearly the usual choice of bandwidth $B = 1/\tau$. The sampling period for the complex video signal with I and Q components is usually chosen as τ. We recognise that by applying this type of a matched filter for the subpulses there are sidelobes outside the nominal bandwidth and this part of the spectrum is not represented by the usual sampling period τ but may be extracted only by oversampling.

7.10.1   Calculation of the filter function for pulse compression

The compression filter function **w** is computed as in section 7.5.2.

With conventional pulse compression the elements of vector **s** are assigned to the individual subpulses. The centre value 1 represents one subpulse or range bin. But here the value 1 represents, for the achievement of a higher resolution, only an individual value of the oversampled signal sequence. The algorithm must thus try to concentrate the compressed signal on only one clock position of the oversampled sequence. All remaining clock places correspond to sidelobes and are minimised in the mean-square sense. To what extent that is possible is to be examined in the following with simulated and also with measured radar echo signals. The application of the filter vector **w** in a transversal filter represents linear processing. The linear superposed echo sequences of neighbouring targets can thus be separated according to the response function of the compression function calculated after equation 7.17.

7.10.2   Subpulse filter

The transmitter emits a pulse containing a finite sequence of subpulses each of the length τ with the individual phases of the selected phase code. Each subpulse is filtered at the receiver by a filter matched to the subpulse, i.e. convolved with its filter function in the time domain. For the ELRA system in the long-range search mode the subpulse length is chosen as $10\,\mu s$. A SAW filter with an approximately rectangular filter function in the time domain is applied as a matched filter for the subpulses. From rectangular received subpulses approximately triangular pulses are created with a length $20\,\mu s$. This triangle function $d(t)$ was measured with amplitude and phase. The sampling points, p, correspond to the oversampling clock rate increased by a factor r compared to the conventional clock rate $1/\tau$:

$$t_p = \frac{\tau}{r} p$$

To prepare the convolution of the original code series $\mathbf{c}_0$ with the function **d** by a matrix multiplication we form a matrix **D** with elements d_{ip}. Each row i contains the response function d_{ip} around the index ir. Matrix $\mathbf{D}$ has therefore the dimension $(N, r(N+1)+1)$. With the code series $\mathbf{c}_0$ according to equation 7.10 or 7.12 written as a $(1, N)$ row matrix, follows:

$$\mathbf{c} = \mathbf{c}_0 \mathbf{D} \tag{7.27}$$

the received signal $\mathbf{c}$ from a point target which is produced after matched subpulse filtering and with the r fold oversampling rate. By equation 7.27 each code element c_{oi} is multiplied with the d_{ip} and all contributions from the subpulses are superposed. This modified or extended signal vector $\mathbf{c}$ has to be applied in the extended matrix $\mathbf{C}$ with equation 7.15. The new $\mathbf{C}$ is then used in equation 7.17 to compute $\mathbf{w}$.

7.10.3   Gain or SNR loss

The price for the LMS compression for higher resolution is, besides the increased sampling rate for the analogue-to-digital converter, also a loss in the signal-to-noise ratio SNR. It is calculated as the ratio of the gain with high-resolution compression (LMS filter) to that with matched filter (MF) compression. For MF compression the oversampling rate is assumed for a fair comparison.

For the gain computation we have to take into account the covariance matrix $\mathbf{R}$ of the noise after the subpulse filter and for the oversampling rate. With an optimal filter for the rectangular subpulse (ELRA, 10 µs) the model for the noise correlation is a triangular function:

$$\rho_t = \rho_0 \left(1 - \left| \frac{t}{r} \right| \right) \quad \text{for } t = -r, \dots, r$$
$$\rho_t = 0 \qquad\qquad\qquad \text{otherwise} \tag{7.28}$$

The covariance matrix $\mathbf{R}$ then contains diagonals with equal values ρ_t at a distance t from the main diagonal. The gain is then with both filter types with y_m as the maximum output signal:

$$g_w = \frac{y_m^2}{\mathbf{w}^* \mathbf{R} \mathbf{w}} \tag{7.29}$$

This gain will be computed for both filter types and compared by their ratio. The resultant loss in gain, expressed in dB, indicates the necessary increase in SNR to achieve high resolution using the LMS filter compared to that when using the MF.

7.10.4   Pulse compression of simulated signals

The ELRA system offers, by virtue of its signal variety, the possibility of practical testing of the suggested compression procedure. For this purpose pulses for high and low resolution are transmitted in a selected beam direction from which we receive echoes from a group of neighbouring fixed targets:

(i)   Short pulses with a duration of 1 or 2 µs. Their reception with the matched bandwidth of 1 MHz resolves the targets with corresponding high resolution.

(ii) Pulses with the code length 16 with Frank code and with a subpulse length of 10 µs. Reception is performed with the optimal subpulse filter (time function 10 µs rectangle). Accordingly, the standard sampling period is 10 µs. These signals are also sampled in the 1 µs period, so that an oversampling with $r = 10$ results. Compression of these signals in the conventional way and with the LMS filter for high resolution then permits a comparison with the echoes after (i).

For an initial investigation the compression is examined comparatively with simulated target signals. Two target signals with Frank code are generated. The range distance of the targets can be selected. The calculation of the LMS filter vector **w** follows equation 7.17. The diagonal noise covariance matrix is set $\mathbf{Q} = q\mathbf{I}$. The factor q is selected to $q = 0.16\,trace(\mathbf{CC^*})$. This choice has proved to be appropriate with the later experiments with measured signals.

Additionally, for comparison, the conventional compression function (MF) is applied to the oversampled signal sequence. The results are presented in Figures 7.21 and 7.22. In Figure 7.21 the target distance is selected as $v = 15$ samples of the oversampling rate. This corresponds to 1.5 conventional range cells. It can be seen that with MF compression (dashed line) the targets are just resolved. The result from the LMS filter is given by the solid line. In Figure 7.22 with target distance only $v = 4$ samples, target resolution is achieved with the LMS filter (solid line) but not with the MF (dashed line). So we take $v = 4$ as the resolution limit for the LMS

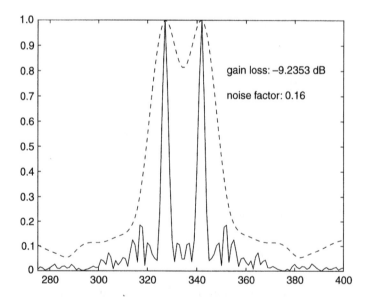

Figure 7.21   *Resolution by MF (dashed line) and LMS compression (solid line), target distance $v = 15$, Frank code $N = 16$, oversampling factor $r = 10$, noise factor $= 0.16$, gain loss $= -9.2353\,dB$, horizontal axis: range samples for oversampling [14]*

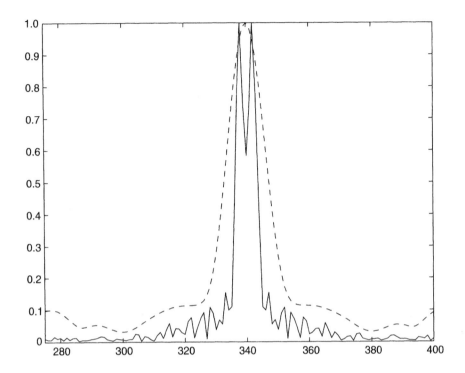

Figure 7.22   *Resolution by MF (dashed line) and LMS compression (solid line), target distance v = 4, Frank code, N = 16, oversampling factor r = 10, noise factor = 0.16, gain loss = 6.3 dB, horizontal axis: range samples for oversampling [14]*

filter. Resolution by the LMS filter is thus improved compared to that by the MF by a factor of 3.75. The gain loss is calculated to 9.23 dB with equation 7.29. So we have to increase the SNR by this gain loss to achieve the higher resolution with the LMS filter compared to compression with the MF. Also, higher sidelobes develop. The demand for high resolution, low sidelobes and moderate SNR loss cannot be achieved together.

In the simulation program a Doppler shift from a velocity of 100 m/s has also been applied to the target signals and no serious resolution degradation was observed. It will thus be possible to estimate the Doppler frequency with sufficient accuracy to apply this high-resolution compression.

7.10.5   Compression of measured echo signals

As a next step the recorded fixed target echoes from the ELRA radar system were processed. With the transmitted Frank code the target reference signal was measured. It was achieved by the reception of the echo pulses of an aircraft. This forms an isolated target with small dimensions in comparison to resolution in space according to the

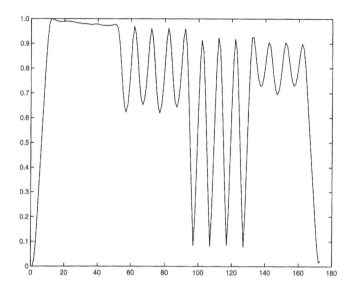

Figure 7.23   Measured reference signal from aircraft [14]

subpulse length. A target was tracked on its tangential flight route, and recorded echo signal blocks from 32 pulses were evaluated. Possible clutter signals were eliminated by elimination of the components' average values. The Doppler frequency was determined by a Fourier transformation, so that flight conditions with very low Doppler frequency could be selected. The received signal function is shown with its amplitude in Figure 7.23. Inaccuracies with the modulation of the transmit signal by the effect of the phase shifter for phase modulation as well as the effect of the transmission amplifier are now covered by the measured reference signal.

The partial decrease of the amplitude results from the effect of the subpulse filter, which converts the originally rectangular subpulses into twice as broad overlapping triangles. The superposition results in an amplitude reduction according to the phase shift between neighbouring code elements.

This signal sequence was used in its complex form as the reference signal for the calculation of **w**. In Figure 7.24 the echo signal from a 1 µs transmit pulse is shown for comparison. In the centre we have the result of conventional pulse compression. The final result at the bottom shows improved resolution from the LMS filter.

A higher range resolution may be achieved without increasing the design bandwidth of the radar system. Because there is no exact bandlimiting in practical radar systems, by oversampling and corresponding pulse compression with a filter function derived with an LMS algorithm the information in the spectral sidelobes of the received signal can be exploited. The price for this additional quality is a higher conversion rate for the ADC, higher processing complexity and a certain loss in signal-to-noise ratio. Especially with phased-array systems, this may be compensated for by a longer dwell time if necessary.

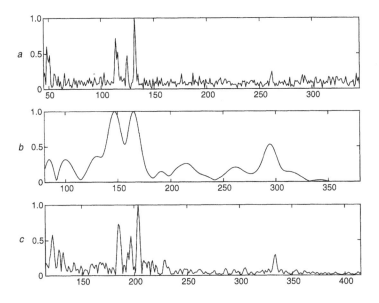

Figure 7.24   *High resolution compression with measured ELRA signals, fixed targets*
      a short pulse echoes from transmit pulse 1 μs
      b transmit subpulse 10 μs, $N = 16$, MF compression
      c transmit subpulse 10 μs, $N = 16$, oversampling $r = 10$, LMS high
      resolution compression, measured reference signal [14]

7.11   Conclusions

Pulse compression with sufficient Doppler tolerance may be achieved using polyphase codes derived from linear and nonlinear frequency modulation.

Low sidelobes in range and Doppler are required, especially for the radar search function. These may be achieved by an LFM-derived phase code together with Hamming weighting or by applying a phase code from nonlinear frequency modulation. A loss in resolution has to be taken into account in both cases.

For a discrete and known Doppler frequency with an expanded and mismatched filter vector a sidelobe reduction is possible. The compression is then achieved without a loss in resolution. A set up for the expanded filter vector results in a least-square minimisation for all range elements. This version may be useful for search in clutter or for target tracking.

Complementary phase codes show increasing sidelobes and responses from unambiguous range from targets with Doppler-shifted echoes.

Pulse eclipsing and the corresponding blind ranges can be tolerated to a certain extent in the search mode. Reduced resolution and signal peak power can be sufficient for target detection at shorter ranges.

Range resolution may be improved by a factor of about four by oversampling and a least-mean-square filter, if the subpulse matched filter is not strictly bandlimiting.

7.12   References

1 WIRTH, W. D.: 'Phased array radar with solid state transmitter'. Proceedings of international conference on *Radar*, Paris, France, 1984, pp. 141–145
2 WIRTH, W. D.: 'Omnidirectional low probability of intercept radar'. Proceedings of international conference on *Radar*, Paris, France, 1989, pp. 25–30
3 COOK, C. E., and BERNFELD, M.: 'Radar signals' (Academic Press, London, New York, 1967)
4 SKOLNIK, M. I.: 'Introduction to radar systems' (McGraw-Hill Book Company, Singapore, 1980, 2nd edn)
5 KRETSCHMER, F. F., and LEWIS, B. L.: 'A new class of polyphase pulse compression codes and techniques', *IEEE Trans. Aerosp. Electron. Syst.*, May1981, **17**, pp. 364–372
6 KRETSCHMER, F. F., and LEWIS, B. L.: 'Doppler properties of polyphase coded pulse compression waveforms', *IEEE Trans. Aerosp. Electron. Syst.*, July 1983, **19**, pp. 521–531
7 FRANK, R. L.: 'Polyphase codes with good nonperiodic correlation properties', *IEEE Trans. Inf. Theory*, January 1963
8 WIRTH, W. D.: 'Compression of polyphase-codes with doppler shift'. IEE international conference on *Radar*, Brighton, UK, October 1992, pp. 469–472
9 FELHAUER, T.: 'Design and analysis of new P(n,k) polyphase pulse compression codes', *IEEE Trans. Aerosp. Electron. Sys.*, July 1994, **30** (3), pp. 865–874
10 GERLACH, K. R., and KRETSCHMER, F. F.: 'New radar pulse compression waveforms'. Proceedings of IEEE national conference on *Radar*, Ann Arbor, Michigan, USA, 1988, pp. 194–199
11 BILLAM, E. R.: 'Eclipsing effects with high-duty-factor waveforms in long-range radar', *IEE Proc. F, Commun. Radar Signal Process.*, December 1985, **132** (7), pp. 598–603
12 HERRMANN, G. E., and KELLY, L. L.: 'Enhanced resolution in simple radars', *IEEE Trans. Aerosp. Electron. Sys.*, 1989, **25** (1), pp. 64–72
13 GABRIEL, W. F.: 'Superresolution techniques in the range domain'. Proceedings of IEEE international conference on *Radar*, May 1990, Arlington, USA, pp. 263–267
14 WIRTH, W. D.: 'High-range resolution for radar by oversampling and LMS pulse compression', *IEE Proc., Radar, Sonar Navig.*, 1999, **146** (2), pp. 95–100

Chapter 8

Detection of targets by a pulse series

The fundamental problems and different aspects of the detection of moving radar targets have long been treated by many authors. We will discuss here those aspects which are especially relevant for a phased-array system.

A series of pulses is generally applied in one beam position. The reason is to provide the possibility for Doppler filtering, clutter suppression, Doppler frequency estimation of targets and eventually for target separation with different Doppler frequencies. Further, the possibility for energy management by adapting the number of pulses to the anticipated target range and cross section is also provided. The first possibility for energy management is in the choice of the pulse length with corresponding modulation, as discussed in chapter 7. Both measures may of course be combined. A special test procedure for phased arrays with a variable number of pulses for each individual test, sequential detection, will be presented and discussed in chapter 9.

For processing all received signals are stored in a memory. Then the signals are read out of this memory according to the individual range cells. For each range cell the period of the signals corresponds to the radar's pulse period T. All Doppler frequency effects of the echoes are considered generally for these signal series corresponding to individual range cells.

A major task for signal processing is the suppression of clutter echoes. We may observe different clutter types:

- Clutter echoes from large fixed targets as buildings and other man made structures. The echoes will be fairly stable in phase and amplitude and will have a large signal-to-noise ratio. The clutter spectrum will be centred around zero Doppler frequency with a narrow bandwidth.
- Clutter echoes from trees. Their fluctuation behaviour depends on the weather conditions, especially on wind strength and season. The clutter spectrum is also centred around zero Doppler frequency but will be broadened to a certain extent.
- Clutter echoes from weather phenomena, especially from rain, hail and snow. Now the clutter spectrum is shifted in the Doppler frequency according to the radial

velocity of the scatterers. The spectral width depends on turbulence phenomena. The shape of the clutter spectrum is changing more or less rapidly with time.

- Clutter echoes from sea waves are also Doppler shifted according to the aspect angle between the radar and the wavefront and wind direction.
- Clutter echoes from moving or flying objects like cars and bird flocks. These clutter echoes may be localised or changing with time.

All types of clutter have been investigated by models and measurement evaluations as a basis for efficient clutter suppression to improve flying target detection within clutter areas [1–3].

Within the radar control system a clutter map will be generated by observing the signals received from all beam directions and range cells. Because with a phased-array radar the beam positions are reproducible by digital steering of the antenna phase shifters, as discussed in chapter 4, the clutter contents of the resolution cells will also be reproducible and depend only on the possible change of the clutter with time.

Clutter types that are relevant for the choice of transmitted signal and the kind of signal processing may be assigned to each resolution cell. We may select for example the following types:

0: no clutter
1: stable fixed clutter at $f_d = 0$
2: fluctuating clutter around $f_d = 0$
3: fluctuating clutter concentrated around a certain Doppler shift f_d
4: fluctuating clutter with a broad spectrum

On the other hand, the echo pulses of flying targets generally show a predominant narrow Doppler spectrum. In other words, they show a fairly constant or regular phase difference between succeeding pulses, determined by the pulse period T and Doppler frequency f_d by $\Delta\varphi = 2\pi f_d T$. This Doppler shift varies only slowly during longer observation times.

Therefore, fortunately, coherent detection procedures, already discussed in chapter 3 in example 6 (page 38) may be applied to achieve optimal detection performance. A test function or test statistic $\lambda(\mathbf{z})$ has to be computed for the received signal series $\mathbf{z}$ and compared with a detection threshold η. If $\lambda(\mathbf{z}) > \eta$ a target is declared detected. The computation of $\lambda(\mathbf{z})$ represents the necessary signal processing.

Equation 3.57 is repeated here for convenience, with the received signal series $\mathbf{z}^* = (z_1, \ldots, z_N)^*$, the covariance matrix $\mathbf{Q} = E\{\mathbf{z}\,\mathbf{z}^*\}$ describing the disturbing clutter signal including receiver noise and the signal vector $\mathbf{s}^* = (s_1, \ldots, s_N)^*$ which is expected at the Doppler frequency f_n:

$$\lambda(\mathbf{z}) = \max_n \left| \mathbf{z}^* \mathbf{Q}^{-1} \mathbf{s}(f_n) \right| \tag{8.1}$$

The radar pulse period T determines the unambiguous range for the Doppler frequency: $f_d = 0, \ldots, 1/T$. With N pulses, a Fourier transform forms N frequency coefficients, which represent N Doppler filters with a separation $1/NT$, forming a filter bank [5–8].

A special condition with phased-array radar for the detection of moving targets is given by the so-called step-scan operation. The beam is steered by the beam-steering computer into a selected direction and is fixed during the complete pulse series. In contrast to conventional mechanical rotating radar there is no modulation of fixed target echoes by the antenna beam pattern. This modulation of mechanical scanning antennas causes a broadening of the clutter spectrum, which is avoided with phased arrays.

The detection test is performed with a finite signal series of length N, selected by the radar control computer. This is also in contrast to conventional radar where for each range cell during the antenna rotation an infinite signal series is fed through the moving-target indication (MTI) filter or the Doppler filter bank. The usual treatment of the MTI filter in the frequency domain is appropriate only for this conventional radar case but not for the step-scan operation. This follows immediately from the following: the rectangular envelope of fixed target echo series in the case of a phased-array radar would produce in frequency-domain representation a spectrum with $(\sin x)/x$ form and therefore high sidelobes and poor clutter suppression. On the other hand we could expect, especially with constant clutter echoes from stable targets, very effective clutter suppression. The explanation of this contradiction lies in the fact that in the phased-array case we have no continuous passing of a signal series through our filter, but have one definite point of time at the end of our signal series in the respective beam position when we have to take the output signal for threshold comparison. That is, we would have to add a switch and consider its effect in the frequency domain. But this turns out to be a very complicated consideration. Therefore, conventional MTI treatments in the frequency domain are not suitable for phased-array radar step-scan operations.

8.1   Filter against fixed clutter

In equation 8.1 the operation $\mathbf{v}^* = \mathbf{z}^*\mathbf{Q}^{-1}$ represents the clutter suppression or filtering part, as is discussed in chapter 3 section 3.5.3. We consider here the suppression of clutter with predominating type 1 (stable and at $f_d = 0$) and only minor parts of type 2 (fluctuating around $f_d = 0$). From fixed targets each echo signal component will be constant in amplitude and phase during the pulse series, because a phased-array radar is generally coherent and stable in frequency. So it would make sense to estimate the mean of the signal series $\hat{m}$ and to subtract it from all individual signals: $v_i = z_i - \hat{m}$.

This may be explained using a simple model for the fixed clutter signal and using the inverse of covariance matrix $\mathbf{Q}$.

Fixed target echoes are given by some $|s_n| \exp(j\varphi_n)$. They are constant in amplitude and phase and therefore independent of index n. All correlation values q are given by $q = s_n^* s_n = |s|^2$. The values q may be normalised to 1. The accompanying receiver noise may then be represented by power r, appearing only at the main diagonal elements, because the noise is uncorrelated from pulse to pulse. For the inverse matrix of $\mathbf{Q}$ we choose to attempt the same structure with unknown values a and b:

$$\mathbf{Q} = \begin{bmatrix} 1+r & 1 & \cdots & 1 \\ 1 & \ddots & \ddots & \vdots \\ \vdots & \ddots & \ddots & 1 \\ 1 & \cdots & 1 & 1+r \end{bmatrix}$$

and

$$\mathbf{Q}^{-1} = \begin{bmatrix} a & b & \cdots & b \\ b & \ddots & \ddots & \vdots \\ \vdots & \ddots & \ddots & b \\ b & \cdots & b & a \end{bmatrix} \tag{8.2}$$

with $\mathbf{Q}\mathbf{Q}^{-1} = \mathbf{I}$ (identity matrix) we get two equations to determine a and b:

$$a(1+r) + (N-1)b = 1$$

for the resulting matrix elements on the main diagonal of $\mathbf{I}$ and:

$$a + (N-1+r)b = 0$$

for all the other elements. From this follows:

$$a = \frac{N-1+r}{r(N+r)}, \qquad b = \frac{-1}{r(N+r)} \tag{8.3}$$

If we normalise to $a = 1$ it follows that $b = -1/(N-1+r)$ and for $\mathbf{Q}^{-1}$ we get:

$$\mathbf{Q}^{-1} = \begin{bmatrix} 1 & \dfrac{-1}{N-1+r} & \cdots & \dfrac{-1}{N-1+r} \\ \dfrac{-1}{N-1+r} & \ddots & \ddots & \vdots \\ \vdots & \ddots & \ddots & \dfrac{-1}{N-1+r} \\ \dfrac{-1}{N-1+r} & \cdots & \dfrac{-1}{N-1+r} & 1 \end{bmatrix} \tag{8.4}$$

By multiplying $\mathbf{z}^*\mathbf{Q}^{-1}$ and for $r \to 0$ we recognise that for each signal element z_i the mean, which is estimated from all the other signal elements is subtracted. This mechanism we expected above. For this procedure first all signals z_i have to be received and stored and then the mean may be computed and subtracted.

A simplified procedure is the following [9,10]. After the beam has been steered into a new direction, the mean is formed from the already received signals for each range cell and subtracted from the current signal. The first filtered output is available after the reception of the second pulse. The estimate of the mean signal is formed by a simple recursive algorithm. After n pulses the filter input is z_n and the output v_n and the available estimated mean from the previous radar period is $\hat{m}_{n-1}$:

$$
\begin{aligned}
v_n &= z_n - \hat{m}_{n-1} \\
\hat{m}_n &= \hat{m}_{n-1} + \alpha_n z_n \\
v_1 &= 0 \\
\hat{m}_1 &= z_1
\end{aligned}
\tag{8.5}
$$

For the recursion factor we choose $\alpha_n = 1/n$. By applying this recursive algorithm, for $n \geq 2$, we arrive as wished:

$$
v_n = z_n - \frac{1}{n-1}(z_1 + \cdots + z_{n-1})
$$

The clutter notch in frequency response becomes narrower with increasing n, and the fluctuating part of the ground clutter may pass the filter more and more. This tendency is limited by substituting the variable factor $\alpha_n = 1/n$ by the constant factor $\alpha_n = 1/n_0$ for $n > n_0$.

The *SNIR* (signal-to-noise-plus-interference ratio) after the clutter suppression is expressed with $\mathbf{s}(f)$ as the Doppler signal for frequency f according to equation 3.80 by:

$$
g(f) = \mathbf{s}^*(f)\mathbf{Q}^{-1}\mathbf{s}(f)
\tag{8.6}
$$

g describes the *SNIR* for a signal at frequency f if the clutter suppression by $\mathbf{Q}^{-1}$ is followed by a Doppler filter at frequency f. Figure 8.1 shows an example for the normalised *SNIR* of the signal-to-clutter ratio for the three types of simplified method of fixed clutter suppression. The curve with markers is valid for the application of the matrix given by equation 8.4. This function is approximated by the recursive mean estimation and subtraction with and without limiting of the recursion factor α_n.

By this notch-filter characteristic the fixed clutter signals will be mostly suppressed and the following Doppler filters will gather much less clutter power by their sidelobes in the clutter frequency region. This reduces the dynamic range of the signals and correspondingly, also, the necessary word length of the Doppler filter processor following clutter suppression.

This method of clutter suppression is suitable for pulse series with the pulse number a variable or unknown in advance. Therefore, it may be applied to flexible Doppler processing with the pulse number adapted to the required Doppler resolution, as discussed in the next section. It may also be applied in combination with sequential detection for target search (chapter 9). Incoherent sequential detection with its advantages may then also be applied within areas containing stable fixed clutter.

For longer observation times we have to apply an integration with some robustness against change of the targets' Doppler frequency. The so-called ACE test procedure, discussed in section 8.5, relies also on previous suppression of fixed clutter signals.

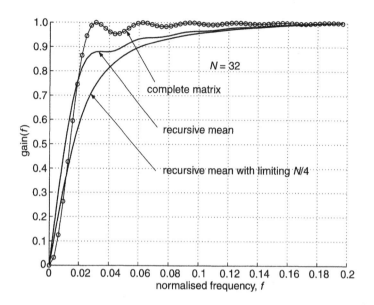

Figure 8.1   Frequency response of gain against fixed clutter

8.2   Doppler filter processor

Doppler processing is the basic method for the detection of moving targets in search and tracking mode. For search it is applied to the output signals of the sum beam; pulse compression and fixed clutter suppression may already have been performed. The aims of Doppler processing are coherent integration of the target echoes for optimal detection out of noise and clutter, separation from clutter signals and estimation of the Doppler frequency f_d as an important target parameter which is used for tracking.

The processing is performed as shown in the block diagram in Figure 8.2. In each radar pulse period the echo signals for all range cells, determined by the subpulse length τ according to chapter 7, are written into the memory in direction write or R (range). When the signals from all pulse periods are stored, the signals can be read out for each range cell in the direction read or pulse periods. By this memory operation a so-called corner turn is achieved. The signals from each range cell are then processed by a Fourier transform, generally realised with the FFT algorithm. Each Fourier coefficient represents a Doppler filter, all coefficients together are equivalent to a filter bank.

The filters are separated by the frequency difference $\Delta f = 1/NT$. As indicated in Figure 8.2, by increasing the pulse number N the resolution in frequency is improved. Also the gain in signal-to-noise ratio (SNR) is increased nearly proportionally to the factor N. By increasing the pulse-recurrence frequency (PRF) the unambiguous frequency range is also increased. So we gain high flexibility by the choice of the parameters N and PRF.

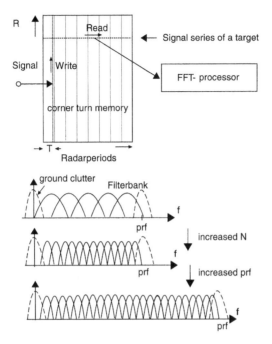

Figure 8.2 Block diagram for Doppler processing

Further processing of the outputs from the individual filters is given by equations 3.56 and 3.57, repeated here:

$$\lambda(\mathbf{z}) = \frac{1}{N} \sum_{n=0}^{N-1} I_0 \left(2 \left| \mathbf{z}^* \mathbf{Q}^{-1} \mathbf{s}(f_n) \right| \right) \quad \text{with } f_n = \frac{1}{NT} n \tag{8.7}$$

In practise, this inconvenient test function is replaced by:

$$\lambda(\mathbf{z}) = \max_n \left(\left| \mathbf{z}^* \mathbf{Q}^{-1} \mathbf{s}(f_n) \right| \right) \tag{8.8}$$

which means, we have to look for that Doppler filter or Fourier coefficient number n with the maximum output signal. This value is then compared with the detection threshold η.

Applying equation 8.8 represents nonlinear processing. This has to be taken into account if further processing steps are planned, e.g. target location, range estimation or coherent Doppler clutter suppression in the frequency domain. In these cases the coherent filter outputs have to be used for further processing.

Because the Doppler frequency is unknown, we cannot achieve the gain of coherent integration given by the factor N, as discussed in chapter 3 in connection with example 5. All N filters give a certain noise output, which is mutually uncorrelated because each filter passes only the noise in its frequency band. We have to apply a somewhat higher detection threshold η compared to that for a single filter (matched to a known Doppler frequency) to achieve the required false-alarm probability P_f. Then the detection probability P_D will be lower and we have to assume or require a higher SNR for the target signal. This increase in SNR is a loss L which we have to pay for ignorance of the Doppler frequency. It has been determined by a series of simulation studies [11] to be:

$$L = 0.26 \log_2(N) \; \text{[dB]} \tag{8.9}$$

$\log_2(x)$ means the logarithm of variable x to the base 2. Doubling N adds 0.26 dB to the loss L.

For a target Doppler frequency halfway between the centre frequencies of adjacent filters there is an additional straddling loss of about 4 dB. This can be avoided by increasing the filter number N, for example doubling to $2N$. With the FFT algorithm this may be achieved by doubling the formal N by zero padding. That means the received signal series of N echoes is completed by an additional N zeros. The result is shown in Figure 8.3: the filter-bank response is computed for the usual FFT response above the frequency, normalised to $1/T$. With zero padding the doubled number of filters reduces the straddling loss to about 1.7 dB.

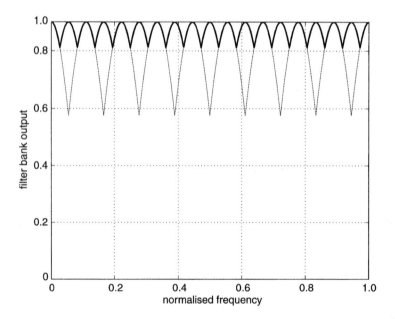

Figure 8.3   Doppler filter-bank response for usual (dashed line) and with double length of signal series (thick line), $N = 8$

8.3 Adaptive suppression of weather clutter

Adaptive clutter suppression is necessary especially against nonstationary disturbing signals, for example weather echoes, echoes from sea waves or chaff, to maintain target detection as far as possible in these situations. The clutter echoes show a Doppler frequency shift depending on the radial velocity of the scatterers. Depending on the PRF (which governs the unambiguous Doppler range), and the velocity profile of the scatterers a considerable part of the Doppler frequency range may be occupied.

By the Doppler filter bank (FFT) a certain separation of frequency regions with and without clutter disturbance is achieved. By a detection threshold individually adapted to the Doppler channels the false-alarm probability could be kept constant for all filter outputs. The threshold could be derived by forming the mean of the outputs from neighbouring range cells.

Parts of the clutter spectrum pass the sidelobes of the individual filters, deteriorating the target detection. For improved clutter suppression one could estimate and apply the covariance matrix $\mathbf{Q}$ of equation 8.8 for the respective clutter situation. The problem is getting the necessary sample number of clutter signals for estimating the covariance matrix with dimension $[N, N]$. There should be about $3N$ sampled data sets or $3N^2$ signals to achieve a reasonable matrix estimation without too high a fluctuation of the matrix elements, as will be discussed in chapter 10. For $N = 64$ and a PRF $= 1\,\text{kHz}$ the necessary time for gathering the signals would amount to 12.28 s.

On the other hand, the clutter output signals of the filters are correlated. This may be used in mutual clutter elimination according to the discussion in chapter 3 section 3.5.2. Instead of applying the covariance matrix between all filters we may construct suboptimal processing using only some neighbour filter outputs around the actual Doppler channel. This was suggested in References 12 and 13 under the name DDL-GLR (Doppler-domain localised generalised likelihood-ratio processor). At the output of these filters the clutter signals are correlated, particularly with the clutter signal at the actual target filter output. This correlation may, to a certain extent, be used for clutter suppression.

We start from equation 8.8 for forming the test function from the measured signal vector $\mathbf{z}$ and its transformation into the frequency domain by applying the Fourier transform. We introduce a Fourier matrix:

$$\mathbf{F} = \frac{1}{\sqrt{N}} \begin{bmatrix} \exp\left(-\lambda_{11}\right) & \dots & \exp\left(-\lambda_{1N}\right) \\ \vdots & \ddots & \vdots \\ \exp\left(-\lambda_{N1}\right) & \dots & \exp\left(-\lambda_{NN}\right) \end{bmatrix} \quad \text{with } \lambda_{ik} = \frac{2\pi j}{N} ik \quad (8.10)$$

According to equation 2.20, we transform the signal $\mathbf{z}$ into the frequency domain $\mathbf{Z}$ with elements Z_n for frequency n/NT $(n = 1, \dots, N)$ by:

$$\mathbf{Z} = \mathbf{Fz} \quad (8.11)$$

The clutter signal (including noise) in the time domain may be denoted by $\mathbf{c}$. After transformation into the frequency domain we get the vector $\mathbf{C}$. For the relationship between the correlation matrices in the time and frequency domain we get:

$$\mathbf{P} = E\{\mathbf{CC}^*\} = E\{\mathbf{Fcc}^*\mathbf{F}^*\} = \mathbf{F}E\{\mathbf{cc}^*\}\mathbf{F}^* = \mathbf{FQF}^*$$

and with $\mathbf{F}^*\mathbf{F} = \mathbf{I}$:

$$\mathbf{P}^{-1} = \mathbf{FQ}^{-1}\mathbf{F}^*$$

This can be seen by $\mathbf{PP}^{-1} = \mathbf{I}$. $\mathbf{P}$ is the correlation matrix of the spectral components and is denoted also as a spectral matrix.

The transformed test function now becomes:

$$\lambda(\mathbf{Z}) = \max_n \left(\left| \mathbf{Z}^*\mathbf{P}^{-1}\mathbf{S}_n \right| \right) = \max_n \left(\left| \mathbf{z}^*\mathbf{F}^*\mathbf{FQ}^{-1}\mathbf{F}^*\mathbf{Fs}(f_n) \right| \right)$$
$$= \max_n \left(\left| \mathbf{z}^*\mathbf{Q}^{-1}\mathbf{s}(f_n) \right| \right)$$
$$= \lambda(\mathbf{z})$$

Both versions for λ are therefore equivalent. In the frequency domain the clutter will be concentrated usually to few components Z_n of $\mathbf{Z}$, but in the time domain the clutter is contained in all signal components.

The suggestion is to construct a suboptimal test by using only a relatively small number of Doppler filter outputs $M < N$ for the purpose of clutter suppression after the Doppler filter bank. This would be a postfiltering technique and the processing scheme is sketched in Figure 8.4. For a Doppler filter with index i a symmetric surrounding of length M is chosen. The local spectral matrix $\mathbf{P}_{M_i}$ is estimated and $\mathbf{P}_{M_i}^{-1}$ computed. We define an ideal expected signal vector in frequency domain $\mathbf{e}$ of length M, with a 1 in the centre and 0 otherwise. Then the filter vector $\mathbf{w}$ of length M becomes equal to the centre column of $\mathbf{P}_{M_i}^{-1}$:

$$\mathbf{w} = \mathbf{P}_{M_i}^{-1}\mathbf{e}$$

For each i $(i = 1, \ldots, N)$ this filter vector is computed and applied to all Doppler filters together with those within their neighbourhood range of length M. This processing may be applied only to the Doppler region disturbed by clutter.

The number of complex multiplications is reduced to $(M^3 + M)N$ in contrast to the optimal processing with $N^3 + N^2$ operations. This number of operations is shown above N with curve 2 in Figure 8.5 for $M = 5$. For comparison curve 1 is given for optimal processing. It is further possible to perform suboptimal clutter suppression for complete sections of length M by multiplying by the matrix $\mathbf{P}_{M_i}^{-1}$. Then we have to estimate and invert $\mathbf{P}$ only once for a Doppler section of length M. The number of operation reduces to $(M^3 + M)N/M$, shown in Figure 8.5 with curve 3. For N above 5 one can see the advantages of this suboptimal procedure.

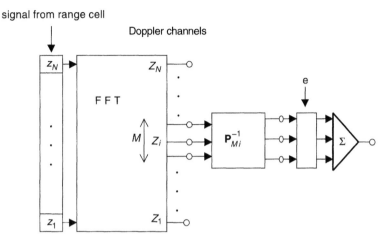

Figure 8.4   Block diagram for suboptimal adaptive clutter suppression in the frequency domain by postfiltering

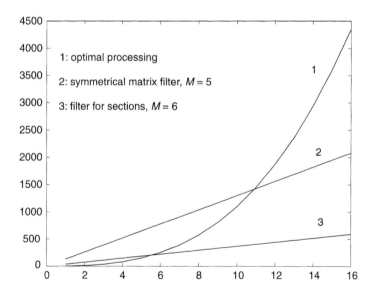

Figure 8.5   Number of complex multiplications for adaptive postfiltering above N

8.3.1   Computation of the gain

The clutter power at the input of the Doppler filter bank is given by:

$$trace(\mathbf{P}) = E\{\mathbf{C}^*\mathbf{C}\} = E\{\mathbf{c}^*\mathbf{F}^*\mathbf{F}\mathbf{c}\} = E\{\mathbf{c}^*\mathbf{c}\} = trace(\mathbf{Q})$$

trace means forming the sum along the main diagonal of a matrix. The gain at Doppler frequency i is the ratio of the signal-to-clutter-plus-noise ratios at the output and the input:

$$G_i = \frac{P_s}{P_c}\bigg|_{out} \bigg/ \frac{P_s}{P_c}\bigg|_{input} \tag{8.12}$$

The target signal with amplitude 1 produces with:

$$\mathbf{e}_i = \begin{bmatrix} 0 \\ \vdots \\ 1 \\ \vdots \\ 0 \end{bmatrix}$$

the one at row i, all other elements are 0, the transformed signal:

$$\mathbf{S}_i = \mathbf{F}\mathbf{s}_i = \sqrt{N}\mathbf{e}_i$$

With this we get:

$$\frac{P_s}{P_c}\bigg|_{i_{in}} = \frac{N}{trace(\mathbf{Q})} = \frac{N}{trace(\mathbf{P})} \tag{8.13}$$

and:

$$\frac{P_s}{P_c}\bigg|_{i_{out}} = \frac{S_i^* \mathbf{g}_i \mathbf{g}_i^* S_i}{E\{(\mathbf{g}_i^*\mathbf{C})(\mathbf{C}^*\mathbf{g}_i)\}}$$

with:

$$\mathbf{g}_i = \mathbf{P}_{M_i}^{-1}\hat{\mathbf{e}}_i \tag{8.14}$$

$\mathbf{P}_{M_i}$ is a $M \times M$ section out of $\mathbf{P}$ around the diagonal position ii. For the output we get:

$$\frac{P_s}{P_c}\bigg|_{i_{out}} = \frac{\sqrt{N}(\hat{e}_i^*\mathbf{P}_{M_i}^{-1}\hat{e}_i)(\hat{e}_i^*\mathbf{P}_{M_i}^{-1}\hat{e}_i)\sqrt{N}}{\hat{e}_i^*\mathbf{P}_{M_i}^{-1}\mathbf{P}_{M_i}\mathbf{P}_{M_i}^{-1}\hat{e}_i} = N(\mathbf{P}_{M_i}^{-1})_{ii} \tag{8.15}$$

$(\mathbf{P}_{M_i}^{-1})_{ii}$ is the element at position ii in matrix $\mathbf{P}_{M_i}^{-1}$. The final result for the gain is:

$$G_i = (\mathbf{P}_{M_i}^{-1})_{ii}\,trace(\mathbf{P}) \tag{8.16}$$

This equation is valid for the symmetrical post filter: the respective central element has to be taken. For postfiltering in sections the diagonal elements corresponding to the respective Doppler channel have to be taken.

8.3.2 Evaluation of experimental signals

With the experimental phased-array system ELRA of FGAN, described in chapter 17, it was possible to record clutter signals from fixed ground targets and rain clouds. With these data the suggested postfiltering could be evaluated for some examples.

For a fixed beam position out of a range interval with 32 range cells a series of 64 pulses was transmitted and the echoes stored in a memory. This was repeated 64 times. First, the echoes are transformed into the Doppler domain. In Figure 8.6 the range Doppler matrix is shown, that is the 64 Doppler output amplitudes for 32 range cells. We recognise some range cells with Doppler-shifted weather clutter and also some fixed clutter. The matrix estimation could be performed with 64 data-sample vectors for a selected range cell. For matrix estimation and inversion up to the dimension $M = 7$ this sample number is sufficient and fulfils the condition $64 > 3M$.

In Figure 8.6 we selected range $r = 26$. The corresponding estimated correlation matrix $\mathbf{P}$ is shown in Figure 8.7. From this matrix $\mathbf{P}$ are selected the submatrices $\mathbf{P}_{Mi}$ as described above and the gain dependent on the Doppler index i is computed with equation 8.16. The result is given in Figure 8.8.

Using the Doppler filter bank a certain improvement against the clutter has already been achieved, especially for those filters with low sidelobes within the clutter frequency area. The gain is computed as a differential gain, that is, the gain which would be achieved by the single Doppler filters alone is subtracted from the result of equation 8.16. We show only the additional gain by applying the matrix filter with

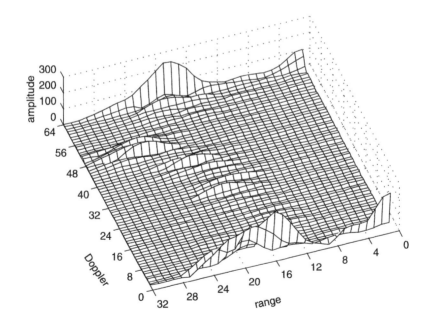

Figure 8.6   Range Doppler matrix with ground and weather clutter, range cells =
32, Doppler filters = 64, 64 samples

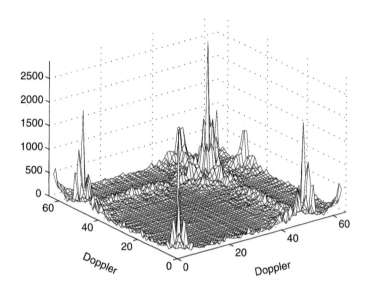

*Figure 8.7   Doppler correlation matrix **P** for N = 64, range cells = 26, 64 samples*

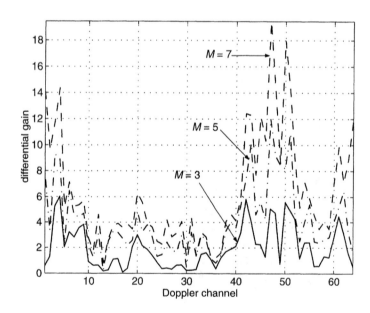

Figure 8.8   Differential gain for weather and ground clutter postfiltering, pulses =
64, M = 3, 5, 7

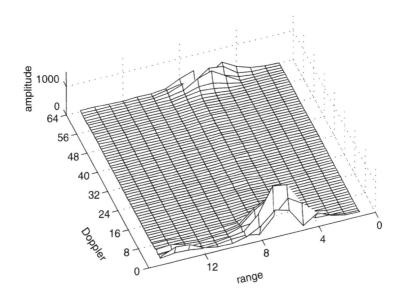

Figure 8.9 Range Doppler matrix for weather clutter, range cells = 15, Doppler filters = 64, no. of measurements = 64

dimension M. For this example the differential gain achieves values up to 18 dB for $M = 7$; with $M = 3$ the filter would not be efficient. In another example we recorded weather clutter alone. These results are given by Figures 8.9–8.11. Now even the matrix filter with $M = 3$ achieves good results. We recognise in the range Doppler matrix Doppler-shifted weather clutter with frequencies near the Doppler frequency equal to zero. We selected range $r = 6$ for the evaluation. The result for the differential gain in Figure 8.10 shows values up to 13 dB.

Next we show in Figures 8.12 and 8.13 the corresponding results for postfiltering in sections of M with a corresponding matrix $\mathbf{P}_M$, as mentioned in section 8.3.1. The gain is found as the diagonal elements of $\mathbf{P}_{M_i}^{-1}$ corresponding to the Doppler channel. The results for the differential gain are comparable with the results for the sliding application of $\mathbf{P}_{M_i}^{-1}$, described before. The saving in computing complexity, as shown with curve 3 in Figure 8.5, is considerable.

8.4 Suppression of sea clutter

Special clutter condition is given for the detection of small targets, such as buoys, growlers or periscopes, in sea clutter. According to measurements the coherent clutter spectrum has a width of about 100 Hz, whereas the spectrum of targets will have only a spectral width of 5 Hz, caused by target movement owing to water waves [14]. To

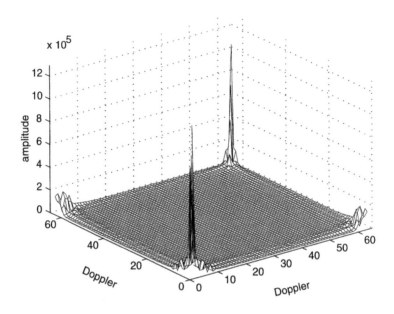

*Figure 8.10   Doppler correlation matrix **P** for weather clutter, range cell number 6,
samples = 64*

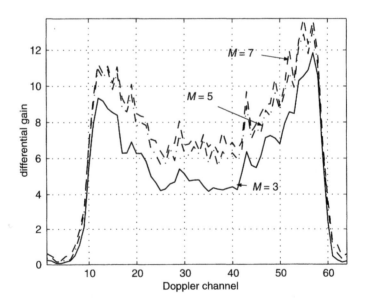

*Figure 8.11   Differential gain against weather clutter by postfiltering, pulses = 64,
M = 3, 5, 7*

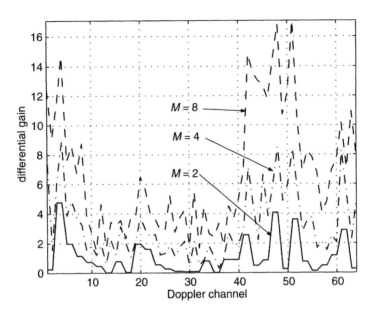

Figure 8.12 Postfiltering in sections equal to M, pulses = 64, M = 2, 4, 8

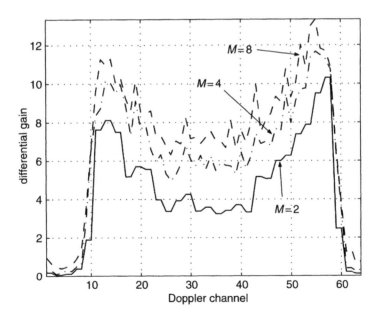

Figure 8.13 Post-filtering in sections equal to M, pulses = 64, M = 2, 4, 8

resolve such a narrow spectrum out of the broader clutter spectrum the dwell time for each beam direction has to be about 200 ms. For a horizon search with about 100 beam positions the radar would need a time of 20 s. This is an unacceptable and unnecessarily long time.

A solution is possible with a phased-array radar using beam agility to establish a direction multiplex operation [15].

The pulse-recurrence frequency is reduced substantially for each beam direction. The number of pulses necessary for target detection are distributed within a longer time interval. After the possible echo delay time corresponding to the selected range there is time left for steering the beam into another direction, transmitting a pulse and receiving those echoes, and so on. Then the beam returns to the first direction for the second pulse. By this scheme the direction-time multiplex operation is established. This would not be possible with a mechanical scanning radar.

All echoes have to be stored in a memory. After all echoes of the selected set of beam directions are received the signals series are read out of the memory according to the respective beam positions and evaluated with a Fourier transform (FFT). In Figure 8.14 an example for the time schedule with blocks of 35 beam positions (BP) and 42 pulses for each direction is given. Each little vertical bar represents the radar transmit pulse and the corresponding receive interval.

Because the pulse period in each direction is longer than usual the second time around echoes of big targets like ships are avoided and small targets at short range are not masked. Second time around echoes from other directions are attenuated according to the respective sidelobe level.

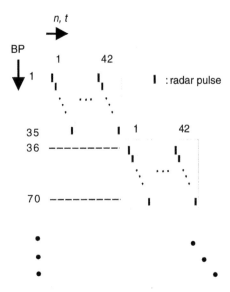

Figure 8.14   Time schedule for sea-clutter suppression

8.5 Estimation of Doppler frequency

The estimate of the target Doppler frequency may be of interest, for example, for target tracking [16]. We apply the estimation procedure by maximising the probability density dependent on the parameter sought, as discussed in chapter 3 section 3.4. The received target signal may be given by:

$$a = |a| \exp(j\varphi)$$

where T = radar pulse period, f = frequency to be estimated:

$$\mathbf{s} = a \begin{bmatrix} \vdots \\ \exp[2\pi j T f n] \\ \vdots \end{bmatrix} = a\mathbf{d} \quad n = -N/2, \dots, N/2 \qquad (8.17)$$

$\mathbf{d}$ is the Doppler vector. The received signal including noise is $\mathbf{z} = \mathbf{s} + \mathbf{n} = a\mathbf{d} + \mathbf{n}$ and its probability density is given with equation 3.41 by:

$$p(\mathbf{z}/a, f) = \frac{1}{(2\pi)^{(N+1)}} \exp\left[-\frac{1}{2}(\mathbf{z} - a\mathbf{d})^*(\mathbf{z} - a\mathbf{d})\right]$$

The maximum of p is given by the minimum of $Q = (\mathbf{z} - a\mathbf{d})^*(\mathbf{z} - a\mathbf{d})$:

$$\frac{\partial Q}{\partial a} = -\mathbf{d}^*(\mathbf{z} - a\mathbf{d}) + (\mathbf{z} - a\mathbf{d})^*(-\mathbf{d})$$

$$\frac{\partial Q}{\partial a} = 0$$

is the condition for the estimate of a:

$$\hat{a} = \frac{1}{N+1}\mathbf{d}^*\mathbf{z} \qquad (8.18)$$

Now only $\mathbf{d}$ is left unknown. For Q then results:

$$Q = \left(\mathbf{z} - \frac{1}{N+1}\mathbf{d}\mathbf{d}^*\mathbf{z}\right)^*\left(\mathbf{z} - \frac{1}{N+1}\mathbf{d}\mathbf{d}^*\mathbf{z}\right) = \mathbf{z}^*\mathbf{z} - \frac{1}{N+1}(\mathbf{z}^*\mathbf{d})(\mathbf{d}^*\mathbf{z})$$

$\mathbf{d}^*$ may be seen as the filter at frequency f. $\mathbf{z}\mathbf{z}^*$ is independent of f. $S = \mathbf{d}^*\mathbf{z}$ is the filter output for signal $\mathbf{z}$. The minimum of Q means maximum of $P = SS^* = |S|^2$.

We look for the maximum of P dependent on f: we use $F = \ln P(f)$ instead of P [18]. From detection by the filter bank we know the corresponding filter at f_0 near f. The estimate $\hat{f}$ will also be near f_0. We look for the zero point of the derivative of F with respect to f.

We abbreviate $dF(f)/df = F_f$. We then apply a Taylor series of F_f around $\hat{f}$:

$$F_f(f) = F_f(\hat{f}) + F_{ff}(\hat{f})(f - \hat{f}) + \cdots$$

We set $f = f_0$ and take into account $F_f(\hat{f}) = 0$ and we arrive at the general equation:

$$\hat{f} - f_0 = -\frac{F_f(f_0)}{F_{ff}(\hat{f})} \tag{8.19}$$

For the numerator we get:

$$F_f = \frac{d\ln P}{df} = \frac{1}{P}\frac{dP}{df} = \frac{(\mathbf{d_f^* z})(\mathbf{z^* d}) + (\mathbf{d^* z})(\mathbf{z^* d_f})}{(\mathbf{d^* z})(\mathbf{z^* d})}$$

$$= 2\,\text{Re}\left\{\frac{\mathbf{d_f^*}(f_0)\mathbf{z}}{\mathbf{d^*}(f_0)\mathbf{z}}\right\} \tag{8.20}$$

The denominator of equation 8.20 is with equation 8.18 proportional to $\hat{a}$. Further we get:

$$\mathbf{d_f^*} = -2\pi jT\mathbf{d^*}diag(n)$$

with the diagonal matrix:

$$diag(n) = \begin{bmatrix} \ddots & 0 & 0 \\ 0 & n & 0 \\ 0 & 0 & \ddots \end{bmatrix} \quad n = -N/2, \ldots, N/2$$

For the computation of F_{ff} at frequency $\hat{f}$ we get:

$$F_{ff} = \left(\frac{P_f}{P}\right)_f = \frac{P_{ff}P - P_f^2}{P^2} = \frac{P_{ff}}{P} \quad \text{because } P_f(\hat{f}) = 0$$

With $S = \mathbf{z^* d}$ (as above) we get:

$$P_f = S_f S^* + SS_f^*$$

$$P_{ff} = S_{ff}S^* + S_f S_f^* + S_f S_f^* + SS_{ff}^* = S_{ff}S^* + SS_{ff}^* \quad \text{because } S_f(\hat{f}) = 0$$

$$F_{ff} = \frac{S_{ff}S^* + SS_{ff}^*}{S^*S} = 2\,\text{Re}\left\{\frac{S_{ff}}{S}\right\}$$

For a high signal-to-noise ratio (SNR) we have at $\hat{f}$ for the received signal $\mathbf{z} = \hat{a}\mathbf{d}$, that is we neglect noise, and $S = (\mathbf{d^* d})\hat{a} = (N+1)\hat{a}$. With this assumption we get:

$$S_{ff} = \mathbf{d_{ff}^* z} = \mathbf{d_{ff}^*}\mathbf{d}\hat{a} = -\hat{a}(2\pi T)^2 \sum_{n=-N/2}^{N/2} n^2$$

and:

$$F_{ff} = -\frac{2}{(N+1)}(2\pi T)^2 \sum_{n=-N/2}^{N/2} n^2 = -\frac{2}{k}$$

with:

$$k = \left(\frac{1}{(N+1)}(2\pi T)^2 \sum_{n=-N/2}^{N/2} n^2\right)^{-1} \tag{8.21}$$

and finally as the estimation processing rule:

$$\hat{f} - f_0 = k \operatorname{Re}\left\{\frac{\mathbf{d}_f^* \mathbf{z}}{\mathbf{d}^* \mathbf{z}}\right\} \tag{8.22}$$

The signal series has to be weighted additionally with n and with the Doppler vector $\mathbf{d}$ at frequency f_0. It results in a differential filter function with its zero at f_0. An example for $N = 64$ is given in Figure 8.15. The abscissa x is the frequency difference from f_0, normalised to the filter bank separation frequency difference $1/NT$. The vertical dashed lines indicate this frequency separation, which is approximately the half-power bandwidth of the filters.

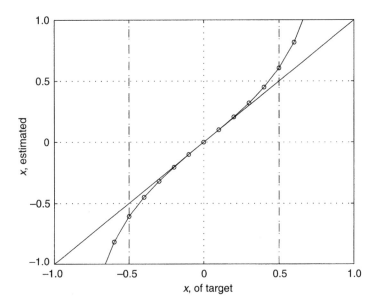

Figure 8.15   *Estimation of Doppler frequency, x: frequency difference to f_0, normalised to filter separation $1/NT$*

8.5.1   Accuracy of Doppler estimation by Cramér-Rao bound

The unknown parameters are the components of the complex signal amplitude $a = g + jh$ and the Doppler frequency f. Our parameter set is $\vartheta = (\vartheta_1, \vartheta_2, \vartheta_3) = (g, h, f)$.

The basic relations are given by equations 3.71 and 3.74 in chapter 3. For a parameter set we have to compute the matrix $\mathbf{G} = \mathbf{F}^{-1}$. The elements of the matrix $\mathbf{F}$ (dimension $[3, 3]$) are given by:

$$F_{ij} = -E\left\{ \frac{\partial^2 \ln p}{\partial \vartheta_i\, \partial \vartheta_j} \right\} \tag{8.23}$$

The matrix elements G_{11}, G_{22}, G_{33} will be the desired lower limits for the variances of g, h, f, respectively. We start from equation 3.41 with $\mathbf{s}$ given by equation 8.17 by $\mathbf{s} = a\mathbf{d}$:

$$p(\mathbf{z}/a,\, f) = \frac{1}{(2\pi)^{(N+1)}} \exp\left[-\frac{1}{2}(\mathbf{z} - \mathbf{s})^*(\mathbf{z} - \mathbf{s}) \right]$$

$$\ln p(\mathbf{z}/f) = const - \tfrac{1}{2}[(\mathbf{z} - \mathbf{s})^*(\mathbf{z} - \mathbf{s})] = const - Q$$

Equation 8.23 becomes:

$$F_{ij} = E\left\{ \frac{\partial^2 Q}{\partial \vartheta_i\, \partial \vartheta_j} \right\} \quad i, j = 1, 2, 3$$

The first derivative results in:

$$Q_{\vartheta_i} = \tfrac{1}{2}\left[-\mathbf{s}_{\vartheta_i}^*(\mathbf{z} - \mathbf{s}) + (\mathbf{z} - \mathbf{s})^*(-\mathbf{s}_{\vartheta_i}) \right]$$

and from this the second derivative:

$$Q_{\vartheta_i \vartheta_j} = \tfrac{1}{2}\left[-\mathbf{s}_{\vartheta_i \vartheta_j}^*(\mathbf{z} - \mathbf{s}) + \mathbf{s}_{\vartheta_i}^* \mathbf{s}_{\vartheta_j} + \mathbf{s}_{\vartheta_j}^* \mathbf{s}_{\vartheta_i} - (\mathbf{z} - \mathbf{s})^* \mathbf{s}_{\vartheta_i \vartheta_j} \right]$$

Both first terms with $(\mathbf{z} - \mathbf{s})$ have an expectation equal to zero and we get:

$$Q_{\vartheta_i \vartheta_j} = \mathrm{Re}\left[\mathbf{s}_{\vartheta_i}^* \mathbf{s}_{\vartheta_j} \right] \tag{8.24}$$

With equation 8.24 we compute all elements of $\mathbf{F}$:

$$F_{11} = \mathrm{Re}\left(\mathbf{s}_g^* \mathbf{s}_g \right) = \mathrm{Re}\,(\mathbf{d}^*\mathbf{d}) = N + 1$$

$$F_{12} = F_{21} = \mathrm{Re}\left(\mathbf{s}_g^* \mathbf{s}_h \right) = \mathrm{Re}\,(j\mathbf{d}^*\mathbf{d}) = 0$$

$$F_{22} = \mathrm{Re}\left(-j\mathbf{d} \cdot j\mathbf{d} \right) = N + 1$$

$$F_{33} = \mathrm{Re}\left(\mathbf{s}_f^* \mathbf{s}_f \right) = \mathrm{Re}\left(a^* a (2\pi T)^2 \mathbf{d}^* diag(n)\, diag(n)\mathbf{d} \right)$$

$$= 4\pi^2 |a|^2 T^2 \sum_{n=-N/2}^{N/2} n^2$$

$$F_{13} = F_{31} = F_{23} = F_{32} = 0$$

because these expressions contain the sum:

$$\sum_{n=-N/2}^{N/2} n = 0$$

As a result we get for $\mathbf{F}$ a diagonal matrix. This means that the errors are independent and we could also have applied equation 3.67 for a single parameter. The matrix $\mathbf{G}$ is then simply a diagonal matrix with reciprocal elements of $\mathbf{F}$. For the desired lower limit for the variance of the frequency estimate $\hat{f}$, $1/F_{33}$ applies. We normalise the standard deviation to the frequency separation of the filters within the filter bank $\Delta f = 1/T(N+1)$:

$$\sigma_{\hat{f}}^2 \geq \frac{(N+1)^2}{4\pi^2 |a|^2 \sum_{n=-N/2}^{N/2} n^2}$$

and with $\sum_{i=1}^{K} i^2 = \frac{1}{6}K(K+1)(2K+1)$ and the signal-to-noise ratio at the output of the filter:

$$\text{SNR} = \frac{|a|^2}{2}(N+1)$$

we get finally:

$$\frac{\sigma_f}{\Delta f} \geq \approx \frac{\sqrt{6}}{2\pi} \frac{1}{\sqrt{\text{SNR}|_{out}}} = \frac{0.39}{\sqrt{\text{SNR}|_{out}}} \tag{8.25}$$

As expected, the standard deviation is inversely proportional to the SNR $= |a|^2/2$ of the received target signal.

Example: for SNR $= 13$ dB or SNR $= 20$ we get $\sigma_f/\Delta f \geq 0.087$.

For a reasonable SNR the Doppler frequency may be estimated to better than $1/10$ of the filter separation. This result compares well to the monopulse estimation discussed in chapter 11.

8.5.2 Simplified doppler estimator

A simplified estimation is possible by using directly the filter outputs. We assume that a target is detected by the filter within the filter bank with frequency index i. Then we apply the suboptimal estimator:

$$\hat{f} = f_i + k_1 \text{ Re} \left\{ \frac{(\mathbf{d}_{i+1} - \mathbf{d}_{i-1})^* \mathbf{z}}{\mathbf{d}_i^* \mathbf{z}} \right\} \tag{8.26}$$

That is we use the neighbouring filters with indices $i+1$ and $i-1$ to produce a difference pattern and use the actual filter with index i for normalisation. The resultant estimation characteristic in shown in Figure 8.16, which looks very similar

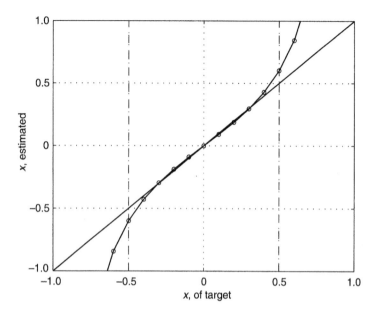

Figure 8.16   *Suboptimal Doppler estimator,* x: *frequency difference to* f_0, *normalised to filter separation* $1/NT$

to the optimum characteristic given in Figure 8.15. The factor k_1 has been found by simulation trials:

$$k_1 = 0.9$$

The price for this simplified estimator without additional weighting with n is an increase in the standard deviation of $\hat{f}$. This has been found by simulation to be about a factor of only 1.12 or an increase of 12 per cent.

8.6   Coherent detection with long pulse series

For the purpose of energy management in favour of the detection of targets with very low radar cross section a much longer pulse series may be applied. Dependent on the PRF, the target dwell time may then also be relatively long. For these applications a change of the Doppler frequency during the dwell time has to be taken into account.

8.6.1   Variable Doppler frequency of the target signal

During longer dwell times a target flying on a tangential course shows a varying radial velocity and Doppler frequency. The range, $r(t)$, at time t for a target with velocity

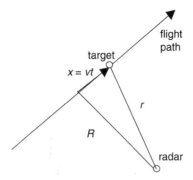

Figure 8.17   Tangential flight

v is given by (R = range at boresight for $t = 0$) (Figure 8.17):

$$r^2 = R^2 + x^2 \quad x = vt$$

The phase changes with:

$$\Phi(t) = \frac{4\pi}{\lambda} r(t)$$

where λ is the radar wavelength. The Doppler frequency is given by the time derivative:

$$\Phi(t) = \frac{4\pi}{\lambda} R \left\{ 1 + \left(\frac{vt}{R} \right)^2 \right\}^{1/2}$$

$$\omega_d(t) = \frac{4\pi v^2}{\lambda R} \frac{t}{\{1 + (vt/R)^2\}^{1/2}} \approx \frac{4\pi v^2 t}{\lambda R} \tag{8.27}$$

The approximation is valid only for small values t. The course of $\omega_d(t)$ is given in Figure 8.18. The fastest change in Doppler frequency $\omega_d(t)$ occurs at the tangential point of the flight path with respect to the radar and may be given by the linear approximation around $t = 0$.

This linear frequency modulation (LFM) given by equation 8.27 results in a problem for target detection with a filter bank, particularly for high values of N, because the filter bandwidth, given by $1/NT$, becomes very narrow and the signal energy would be distributed into several filters. The maximum filter output would therefore decrease and the detection probability be reduced.

The frequency sweep width is characterised by a sweep factor f, which gives the number of filters that are passed by the signal frequency during the target dwell time. The filter spacing is given by $\Delta\omega = 2\pi/NT$. With equation 8.27 follows for the Doppler frequency change:

$$\Delta\omega_d = \frac{4\pi v^2 NT}{\lambda R}$$

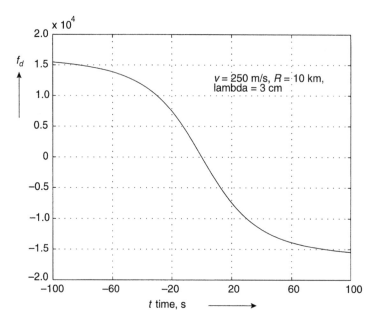

Figure 8.18   Change of Doppler frequency for tangential flight with time

and for the sweep factor f:

$$f = \frac{2v^2(NT)^2}{\lambda R} \tag{8.28}$$

An example is given for illustration: for $\lambda = 0.03$ m, $R = 10^4$ m, PRF $= 1$ kHz, $N = 100$ and $v = 300$ m/s ($NT = 0.1$ s) the result is $f = 6$.

Flying targets may have even higher velocities and a corresponding higher number of filters will be passed.

In the case of an airborne phased-array radar, for the high PRF (HPRF) mode with a PRF of 150 kHz, there are a high number of pulses with $N = 2000$. But the dwell time is only $NT = 0.0133$ s and for the other parameters as above follows $f = 0.1$. The Doppler signal will remain within one filter.

8.6.2   Coherent test function for long echo series

In the following a test function suitable for longer pulse series will be discussed. This test function is derived from the likelihood ratio for coherent detection. The test function corresponding to one range cell for a signal series $\mathbf{z}$ of length N, amplitude a, unknown starting phase and an unknown Doppler phase difference $\varphi = \omega_d t$ may be written according to chapter 3 equation 3.55 as:

$$T(\mathbf{z}) = \frac{1}{2\pi} \int\limits_0^{2\pi} I_0(2a|\mathbf{z}^*\mathbf{s}(\varphi)|)\,d\varphi$$

After Selin [17] this expression may be approximated for small values of a by a Taylor series for the function I_0. The result is:

$$T = \frac{a^2}{64}\left[\frac{16}{a^2}S_0 + S_0^2 + 2\sum_{k=1}^N |S_k|^2\right] \tag{8.29}$$

with:

$$S_k = \sum_{n=k+1}^N z_n z_{n-k}^* \tag{8.30}$$

The sum with S_k is the only term dependent on the signal's Doppler phase. The other sums are only dependent on the signal power which could result also from noise and interference. The meaning of the partial sums S_k may be explained as follows. With an assumed Doppler signal:

$$z_n = a\exp(j\varphi n)$$

the products:

$$z_n z_{n-k}^* = a^2 \exp(j\varphi k)$$

show a phase argument independent of n and therefore the summation adds coherently all the signal contributions $z_n z_{n-k}^*$.

On the other hand, the following expression for ρ_k gives the autocorrelation estimate (ACE) of the signal series z_n for the time lag kT:

$$\rho_k = \frac{1}{N-k}\sum_{n=k+1}^N z_n z_{n-k}^* = \frac{1}{N-k}S_k \tag{8.31}$$

The test function may therefore be expressed with S_k given by equation 8.30 as:

$$T_K = \sum_{k=1}^K |S_k|^2 = \sum_{k=1}^K (N-k)^2 |\rho_k|^2 \tag{8.32}$$

The parameter K is at most $N-1$. The detection performance will improve with increasing K. This has been investigated by simulation studies [11], which have been performed as follows: as a first step for selected pairs of (N, K) the detection thresholds corresponding to the false-alarm probabilities P_F (10^{-3}, 10^{-4}, 10^{-5}, 10^{-6}) have been determined. With these thresholds the functions for the detection probabilities

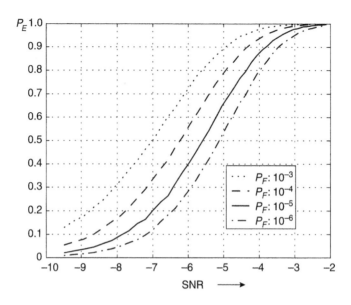

Figure 8.19 P_D *curves for ACE-test function,* $N = 64$, $K = 16$, *20 000 trials,*
$P_F = 10^{-3}, 10^{-4}, 10^{-5}, 10^{-6}$ © *IEEE 1995 [11]*

P_D dependent on the signal-to-noise ratio SNR have been achieved. An example for
the set of curves is given in Figure 8.19.

All sets of P_D curves have approximately the same form, but differ in their position
above the SNR axis. For each individual set a central point has been selected for
$P_D = 0.5$ between $P_F = 10^{-4}$ and 10^{-5}. Each set of curves has been characterised
by the respective value of SNR. Therefore it is convenient to compare the detection
performance for different sets (N, K) by the differences in SNR for these central
points. For N the values 4, 8, 16, 32, 64, 128, 256 and for K values up to $N/2$ have
been used. $K = 0$ corresponds to incoherent integration. The results are given in
Figure 8.20. Here the additional SNR when compared to detection with a filter bank
is shown. For $K = N/2$ the loss compared to a filter bank is only 0.5 dB. Because
there is no filter straddling loss assumed, which is about 1 dB, there is even a gain of
0.5 dB compared to a filter bank.

Processing for the ACE test is nonlinear, and only one signal should be present.
A weak signal would be suppressed by a stronger signal, therefore clutter suppression
has to be performed in advance.

8.6.3   Comparison of detection performance of the filter bank and
ACE test for LFM Doppler signals

The comparison studies have been performed by assuming an LFM signal for simulations and computations. For a sweep factor f the signal around e.g. the frequency

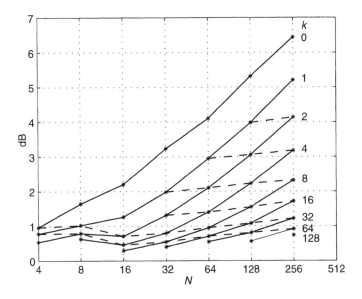

Figure 8.20   ACE test, additional SNR compared to a filter bank © IEEE 1995 [11]

$1/(2T)$ is given by:

$$z_n = a \exp\left[j\left(\pi + \frac{\pi(n+1)f}{N^2} - \frac{\pi f}{N}\right)n\right] \tag{8.33}$$

For a filter bank for $N = 64$ the decrease of the maximum output was first computed dependent on f. The loss in the signal-to-noise ratio (SNR) is shown in Figure 8.21 (dotted curve). Because with increasing f an increasing number of filters show a decreasing output signal, their chance to contribute to the detection should also be covered. A corresponding simulation study was performed. The result of the simulation is also given in Figure 8.21, by the dashed curve.

To compute the loss in the ACE test for LFM signals the test value of equation 8.32 was used. The dependence of:

$$Q_K = \left[\sum_{k=1}^{K}\left|\sum_{n=k+1}^{N} z_n z_{n-k}^*\right|^2\right]^{1/2} \tag{8.34}$$

on f was computed with a signal, following equation 8.33. The exponent 1/2 in equation 8.34 results in Q_K being a quadratic function of the signal amplitude, a. This allows us to consider the decrease of Q_K as a loss in SNR. The result as a function of the sweep factor f is also shown in Figure 8.21 for $N = 64$ and with K as a parameter for the individual curves (solid curves). A certain starting loss for $f = 0$ for different values of K follows from Figure 8.20. It can be seen that the

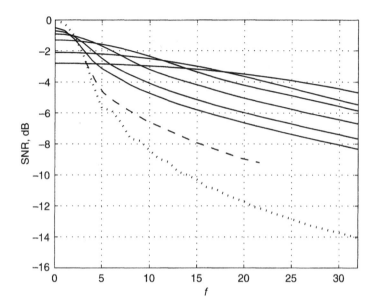

*Figure 8.21   SNR loss for LFM signals dependent on frequency variation factor
f, ACE test N = 64, K = 1, 2, 4, 8, 16, 32, Filter bank: N = 64,
computed maximum (dotted line) simulation (dashed line) © IEEE
1995 [11]*

ACE test is clearly more robust against a frequency variation described by the sweep
factor f.

8.7   Conclusions

Generally, a series of echo signals is the basis for target detection. An arbitrary length
of this series is applicable by using the beam agility, which enables adaptation of the
transmit energy to target strength and of the required Doppler resolution. The step
scan operation of a phased array provides an important difference when compared
with mechanical scanning radars with respect to the detection procedure and clutter
suppression. Clutter suppression may be divided between strong fixed clutter and
varying Doppler-shifted clutter, the first of which may be suppressed by a simple
recursive filter. Using post processing the latter can be arranged advantageously, in
an adaptive manner, after the filter bank. Sea clutter suppression can be improved
by a special multiplex operation of the beam, resulting in elongated dwell times in
each beam position. For a longer series of pulses, a possible Doppler shift during the
dwell time has to be considered. The usual filter bank shows a degraded detection
performance because the target energy is distributed into several filters. An autocorrelation estimation test will give a more robust performance in this case. The detection
performance is evaluated and compared on the basis of simulation studies.

8.8 References

1 NATHANSON, F. E.: 'Radar design principles' (McGraw-Hill, New York, 1996)
2 'Models of clutter' in GALATI, G. (Ed.): 'Advanced radar techniques and systems' (Peter Peregrinus Ltd., 1993) chapter 2
3 WIRTH, W. D.: 'Clutter- und Signalspektren aus Radicord-Aufnahmen und die Bewegtzielentdeckung', *Nachrichtentechnische Zeitschrift (Communications Journal)*, December 1968, (12), pp. 759–765 (in German)
4 VON SCHLACHTA, K.: 'A contribution to radar target classification'. IEE international conference on *Radar*, London, UK, 1977, pp. 135–139
5 WIRTH, W. D.: 'MTI in pulse radar systems', *Nachrichtentechnische Zeitschrift (Communications Journal)*, 1966, (5), pp. 279–287
6 WIRTH, W. D.: 'Detection of Doppler shifted radar signals with clutter rejection'. AGARD *Radar* symposium, Istanbul, Turkey, 1970, pp. 30.1–30.19
7 ROHLING, H.: 'Zur Berechnung von Dopplerfilterbänken für die Radarsignalverarbeitung', *Wiss. Ber. AEG-Telefunken*, 1980, **53** (4–5), pp. 145–153 (in German)
8 LUDLOFF, A.: 'Handbuch Radar und Radarsignalverarbeitung', (Vieweg, 1993) chapter 7
9 WIRTH, W. D.: 'Signal processing for target detection in experimental phased array radar ELRA', *IEE Proc. Commun. Radar Signal Process.*, 1982, **128** (5), pp. 311–316
10 BÜHRING, W.: 'Fixed clutter suppression by a pre-filter for constant beam direction'. FGAN-FFM report no. 257, 1977
11 WIRTH, W. D.: 'Long term coherent integration for a floodlight radar'. Proceedings of the IEEE international conference on *Radar*, Washington, USA, 1995, pp. 698–703
12 KLEMM, R.: 'Adaptive clutter suppression in step scan radars', *IEEE Trans. Aerosp. Electron. Syst.*, 1978, **14** (4), pp. 685–687
13 WANG, H., and CAI, L.: 'A localized adaptive MTD processor', *IEEE Trans. Aerosp. Electron. Syst.*, 1991, **27** (3), pp. 532–539
14 HAYKIN, S., and NOHARA, T. J.: 'Growler detection in sea clutter with coherent radars', *IEEE Trans. Aerosp. Electron. Syst.*, 1994, **30** (3), pp. 836–847
15 WIRTH, W. D.: 'Verfahren für kohärente Radarmessung mittels elektronisch gesteuerter Antennen sowie Vorrichtung zur Durchführung des Verfahrens'. Patent DE 196 49 838 C1, 1998
16 FLESKES, W., and VAN KEUK, G.: 'On single target tracking in dense clutter environment'. IEE international conference on *Radar*, London, UK, 1987, pp. 130–134
17 SELIN, I.: 'Detection of coherent radar returns of unknown Doppler shift', *IEEE Trans. Inf. Theory*, July 1965, p. 39
18 NICKEL, U.: 'Radar target parameter estimation with array antennas', *in* HAYKIN, S., LITVA, J., and SHEPHERD, T. J. (Eds.): 'Radar array processing' (Springer Verlag, Berlin-Heidelberg-New York, 1993) chapter 3, pp. 47–98

Chapter 9

Sequential detection

Sequential detection is based on the theories of sequential analysis developed and published by A. Wald [1] in 1947. The basic problem is the decision between two hypotheses, in the radar case H_0 for 'no target present' and H_1 for 'target present'. This has been discussed already in chapter 3 section 3.3. The decision has to be based on a series of received signals, which always contain noise and perhaps a target signal. The basic idea and characteristic feature of sequential analysis is that the number of necessary signals or radar pulses is not determined in advance. Instead, after each new signal a decision is derived to terminate the test with a decision or to continue with a further pulse. The advantage of sequential tests is that on average a substantially smaller number of pulses is needed compared to equally reliable tests with a predetermined fixed number of pulses. Equally reliable means the same pair of error probabilities α, β for false alarms and target missing, respectively.

Because of the variable test length of the individual tests only a phased-array radar with its beam agility is able to apply sequential detection. Because of its potential advantages there was early interest, and the application of sequential detection for radar was studied by several authors [2–7]. First, considerable gain in mean test length or signal-to-noise ratio (SNR) of about 8 dB was discovered, but this was valid for only one range element. In the search mode a radar has to test up to thousands of range elements or cells. In this case the gain of the suggested sequential procedures dropped to only 1–2 dB and the general interest decreased.

With the introduction of a suitable range-dependent test, adapted to a radar target with a certain cross section, and a weighted combination of range element tests for detection within all ranges, the gain may be recovered to almost the original value [8,9].

Practical applications were first restricted to incoherent processing, because of the complexity and cost for realisation of a coherent test (the gain of coherent tests would be considerably higher). Recently, suggestions have been made for a practical coherent test [16]. These show a robustness against Doppler frequency variations of flying targets during the test. Therefore sequential detection may also be applied in coherent realisation for special applications with a longer mean test length to achieve a high detection performance for targets with a low radar cross section.

In a selected beam direction a series of pulses is transmitted. Out of the received echo signals for each range element a decision has to be derived with a suitable test procedure on whether there is only noise (hypothesis H_0) or a target present (hypothesis H_1). The range elements are matched in size to the radar's range resolution given by the subpulse length.

This test operates as follows. For a certain range element the echo signals already received are denoted by $x_1, \ldots, x_n$ and put into the test function:

$$\lambda_n = \frac{p_1(x_1, \ldots, x_n)}{p_0(x_1, \ldots, x_n)} \tag{9.1}$$

The functions p_0 and p_1 are the joint probability densities for the pulse series $x_1, \ldots, x_n$ for the hypothesis H_0 (no target, only noise) and H_1 (target present), respectively. Therefore λ_n is the likelihood-ratio function for $x_1, \ldots, x_n$. The test value λ_n is compared with two decision thresholds A and B after each new pulse, starting after the first pulse x_1. Three decisions are possible:

$B < \lambda_n < A$:   there is no decision for one of the two hypotheses H_0 or H_1; a new pulse is transmitted and the signal series is extended by one further echo signal

$\lambda_n \leq B$:   the test is terminated with the decision for H_0

$\lambda_n \geq A$:   the test is terminated with the decision for H_1

This arrangement for a test is called a sequential probability ratio test. The value n is the test length of the actual test. n is a random variable because of receiver noise influence. Extreme dwell times in a beam position are avoided by a preset maximum test length (truncation length). The length of the overall search time shows a certain variance which depends also on the truncation length.

The principal test development is qualitatively illustrated in Figure 9.1: the test function λ randomly goes down in the case of noise alone. It goes upwards in the case of a target present. If there is a high target signal the upper threshold A is crossed with few pulses, if there is only a weak target signal the test needs more pulses to reach a decision.

Both decision thresholds A and B may be selected independently. They are related to the decision error probabilities α for false alarms and β for missing a target after Wald [1] by the fundamental inequalities:

$$A \leq \frac{1 - \beta}{\alpha} \quad \text{and} \quad B \geq \frac{\beta}{1 - \alpha} \tag{9.2}$$

For $\alpha \ll 1$ follows $B \geq \beta$, which means that B determines the maximum value for the target missing probability β. Because most of the beam positions contain noise only, the test is terminated in almost all cases by passing the lower threshold B and therefore this case determines the overall mean test length for the radar's search operation.

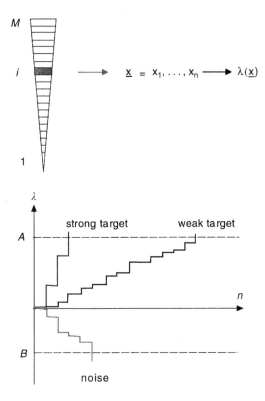

Figure 9.1 Principle of sequential detection

A determines the false-alarm probability α. The test length in the case of the detection of a target by crossing the upper threshold *A* depends on the value of *A* and the signal strength of the target present. Because there are always only a few targets compared to the high number of beam positions which have to be tested there is practically no cost associated with an increased overall mean test length by specifying a low α (high threshold *A*).

The advantage of sequential detection over fixed sample size tests therefore increases with decreasing α.

9.1 Incoherent detection

For the detection of a nonfluctuating target by incoherent processing the well-known test function in the logarithmic domain for the signal amplitudes, x_i, after *n* pulses, is given by [3]:

$$l_n = \ln \lambda_n = -n\frac{a^2}{2} + \sum_{j=1}^{n} \ln I_0(a_0 x_j) \tag{9.3}$$

This is approximately the incoherent sum of amplitudes combined with a bias which reduces with increasing n the value of the test function in the case of only noise. I_0 is the modified Bessel function of order zero and a_0 is a design signal. The signals x_j and a_0 are normalised on the rms value of the noise. If a target signal with amplitude $a = a_0$ is present, the resulting detection probability will be $P_D = 1 - \beta$.

Pairs of (a_0, B) determine the so-called operating characteristic function (OCF) that is $P_D = f(\mathrm{SNR})$. Different suitable pairs all resulting in the same value for the mean test length in noise alone, n_n, lead in practice to the same function $P_D = f(\mathrm{SNR})$. We can therefore say that sequential detection results in a uniformly most powerful test. The important parameter which determines the search energy is also n_n. This important result was not achievable by a mathematical derivation but by extensive simulation studies.

Because the choice of a_0 is uncritical, the target signal need not be known in advance. It follows moreover that the detection performance is also optimal for slowly fluctuating targets (Swerling case 1) with an amplitude variation from dwell to dwell.

For rapidly fluctuating targets (Swerling fluctuation model [11,12] case 2) a corresponding test function may be set up. By comparative simulation studies it was shown that the test function of equation 9.3 performs equally well.

The test length for a target present, n_t, is dependent on the value of amplitude a. For small signals n_t may be large, therefore a truncation test length n_{tr} may be applied. After simulation studies the detection performance is not deteriorated if $n_{tr} \geq 3n_n$ is chosen.

The incoherent sequential test has been compared with the conventional fixed sample size test FSST. It is assumed that the mean test length in noise for the sequential test, n_n, equals the number of pulses, N, of the FSST. The gain for testing one range cell expressed as a saving in SNR is about 6–7 dB for $\alpha = 10^{-8}$ [3]. The gain decreases roughly by 1 dB for each factor 10^2 by which α increases.

9.2   Multiple range elements

For the detection of radar targets in the search mode the test has to be applied for each beam position to all multiple range cells. For this extension of the test different suggestions have been made; the most important proposals will be discussed in the following.

9.2.1   Independent test of all range elements

All M range elements are tested independently at the same time using the test function of equation 9.3. The test for the current beam position is terminated after all independent individual tests have been finished. The resultant test length is given by the maximum individual test length. A loss is apparent, because in most range elements a dead time occurs until the last or longest individual test in a certain range element is finished. The increase in the test length has been computed dependent on the number of range elements M. For $M = 266$ the test length increases by a loss factor 6 and for $M = 1000$ this factor is about ten. Therefore the advantage or gain factor of 8 dB for testing a single range element disappears [8].

9.2.2   Common test for all range elements

A reduction of the mean test length for the test of M multiple range elements is achieved by setting up a common test function for all range elements together. Marcus and Swerling [3] introduced a test to address the question: is there for a beam position at least one target in any range element with range index number i? For the test step n the test function is:

$$\lambda_n = \frac{1}{M} \sum_{i=1}^{M} \lambda_{in} \qquad (9.4)$$

With this set up an averaging is performed over the unknown parameter range i. For all range values i the same probability $1/M$ is assumed for the presence of a target. The test function λ_{in} is given by equation 9.3 and the design signal a_0 has been chosen the same for all i. For higher values of M, the advantage in saving of time or energy also decreases with this test: for $M = 1000$ there is a saving left of only about 2 dB. But a certain improvement is achieved compared to an independent test, discussed above.

9.2.3   Combined test with range-dependent design signal

With the introduction of a combined test for all range elements together with a range-dependent design signal a_i according to the radar range equation the gain given in section 9.1 for a single range element could be almost recovered. a_i increases with descending range index i beginning with a_0 at the maximum range $i = M$ according to the radar equation:

$$a_i = a_0 \left(\frac{M}{i} \right)^2 \qquad (9.5)$$

The combined test, suggested by the author [9], is defined by a weighted sum of test functions λ_{in} from the individual range elements i:

$$\lambda_n = \sum_{i=1}^{M} g_i \lambda_{in} \qquad (9.6)$$

Each λ_i is computed with the design signal a_i instead of a_0. By the range-dependent design signal a_i a parameter related to the signal power is introduced. By simply averaging as in equation 9.4 the required detection probability P_D would be achieved only in the mean over all i, low at maximum range and high for near ranges. But we require a certain detection probability given by $P_D = 1 - \beta$ independent of range. This can be achieved only with the introduction of a weighting function g_i in equation 9.6. An appropriate weighting function has been found by simulation studies [9] with:

$$g_i = \frac{1}{S} \exp\left(- a_i^2\right) \tag{9.7}$$

and the term S for normalisation:

$$S = \sum_{i=1}^{M} \exp\left(- a_i^2\right)$$

so the sum is:

$$\sum_{i=1}^{M} g_i = 1$$

This approach provides a $P_D = 1 - \beta$ nearly independent of range for a certain radar cross section σ of the target giving a receiving amplitude a_0 at maximum range $i = M$ and a_i at range i following equation 9.5.

With equations 9.3, 9.6, and 9.7 we get for the combined test function:

$$\lambda_{Mn} = \frac{1}{S} \sum_{i=1}^{M} \exp\left(- a_i^2 + l_{in}\right) \tag{9.8}$$

According to simulation studies [10] this test function may be replaced by a simpler one:

$$\tilde{\lambda}_{Mn} = \max\left(- a_i^2 + l_{in}\right) \quad \text{out of all } i = 1, \ldots, M \tag{9.9}$$

For a beam position this test function leads to target detection if there is at least one target present. In the case of target detection the index i of the maximum of all the stored λ_i values determines the range cell i and therefore the target's range.

Limitation of the value of the increasing a_i by $\sqrt{2}a_0$ avoids an unnecessary loss in sensitivity for targets at lower ranges, which means that a_i is increased with descending range i only up to 3 dB. This limitation does just not increase the mean test length in noise for the whole beam position. This was a result of corresponding simulation studies [10].

The test length in the case of a target depends on the signal amplitude: high amplitudes result in a fast detection with few pulses. This is a form of signal-dependent radar energy management.

This test function for incoherent processing has already been implemented and tested in the experimental electronic steerable radar system ELRA, described in chapter 17.

9.2.4 Comparison of the test for multiple range cells with the fixed sample size test

With available detection performance results it is possible as an illustration to compare the conventional fixed sample test (FSST) and the sequential test by their required SNR dependent on range (Figure 9.2). The comparison is made for $M = 64$ range cells. The required SNR of the sequential test for achieving $P_D = 0.5$ is 0 dB for the maximum range $i = 64$ and a design signal $a_0 = \sqrt{2}$ (0 dB).

From simulation resulted a mean sequential test length in noise of $n_n = 6.23$. From [11] follows for the FSST for $\alpha = 10^{-8}$ a required SNR of 6 dB. So the gain at this range is 6 dB.

The SNR increasing with descending range i required by the combined sequential detection is given by the ascending function up to 3 dB. The required SNR by the FSST is independent of range and is given by the horizontal line. So there is a gain of 6 dB at maximum range, descending to only 3 dB for all lower ranges.

The saving at maximum range is practically nearly the same as with a single element test [3]. The gain is lower for higher values of α, also indicated in the figure by dotted horizontal lines, or higher with lower values of α.

The influence of the number of range elements is also indicated in Figure 9.2; because the mean test length in noise increases slightly with the number of range

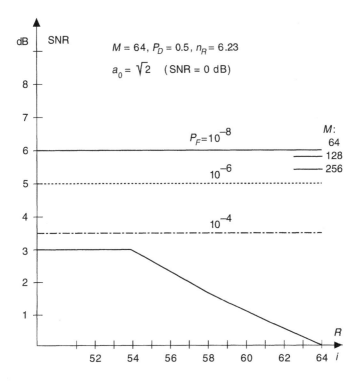

Figure 9.2   SNR gain of sequential test compared to FSST © IEEE 1996 [17]

elements M the SNR of the corresponding FSST decreases as indicated by the short bars on the right-hand side of the diagram. The gain reduces at maximum range $i = M$ to 5.1 dB for $M = 128$ and to 4.7 dB for $M = 256$.

Therefore, to achieve a high energy saving for the long-range search mode of a phased-array radar with sequential detection one should choose only a moderate range resolution. The following acquisition function could then be performed with a higher range resolution combined with monopulse for direction estimation to achieve an accurate target location for track initiation.

Figure 9.2 clearly demonstrates that the saving in required SNR is maximum at maximum range where the target detection is most critical and determines the radar's search performance. But there is also a saving in SNR or an improvement in sensitivity for shorter ranges.

9.2.5   Sequential detection with a filter for the rejection of stationary clutter

The output signals of the filter suppressing echoes from fixed clutter, as discussed in section 8.1, may be used for sequential detection. The I and Q components of the received signal are used to form the amplitudes as the input to the incoherent sequential detection. As the frequency transfer function of this filter varies with n, and as n is a random variable, the detection performance dependent on the Doppler frequency can only be found by simulation. The result for a mean test length in noise $n_n = 10$ is shown as an example in Figure 9.3. The detection probability P_D as a function of the Doppler frequency $f_d T$ (T = radar period) increases rapidly at low Doppler frequencies. The explanation for this effect may be deduced from Figure 9.4; the

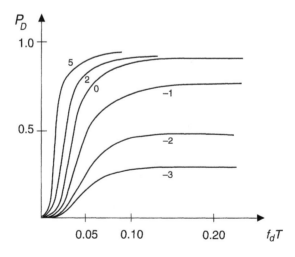

Figure 9.3   *Detection probability of sequential test with clutter filter as a function of Doppler frequency $f_d T$, curve parameter: SNR in dB, © IEE 1981 [10]*

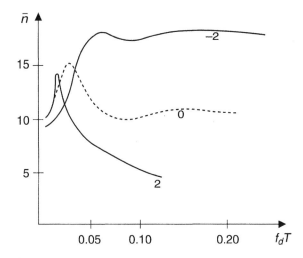

*Figure 9.4   Mean test length of sequential test with clutter filter as a function of
Doppler frequency $f_d T$, curve parameter: SNR in dB, © IEE 1981 [10]*

mean test length for target detection increases for low Doppler frequencies because the
signal amplitude at the output of the clutter filter offered to the test function is relatively
small for the first pulses, giving the clutter filter the chance to develop a favourable
frequency-transfer function. The transfer function of the clutter filter develops from
a single canceller after two pulses to a more and more efficient function matched
to the fixed clutter spectrum with a narrow notch around zero Doppler frequency,
as explained in chapter 8. The combination of this filter and the sequential detector
is therefore recommended for operation in moderate stationary clutter situations,
especially if there is only sidelobe clutter.

9.3   Coherent test function

For a fixed sample size test, there is a gain in detection performance for coherent
compared to incoherent processing. This gain may be expressed by the saving in
SNR required for the same detection and false-alarm probabilities. It is approximately
given by the factor $\sqrt{N}$. It can therefore be expected that by introducing a suitable
coherent test function the so-far incoherent sequential test will be improved likewise.

   A Doppler filter bank is usually used to process coherently and detect N samples
of a signal with an unknown Doppler frequency. This may be achieved efficiently by
an FFT, each coefficient of the transform representing an individual filter within the
filter bank. For sequential detection, the FFT would have to be computed for each
new test step, after transmitting a new pulse, for a value n increased by one. This
leads to an increasing number of resolution cells in the Doppler domain and also
to an increasing computational effort with increasing test length. Another factor to
consider is the passband of the individual filters decreasing with $1/n$. With a long test

length for the detection of targets with low SNR the received signal energy may be distributed into several filters if the Doppler frequency changes during the test and therefore a loss would result for the detection performance. This is especially the case for targets on a tangential flight path, as discussed already in chapter 8, section 8.6.1.

In the following the ACE test function is proposed as being especially suitable for sequential detection.

9.3.1   Test function with autocorrelation estimates (ACE)

The likelihood-ratio coherent test function corresponding to one range element for a signal series of length N and an unknown Doppler phase difference was developed for a small signal approximation in chapter 8, section 8.6.2, to the autocorrelation estimate ACE test with the expression:

$$S_k = \sum_{m=k+1}^{n} z_m z_{m-k}^* \tag{9.10}$$

As the test function the following is suggested:

$$T_{Kn} = \sum_{k=1}^{K} |S_k|^2 \tag{9.11}$$

The parameter K is at the most $n - 1$ for the test step n. Beyond a certain selected value K_{max}, which determines the processing load, K may be set to K_{max}. The detection performance will improve with increasing K_{max}. This has also been studied by simulations for fixed sample size tests, as described in section 8.6.2 [13].

Figure 9.5 gives a possible structure for an implementation which computes T_{Kn}. The shift register is normalised with zeros at the beginning of the test. The sums $S_1, \ldots, S_k$ are accumulated for each test step n and the intermediate results for the $S_1, \ldots, S_k$ are summed magnitude squared for T_{Kn} according to equation 9.11. Because of the stepwise accumulation procedure, the ACE test function seems especially suited for performing sequential testing.

As the summation in equation 9.11 starts at $k = 1$, the minimal test length n_{min} for a sequential test is 2. The parameter K_{max} determines the detection performance but also the processing complexity and has to be selected carefully.

9.3.2   Sequential test function with autocorrelation estimates

The expression T_{Kn} of equation 9.11 is used for the design of the test function [13]. The convenient relationship of equation 9.2 between the thresholds A and B and the error probabilities α, β holds if it is possible to derive the test function λ as the likelihood ratio according to equation 9.1. However, the probability density for the expression T_{Kn} is not available analytically and therefore the likelihood ratio cannot be derived. One cannot therefore expect to find a test function which fulfils both relations of equation 9.2.

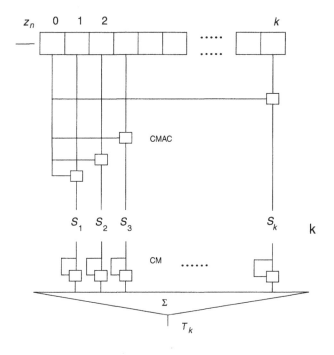

Figure 9.5   Processing structure to compute the test value T_k, © IEE 1995 [13]

As requirements for the construction of a test function the following aims should be achieved:

- for a signal with amplitude $a = a_0$ the detection should be performed with $\beta \approx B$, according to equation 9.2
- this detection performance should be valid for different values of a_0 to achieve, later, the range-dependent tests for multiple range elements according to equations 9.4 to 9.9
- the variance of the test length should be small, especially in the case of noise only
- the mean test length for noise only should be as small as possible

We have to anticipate that the effective false-alarm probability $\hat{\alpha}$ will not be related to threshold A according to equation 9.2 but has to be determined by simulation studies.

For these simulation studies complex signals, z, are generated from pairs (x_1, x_2) with a Gaussian probability density function, described in chapter 3 section 3.2.6. Mean and standard deviation are given by $(0,1)$. A target signal with amplitude, a, is represented by the pair $(x_1 + a, x_2)$. The corresponding SNR is $a^2/2$. The result is a signal with Rayleigh or Rice probability density function as desired.

As a first step we develop the test function for a single range element. The test function which has to be constructed should behave in a way similar to the incoherent test function given by equation 9.3. The signal-dependent last term increases for the noise case linearly with the test length n. Therefore, we need first for comparison the expectation of T_{Kn}.

We look first to the variable S_k, which be written:

$$S_k = M_k + X_k + jY_k$$

where M_k is the mean and X_k, Y_k are real variables with zero mean. The mean and variances for these variables are given in [15]. For noise only and a signal z with components with zero mean and variance equal to 1 we get for the variances of X_k, Y_k:

$$\sigma_x^2 = \sigma_y^2 = 2(n - k)$$

For the expectation of $t_k = |S_k|^2$ follows:

$$E\{t_k\} = E\{(M_k + X_k)^2\} + E\{Y_k^2\} = 4(n - k)$$

For the test function T_{Kn} we get:

$$E\{T_{Kn}\} = E\left\{\sum_{k=1}^{K} t_k\right\} = 4Kn - 2K(K + 1)$$

With the sequential test we have at each test step $K = n$ resulting in:

$$E\{T_{Kn}\} \approx 2n^2$$

Therefore, a factor $1/n$ has to be applied to T_{K_n} to achieve the mentioned increase in noise proportional to n.

This leads to the general approach for the test function l_1 corresponding to equation 9.3:

$$l_1 = C_1 - C_2 n \frac{a_0^2}{2} + C_3 \frac{a_0^2}{n} T_{Kn} \qquad (9.12)$$

The factor a_0^2 gives a dependence on the design signal, a_0. The third term is dependent on the received signal by T_{Kn}. By dividing this term by n the desired linear increase with n is achieved. The constant C_3 allows us to adjust the mean increase of the test function with n. The second term corresponds to the first term in equation 9.3 and effects the approach of l_1 to the threshold B_1 in the case of noise only. The low threshold is now $B_1 = \ln(B)$, because we are, with equation 9.3, in the logarithmic domain. The upper threshold in the logarithmic domain will be denoted by A_1.

The ratio of C_2 and C_3 determines the desired different behaviour in the cases of noise only or signal and noise, respectively. The absolute values of C_2 and C_3 have to be chosen suitable to B_1. The constant C_1 is used for a fine adjustment of the detection performance.

Another requirement is given by the detection probability P_D: the lower threshold B_1 should be related to P_D by:

$$\log (1 - P_D) \approx B_1$$

which is applicable for the test function of equation 9.3, with the effect that signals with the amplitude $a = a_0$ are detected with P_D.

We cannot also expect to achieve the dependence between A_1 and α given by equation 9.2, because we have not constructed the test function with the likelihood ratio. A suitable threshold A_1 for a required design signal a has to be found by simulation. We have to remember that the choice of the threshold A_1 has no influence on the mean test length n_n in noise and therefore may be chosen high for a low α without increasing the mean search time.

The constants are then determined by simulation trials to achieve favourable test behaviour. No general solution for the values of the constants have been found but only values for certain intervals of a. To determine the constants $C_{1,2,3}$ first the development of the test function l_1 with the pulse number n was observed for a certain number of trials, e.g. 100, as represented in Figures 9.6 and 9.7, and compared with the behaviour of the incoherent test function. This gave a certain first selection for the constants. Then by simulation trials with systematic variation of the constants $C_{1,2,3}$ the test performance was improved stepwise. This procedure was performed for different values for a_0 (0.15, 0.25, 0.50, 0.75, 1, 1.5, 2) and $B_1 = 0.1$ or 0.5. By alternating $a = 0$ (noise only) or $a = a_0$ (target and noise) both cases have been tested. Finally, there have been found as an example the following sets for a reasonable test performance in view of the

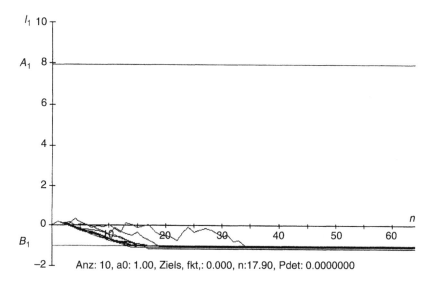

Anz: 10, a0: 1.00, Ziels, fkt,: 0.000, n:17.90, Pdet: 0.0000000

Figure 9.6   Test function realisations for noise only, $a_0 = 1$, $n = 17.9$, $P_{det} = 0$

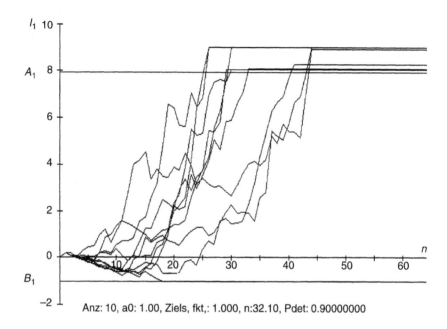

Anz: 10, a0: 1.00, Ziels, fkt,: 1.000, n:32.10, Pdet: 0.90000000

Figure 9.7   *Test function realisations for target and noise, $a_0 = 1$, $n = 32.1$,*
$P_{det} = 0.9$

above-stated requirements:

$$a_0 = 1\ldots 2: \quad C_1 = 0.5, \quad C_2 = 0.5, \quad C_3 = 0.035$$
$$a_0 = 0.25: \quad C_1 = 0.5, \quad C_2 = 0.92, \quad C_3 = 0.4$$
$$a_0 = 0.15: \quad C_1 = 0.5, \quad C_2 = 1.7, \quad C_3 = 0.875$$

The received signal components have to be normalised for a noise variation $\sigma^2 = 1$. This set of constants would be sufficient to achieve, later on, the desired range dependence for the multiple range-element test function. With programmable signal processing the values may be changed according to the desired search mode parameters.

Figure 9.6 shows for illustration the value of l_1 for noise only as a function of n for ten trials. It is seen that l_1 randomly approaches the threshold B_1 as n increases. The test is terminated when the threshold is reached. The last value of l_1 is shown in the Figure until $n = n_{max}$.

Figure 9.7 gives realisations of the test values in the case of signal plus noise. Note that a higher setting of threshold A_1 in favour of a lower α would cause an increased test length in the case of a target.

9.3.3 Simulation studies for a comparison of incoherent and coherent sequential tests

The statistical behaviour of the test function given by equation 9.12 has been evaluated by simulation trials. The outcome of K trials for detections may be given by a series with the variable $y = 0$ or 1. The mean of this series is p and its estimate is the estimate for p. The variance of the estimate is given with equation 3.33 by:

$$\sigma_p^2 = \frac{\sigma_y^2}{K} = \frac{1}{K}E\{(y-m)^2\} = \frac{1}{K}\left(E\{y^2\} - (E\{y\})^2\right) = \frac{1}{K}(p - p^2)$$

and finally

$$\sigma_p^2 = \frac{p(1-p)}{K} \tag{9.13}$$

For example if $p = 0.1$ and the required $\sigma = 0.003$ then follows $K \geq 10\,000$. All the following simulations were performed with at least $10\,000$ trials.

From $\beta = 0.1$ for $P_D = 0.9$ follows with equation 9.2 the low threshold B and $B_1 = \ln(B)$. The high threshold A_1 was selected after trials to achieve an effective $\hat{\alpha} \approx 10^{-5}$.

We first determined by the simulations the mean test length in noise, n_n, for different values of the design signal a_0. Then we obtained with another set of simulations the detection probabilities P_D and the mean test length n_t when the target is present for various design signals a_0 and the target signal $a = a_0$. All results are given in Table 9.1.

Since α can be chosen to be arbitrarily low by the selection of a high threshold A_1 without changing n_n and P_D, the results of Table 9.1 are sufficient to characterise the detection performance of the test. We observe that the detection probability, which is specified by $B_1(\beta)$, is achieved with our test function equation 9.12 fairly well.

The value for $K_{\max}$ determines the processing load and should be chosen to be as low as possible without a loss in detection performance. In Figure 9.7 it is seen that

Table 9.1   Simulation   results   for ACE sequential test

a	P_D	n_n	n_t
0.75	0.909	37.16	77.90
1.00	0.894	17.85	34.11
1.50	0.911	6.90	13.53
2.00	0.887	4.20	7.40

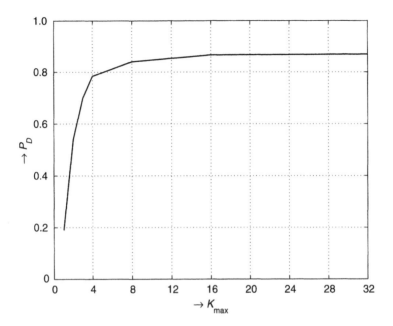

*Figure 9.8   Detection probability P_D as a function of $K_{max}, n_n = 6.9$, © IEE
1995 [13]*

very frequently the curve of the test function over n moves down towards threshold B_1
when the target is present before striving finally upwards to A_1 and leading to target
detection. This is typical behaviour of the sequential test function, also observed with
the incoherent test of equation 9.3.

From this observation it can be concluded that the value of K_{max}, to achieve an
efficient test performance, should be selected at least to a level which supports the
test function beyond this minimum area.

In a special simulation the dependence of P_D from K_{max} was evaluated. An
example is given in Figure 9.8 for $n_n = 6.90$ which shows that K_{max} should be at
least selected to about the value of n_n.

The simulated target signal amplitude, a, may be varied resulting in the operating
characteristic function of $P_D = f(\text{SNR})$, as shown in Figure 9.9. The position of this
typical S-shaped curve in relation to the SNR axis describes the detection performance.
In practise, it is only dependent on the mean test length, n_n, which is determined by
the parameters a_0 and B_1.

Figure 9.10 shows the dependence of the mean test length, n_t, on the actual SNR.
A maximum of n_t appears for a detection with $P_D = 0.5$.

The important consideration now is to evaluate the gain of the new coherent
sequential test using equation 9.12 in comparison to the incoherent test of equation 9.3.
This gain may be expressed by the ratio of the test-length values for the same input
signal a and P_D. Respective mean test lengths in noise were found by simulation

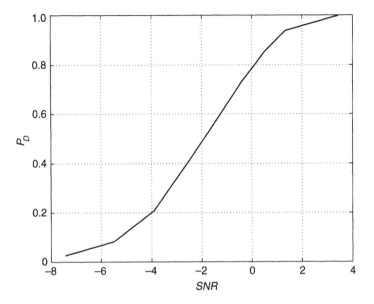

Figure 9.9 P_D dependent on SNR, © IEE 1995 [13]

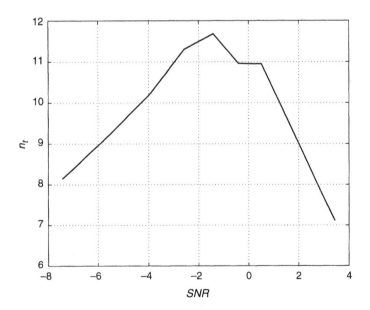

Figure 9.10   Mean test length dependent on SNR, © IEE 1995 [13]

Table 9.2   Comparison of incoherent and coherent sequential test for $\beta = 0.1$

Incoherent test:				Coherent test:			
a	n_n	f_i	P_D	n_n	f_c	P_D	gain (dB)
2.00	4.39	1.00	0.943	4.19	1.02	0.901	0.02
1.50	10.19	0.96	0.911	6.90	1.03	0.903	1.08
1.00	41.70	0.94	0.918	17.85	1.02	0.913	2.98
0.75	152.84	0.92	0.926	37.16	1.01	0.925	5.33

Table 9.3   Comparison of incoherent and coherent sequential test for $\beta = 0.5$

Incoherent test:				Coherent test:			
a	n_n	f_i	P_D	n_n	f_c	P_D	gain (dB)
1.00	14.60	0.90	0.51	7.26	1.15	0.522	0.91
0.75	47.72	0.85	0.54	14.27	1.10	0.529	3.00

for both cases. An additional fine adjustment for equal P_D values in both cases was achieved by the factors f_i or f_c for the incoherent or coherent test, respectively, to change the input signal amplitude from a to $a \cdot f$. The gain factor g for the energy saving is then given by:

$$g = \frac{n_{ni}}{n_{nc}} \cdot \frac{f_i^2}{f_c^2} \qquad (9.14)$$

Table 9.2 shows the results for $\beta = 0.1$. Table 9.3 gives the results for $\beta = 0.5$.

As expected, the gain of the coherent test compared to the incoherent test increases distinctly with n_n. This behaviour is also true for tests with fixed sample sizes.

9.3.4   Gain comparison of the coherent sequential and fixed sample size test

The optimal coherent test function for a fixed sample size, N, is given for unknown initial phase and Doppler frequency as discussed in chapter 8 in section 8.2 and by

Table 9.4   Comparison of coherent fixed sample size and sequential test

Coherent FSST:			ACE sequential test:			
N	P_D	SNR (dB)	n_n	P_D	SNR (dB)	G (dB)
8	0.9	4.75	7.00	0.9	0.511	4.88
16	0.9	1.90	17.85	0.9	−2.840	4.24
32	0.9	−0.25	37.16	0.9	−5.420	4.57
16	0.5	0.20	14.27	0.5	−4.680	5.37
32	0.5	−2.40	31.67	0.5	−7.370	5.01

equations 8.7 and 8.8. It may be achieved with a filter bank of N_1 filters:

$$T = \sum_{m=1}^{N_1} I_0\left[2\left|\mathbf{z}^*\mathbf{f}_m\right|\right] \tag{9.15}$$

$$= \max_m \left(\left|\mathbf{z}^*\mathbf{f}_m\right|\right) \tag{9.16}$$

If $N_1 = N$ then $\underline{\mathbf{f}}_m$ are the Fourier vectors with independent noise outputs. So the filter bank may be realised by an FFT.

Simulation results [17] for the fixed sample size (FSST) test have been selected for $\alpha = 10^{-6}$ which is an appropriate value for a radar in search mode. Some comparable results for the FSST and the autocorrelation estimation sequential test (ACE-ST) for $P_D = 0.5$ and $P_D = 0.9$ are contained in Table 9.4.

The gain achieved by applying sequential detection with the ACE test is obtained by the ratio of the SNR values and the ratio of the test length n_n/N. The mean value of the gain is about 4.8 dB. This compares well with gain values for sequential testing in the incoherent case mentioned in the introduction: for $\alpha = 10^{-6}$ we have to subtract 2 dB from the value of 6–7 dB given in [3], resulting in a gain of 4–5 dB.

We conclude that a comparable gain is achieved for both coherent and incoherent detection by the application of sequential testing instead of the fixed sample size test.

9.3.5   Extension for multiple range elements

In the search mode, we test a range interval up to a maximum range containing multiple range elements $i = 1, \ldots, M$. The approach given by equations 9.3 to 9.9 will be used as a starting point for the extension to multiple range elements for our coherent test function. The problem is again finding a suitable range-dependent weighting function, g_i, or in the logarithmic domain an additional term to the test

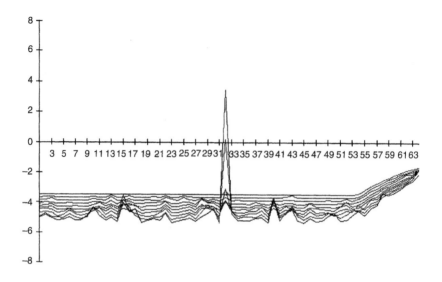

Figure 9.11   Realisations of l_{Mi}, target simulated at $i = 32$, © IEE 1995 [13]

function of equation 9.12. A possible approach is selected as:

$$l_{Mi} = -C_0 a_i^2 + l_{1_i} \qquad (9.17)$$

The range-dependent design signal, a_i, is given by equation 9.5 with the limitation by $a_0\sqrt{2}$.

The combination of the test values from all individual range cells of the actual beam position to one common test function is performed following equation 9.9 by taking the maximum value of all l_{Mi} over i.

To demonstrate the effect of the range-dependent test function in equation 9.17, its application over i is shown in Figures 9.11 and 9.12 in a waterfall representation for increasing n. The value C_0 is chosen to achieve equally good target detections at maximum range $i = M$ and e.g. $i = M/2$.

The test-function values for those range elements containing a target signal should show a maximum over i, to achieve an easy estimation of target range simply by taking that value of i corresponding to the maximum l_{Mi}. Both requirements are fulfilled by the choice of $C_0 = 4$.

The simulation result for the dependence of P_D on range index i is displayed in Figure 9.13, which shows that P_D is nearly constant compared to a_i in the transition region where $M/\sqrt{2} < i < M$, in this example for $54 < i < 64$.

In a further simulation study the choice of the parameter $C_0 = 4$ was tested for validity also for different parameter values of M, e.g. $M = 16, 64, 256$. The results for P_D at $i = M/2$ and $i = M$ are given in the Table 9.5 for $a_0 = 1.5$. As expected the detection performance given by the detection probability P_D is in all three cases practically the same if the target is simulated at $i = M/2$ or $i = M$.

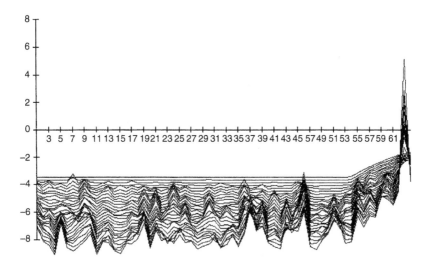

Figure 9.12   Realisations of l_{Mi}, target simulated at i = 63, © IEE 1995 [13]

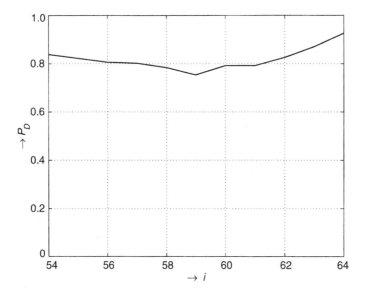

Figure 9.13 P_D for a target in the region i = 54 ... 64

A loss in detection performance is observed when using the test for multiple range elements. This loss increases with the number of range elements, M. It can be overcome by increasing the SNR or the mean test length in noise to achieve the same probability of detection as when $M = 1$.

The results shown in Table 9.6 have been obtained by simulations. Since large values of M should be avoided, one has to chose an adequate range resolution.

Table 9.5 *Sequential detection for multiple range elements*

M	P_D for $i = M/2$	P_D for $i = M$
16	0.956	0.970
64	0.848	0.926
256	0.915	0.852

Table 9.6 *Loss dependent on the number of range elements*

M	n_n	Factor f	P_D	Loss (dB)
1	6.90	1.00	0.8688	—
16	7.03	1.00	0.8770	0.24
64	7.10	1.03	0.8556	0.40
256	7.28	1.10	0.8523	1.06

9.3.6 Variable Doppler frequency of the target signal

The possible Doppler frequency variation, especially of tangential flying targets, is discussed in chapter 8 section 8.6. The results also apply for sequential detection. A sequentially-developed filter bank for coherent processing would suffer with increasing test length from the effect of target energy distribution into several filters.

The ACE test is therefore much more robust against frequency variation, which is especially important for energy management with long dwell times to detect targets with low radar cross section.

9.4 Comparison of detection procedures

For signal detection we may apply incoherent or coherent integration with a fixed sample size test or with a sequential test. These four detection methods are compared with respect to their gain in SNR dependent on the pulse number $N = n_n$ (fixed or mean value for test in noise), see Figure 9.14. The gain is relative to incoherent FSST, it represents the baseline. Dependent on the selected false-alarm probability α a substantial gain by sequential detection results. The gain by applying the coherent ACE sequential test compared to incoherent sequential test

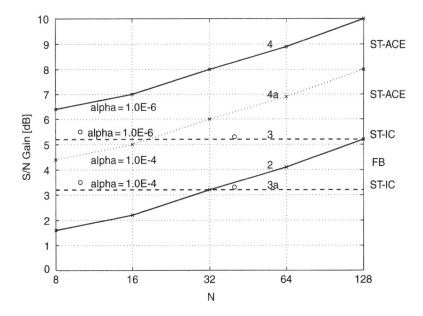

Figure 9.14   SNR gain of detection processing compared to incoherent integration,
© *IEEE 1996 [18]*
FB: filter bank FSST (2)
ST-IC: incoherent sequential test (3, 3a)
ST-ACE: coherent sequential test (4, 4a)

increases from 1.5 dB for mean pulse number in noise $n = 8$ to about 5 dB for $n = 128$.

9.5   Adaptation to the noise level and energy management

Owing to jammer residues despite adaptive jammer-suppression techniques by adaptive beamforming the noise level may be increased above the receiver noise level. A noise estimation from L range cells without clutter may then be used to normalise the received signal to a constant noise level. For Rayleigh-distributed noise amplitudes the estimate of the noise RMS value is given by:

$$\hat{\sigma}_x = \sqrt{\frac{2}{\pi}} \frac{1}{L} \sum_{i=1}^{L} x_i \qquad (9.18)$$

Likewise, the design signal a_0 for the test function is normalised and therefore decreased corresponding to the increased noise level. This effects a constant detection probability and false-alarm rate α despite an increased noise level. The price is an increased test length in noise according to the lower a_0. So we achieve automatic control of the test length (dwell time or pulse number) in the sense of energy management in favour of preserved target detection.

9.6   Sequential test for long signal series

For special radar concepts or functions a long signal series may be necessary. One example is the high PRF mode with an airborne phased-array radar. The PRF is chosen to be about 100 kHz to achieve an unambiguous Doppler measurement. With a mean dwell time of about 20 ms the number of pulses from a target will be $N = 2000$. The second example is the floodlight radar OLPI which will be discussed in detail in chapter 18.

By simulation studies the parameters for the application of the sequential test with ACE have been determined [19]. In the case of the OLPI system the Doppler tolerance of the ACE test function will be of special importance.

The OLPI radar concept applies continuous floodlight illumination for all resolution cells, therefore the test length for each resolution cell may be selected arbitrarily. After the end of a test the next test may be started. Because on average a shorter test length for the same detection performance is achieved, we can also hold the mean test length fixed and enjoy improved detection performance.

First, we estimate the expected signal amplitude which would be necessary with coherent integration by a Doppler filter bank. For N pulses the integration gain is N. For a single pulse the required SNR for $P_D = 0.5$ and $\alpha = 10^{-6}$ is 11 dB [11]. The required input signal-to-noise power ratio SNR for N pulses follows:

$$\text{SNR} = 11 - 10 \log_{10}(N) \, [\text{dB}]$$

For an unknown Doppler frequency the maximum filter output out of the bank with N filters is compared with the detection threshold. The performance loss L by ignorance of the Doppler shift, given in chapter 8 equation 8.9, is:

$$L = 0.26 \log_2(N) \, [\text{dB}]$$

The required input SNR or amplitude a values for the filter bank are then for the application examples, with $\text{SNR} = a^2/2$; for HPRF airborne phased-array radar:

$$N = 2000: \quad \text{SNR} = -18.04 \, \text{dB} \quad a = 0.177$$

for the floodlight OLPI radar:

$$N = 8000: \quad \text{SNR} = -23.52 \, \text{dB} \quad a = 0.0943$$

These values are used in the following as a reference.

We may use again the test function for a sequential test given by equation 9.15, which is repeated here for convenience:

$$\lambda_n = C_1 - nC_2 a_0^2 + C_3 \frac{a_0^2}{n} T_{Kn}$$

T_{Kn} is given by equation 9.13. We selected already $C_1 = 0.5$.

In the case of the airborne radar with HPRF ($a_0 = 0.177$) suitable values for the constants are $C_2 = 1.32$, $C_3 = 0.6725$. From a simulation the mean test length in noise resulted: $n_n = 697$, which compares with $N = 2000$ for the FST with a filter bank. The search time would be reduced by a factor of 2.87 or 4.58 dB.

9.6.1 Sequential test with coherent sections

For higher values of N (e.g. $N > 2000, \ldots, 8000$), for example in the case of the OLPI radar with an overall integration time of about 2 s, the variation of the Doppler shift within the integration time would result in unacceptable losses, even if the ACE test was used.

The possible Doppler variation may be estimated again; for a tangential flight the variation is given by:

$$\Delta f_d = \frac{2v^2 t}{\lambda R}$$

With $R = 40 \, \text{km}$, $\lambda = 10 \, \text{cm}$, $v = 250 \, \text{m/s}$, $t = 2 \, \text{s}$ follows $\Delta f_d = 62.5 \, \text{Hz}$. A filter bank with $N = 8000$ and a radar period of $T_C = 256 \, \mu\text{s}$ would have a single filter bandwidth of $1/N T_C = 1/t = 0.5 \, \text{Hz}$. The signal energy would be distributed over 120 filters and a corresponding high detection loss would result.

Even if the ACE test is applied a high loss would result by the Doppler change. A proposal is to compute ACE-test values for shorter subsections of length N_m. For each section one output signal T_m is derived from the ACE test. These signals T_m are further processed with an incoherent sequential test. Because the test variable T_m is the sum of a high number of contributions it is approximately Gaussian distributed, therefore a likelihood ratio can be set up for this sequential test function. This sequential test will have a mean test length in noise n_n, that is here the mean number of segments of length N_m. For different section length values N_m the detection performance has been compared to find the most efficient performance.

To define the Gaussian probability density function we need its mean and variance. The mean value μ of T_m may be derived analytically, but this is not possible for the standard deviation because the individual terms of equation 9.13 are not statistically independent. Therefore the parameters $(\mu, \sigma)_{0,1}$ have been determined for the two cases, $0 = $ noise and $1 = $ signal and noise, by simulation. The parameters for the signal and noise case are valid for a selected design signal $a = a_0$ used for the signal simulation.

The distribution of the test values $x = T_m$ is given by:

$$p(x) = \frac{1}{\sqrt{2\pi}\sigma} \exp\left[-\frac{(x - \mu)^2}{2\sigma^2}\right] \tag{9.19}$$

The likelihood-ratio test function is then with p_0 the probability density for noise and p_1 for signal and noise:

$$\lambda = \prod_{m=1}^{n} \frac{p_1(x_m)}{p_0(x_m)} \qquad (9.20)$$

With equations 9.19 and 9.20 we get:

$$\ln \lambda = \sum_{m=1}^{n} \ln\left(\frac{p_1}{p_0}\right) = n \ln\left(\frac{\sigma_0}{\sigma_1}\right) + \sum_{m=1}^{n} \left\{ \frac{(x_m - \mu_0)^2}{2\sigma_0^2} - \frac{(x_m - \mu_1)^2}{2\sigma_1^2} \right\} \qquad (9.21)$$

9.6.2   Simulation studies

A program was developed to simulate the complete signal processing, starting from random signals $z = (x + a) + jy$ (components x, y normally distributed with mean 0 and standard deviation $\sigma = 1$), and then computing the ACE test values for signal segments N_m using equation 9.11. An example of simulated probability distribution densities is given in Figure 9.15, with (mn, sign), (ms, sigs) given the resultant values for (μ_0, σ_0) and (μ_1, σ_1) respectively. The final sequential test is simulated according to equation 9.21, using the results for (μ_0, σ_0) and (μ_1, σ_1) from the first simulation step for setting up the test function according to equation 9.21.

In Figure 9.16 is given for illustration an example for the course of test functions of the sequential test in case of noise alone, and in Figure 9.17 for signal plus noise.

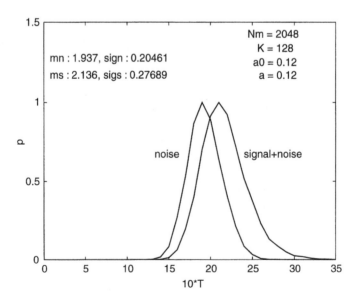

Figure 9.15   *Probability densities of ACE test values T for noise and signal plus noise, originally published in [19]*

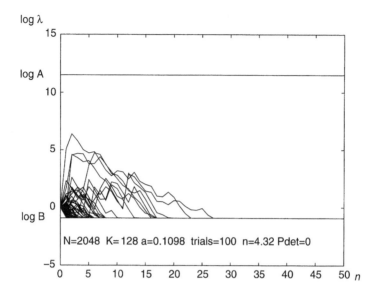

Figure 9.16 Simulation of sequential test for coherent processed segments, noise alone, $\mu_0 = 1.94$, $\sigma_0 = 0.205$, $\mu_1 = 2.14$, $\sigma_1 = 0.28$, $\beta = 0.4$

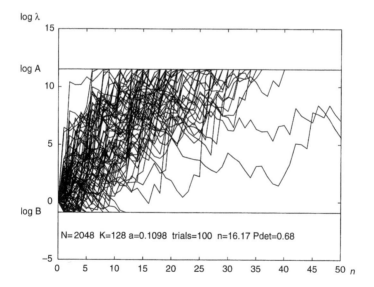

Figure 9.17 Simulation of sequential test for coherent processed segments, signal and noise, $\mu_0 = 1.94$, $\sigma_0 = 0.205$, $\mu_1 = 2.14$, $\sigma_1 = 0.277$, $\beta = 0.4$

The random walk of the test function to one of the two decision thresholds and the resultant random test length is again demonstrated. For the subsection length N_m were selected the values 256, 512, 1024, 2048. As a measure of efficiency the mean overall test length $n_n N_m$ in noise for a detection probability $P_d = 0.5$ and a selected signal amplitude $a = a_0$ were compared.

For the OLPI radar processing with a design amplitude $a_0 = 0.105$, $N_m = 2048$, $K = 512$ and $n_n = 3.9$ results on the one hand in a mean test length of $3.9 \cdot 2048 = 7987 \approx 8000$, which corresponds to the required fixed test length $N = 8000$ for a hypothetical filter bank. On the other hand, $P_d = 0.5$ is now achieved with a simulated signal amplitude $a = 0.09975$ (SNR $= -23$ dB).

This has to be compared with the filter bank concept: the necessary SNR $= -23.52$ dB. There is only a difference of 0.52 dB, so the efficiency of the filter bank (unrealisable because of the missing Doppler tolerance) is practically achieved by this combination of ACE and sequential detection.

The important advantage of the partial coherent test with ACE is the improved tolerance against Doppler frequency changes e.g. for the detection of targets with a tangential velocity component. This is demonstrated by Figure 9.18: the additional SNR with increasing Doppler frequency, normalised to the overall dwell time, is shown for different detection methods. At zero Doppler frequency the filter bank starts with a loss of 0 dB. This loss is soon higher than with incoherent or amplitude integration. Two versions of the combination of ACE test and sequential detection start at somewhat increased losses, but these increase only a little with the increasing Doppler rate. For the parameter example assumed above

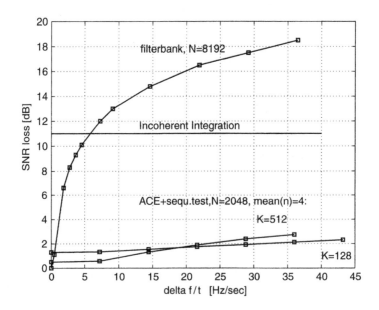

Figure 9.18   S/N detection loss in dB versus Doppler variation

for the OLPI radar a Doppler rate for tangential flying targets with 31.25 Hz/s applies.

9.7 Experimental system

A sequential-detection signal-processing subsystem has been implemented with digital signal processors (DSPs) and combined with the experimental phased-array system ELRA to perform the search function at long ranges. The following figures give a demonstration of the function of the sequential test for multiple range elements. According to equation 9.8 the test function for each range element i is computed and combined with the range-dependent weighting using equations 9.3, 9.6 and 9.7:

$$\lambda_n = \frac{1}{S} \sum_{i=1}^{M} \exp \left[-a_i^2 - n\frac{a_i^2}{2} + \sum_{j=1}^{n} \ln I_0(a_i x_{ij}) \right]$$

or by abbreviating the term in [ ] by $\tilde{l}_{in}$:

$$\lambda_n = \frac{1}{S} \sum_{i=1}^{M} \exp \left[\tilde{l}_{in} \right]$$

The function $\tilde{l}_{in}$ above range i is shown in Figure 9.19 after the first transmit pulse.

The digital value for $\tilde{l}_{in}$ is digital-analogue converted and displayed on an oscilloscope. A weak echo signal (position indicated by the arrow) increases the test function a little for the corresponding range element. The course of $\tilde{l}_{in}$ is mainly determined by the weighting term $-a_i^2$. In Figure 9.20 the signal has been stressed after ten radar

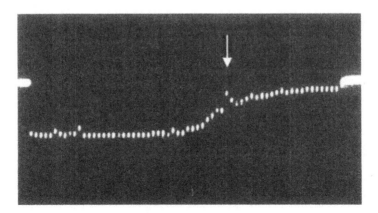

Figure 9.19   Test function $\tilde{l}_{in}$ after the first radar pulse above range i, © IEE 1981 [10]

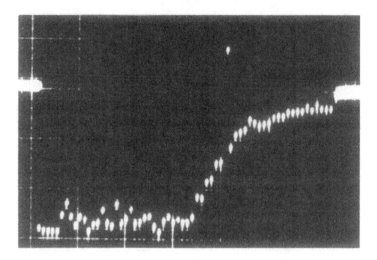

Figure 9.20   Test function $\tilde{l}_{in}$ after 10 radar pulses above range i, © IEE 1981 [10]

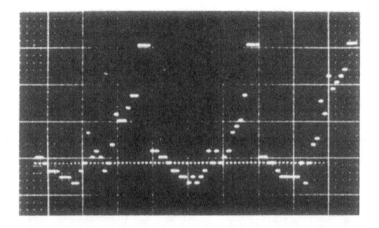

Figure 9.21   Test function λ_n for a beam position, target present, over pulse number n, © IEE 1981 [10]

periods by the integration effect. After a decision for a target for this beam position the corresponding range i can clearly be extracted from the maximum value of the test function $\tilde{l}_{in}$.

The test function for one beam position and for all range elements according to equation 9.8 or 9.9 is displayed in Figures 9.21 and 9.22 for target present and noise alone, respectively. The horizontal axis is now the time in pulse periods or pulse number n. We see the course of several tests with the test function increasing to the upper threshold A (decision: signal present) or to the lower threshold B (decision: noise only) as sketched in Figure 9.1.

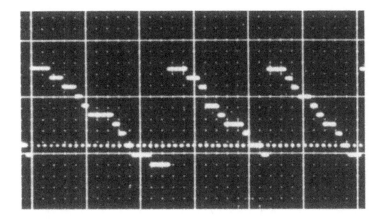

Figure 9.22   Test function λ_n for a beam position, noise only, over pulse number n,
            © *IEE 1981 [10]*

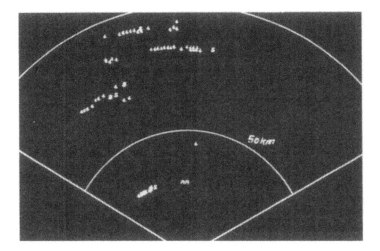

Figure 9.23   Flying targets, detected by sequential detection, © *IEE 1981 [10]*

Aircraft target search plots, produced by sequential detection, are displayed in Figure 9.23.

The plots are integrated for some time to recognise the movement of targets along their flight path.

Adaptation to the noise level as described in section 9.4 is demonstrated with the following figures. In Figure 9.24 plots of some detected targets are shown. Then the noise level is increased by about 10 dB. The resulting numerous false alarms are seen in Figure 9.25. Because the test function has an increased sensitivity for longer ranges the false alarms are concentrated there correspondingly. After activation of noise normalisation, most of the false alarms disappear and the targets are detected

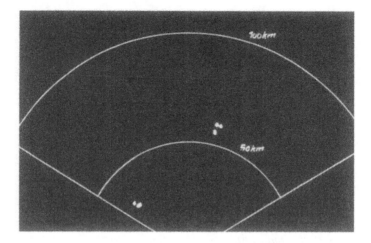

Figure 9.24   Plots of sequentially detected targets, © IEE 1981 [10]

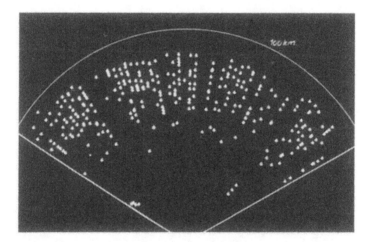

Figure 9.25   Plots of sequential detector for increased noise power, © IEE 1981 [10]

again, as shown in Figure 9.26. However, the mean test length in noise has now increased.

9.8   Conclusions

Sequential detection should be applied for the search mode of phased-array radars. A range-dependent weighting has to be used to combine the test function from range cells for forming one test function for the respective beam position. Then the highest gain, expressed in SNR saving, is achieved at maximum range.

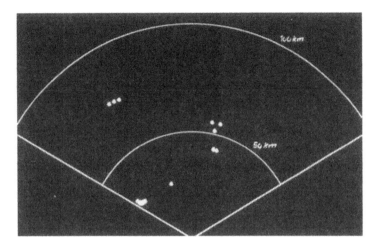

Figure 9.26 *Plots of sequential detector after noise normalisation,* © *IEE 1981 [10]*

The most effective detection performance is achieved with coherent signal processing. As an alternative to a filter bank with an increasing number of filters the necessary test function may be derived from the small-signal approximation by a combination of autocorrelation estimates. This ACE test may be realised by recursive processing and is therefore especially suited to sequential detection. It is also more robust in the case of Doppler signals with varying frequency compared to filter-bank processing. The ACE sequential test function has been devised to achieve the usual probability of detection for a specified design signal and determined by the lower threshold B. The gain compared to fixed sample size coherent processing with a filter bank has been established to about 5 dB for a false-alarm probability of 10^{-6}. This corresponds to the gain of sequential detection in the case of incoherent processing. The test for multiple range elements with a range-dependent design signal has also been proposed and discussed for this coherent case.

For a long signal series in favour of energy management the application of sequential detection results in a considerable improvement, expressed in a reduced mean test length. A combination of ACE test values of signal segments or subsections with an incoherent sequential test achieves the theoretical performance of the optimal filter bank even in the case where, because of possible Doppler variations, the filter bank cannot be applied.

9.9 References

1 WALD, A.: 'Sequential analysis' (John Wiley and Sons, Inc., New York, 1947)
2 BUSSGANG, J. J., and Middleton, D.: 'Optimum sequential detection of signals in noise', *IRE Trans. Inf. Theory*, 1955, **1** (3), pp. 5–18

3 MARCUS, M. B., and SWERLING, P.: 'Sequential detection in radar with multiple resolution elements', *IRE Trans. Inf. Theory*, April 1962, pp. 237–245

4 HELSTROM, C. W.: 'A range-sampled sequential detection system', *IRE Trans. Inf. Theory*, 1962, pp. 43–47

5 ROBERTS, A. R.: 'Sequential detection and composite hypotheses with application to a signal of uncertain amplitude', *IEEE Trans. Inf. Theory*, 1967, **13** (2), pp. 175–183

6 BUSSGANG, J. J.: 'Truncated sequential hypothesis tests', *IEEE Trans. Inf. Theory*, 1967, **13** (8), pp. 512–516

7 BUSSGANG, J. J.: 'Sequential methods in radar detection', *Proc. IRE*, 1970, **58** (5), pp. 731–743

8 WIRTH, W. D.: 'Der Sequentialtest zur Entdeckung von Radarzielen bei einer Vielzahl von Entfernungselementen', *NTZ*, 1972, (2), pp. 72–76

9 WIRTH, W. D.: 'Fast and efficient target search with phased array radars'. Proceedings of the IEEE international conference on *Radar*, Arlington, USA, 1975, pp. 198–203

10 WIRTH, W. D.: 'Signal processing for target detection in experimental phased array radar ELRA', *IEE Proc. F, Commun. Radar Signal Process.*, 1981, **128** (5), pp. 311–331

11 SKOLNIK, M. (Ed.): 'Radar handbook' (Mc Graw Hill, 1990, 2nd edn) p. 2.21

12 SWERLING, P.: 'Probability of detection for fluctuating targets', *IRE Trans. Inf. Theory*, 1960, **6**, pp. 269–308

13 WIRTH, W. D.: 'Energy saving by coherent sequential detection of radar signals with unknown Doppler shift', *IEE Proc., Radar Sonar Navig.*, 1995, **142** (3), pp. 145–152

14 SELIN, I.: 'Detection of coherent radar returns of unknown Doppler shift', *IEEE Trans. Inf. Theory*, July 1965, pp. 396–400

15 LANK, G. W., REED, I. S., and POLLON, G. E.: 'A semicoherent detection and doppler estimation statistic', *IEEE Trans. Aerosp. Electron. Syst.*, March 1973, **9** (2), pp. 151–165

16 WIRTH, W.D.: 'Energy saving by coherent sequential detection of radar signals with unknown Doppler shift'. International conference on *Radar*, Paris, 1994, pp. 133–138

17 WIRTH, W. D.: 'Long term coherent integration for a floodlight radar'. Proceedings IEEE international conference on *Radar*, Alexandria, 1995, pp. 698–703

18 WIRTH, W. D.: 'Sequential detection for energy saving and management'. IEEE international symposium on *Phased Array Systems and Technology*, Boston MA, USA, 15–18 October 1996, pp. 289–292

19 WIRTH, W. D.: 'Detection of long signal series with unknown Doppler shift'. IASTED international conference on *Signal Processing and Communication*, Gran Canaria, February 11–14, 1998, pp. 282–285

Chapter 10

Adaptive beamforming for jammer suppression

Military radar applications must generally take into account hostile countermeasures in the form of irradiation of interference signals from jammers, which emit different waveforms occupying the frequency band of our own radar as an electronic countermeasure (ECM). Usually for the investigation of countermeasures against this threat the assumed interference waveform is noise. The signal-to-noise ratio for target echoes at the output of the radar receiver would be dramatically decreased by this interference and target detection would be impossible or a lot of false alarms generated. Therefore we have to develop adequate electronic counter countermeasures (ECCM) to maintain the operation of our radar [1]. One common technique is to spread our own frequency band by changing the operating frequency using a random pattern for the frequency selection. This technique is named frequency agility. We thus force the enemy also to spread his noise power over a wider frequency band and therefore the spectral power density of the jammer in our receiver is minimised.

Several radar functions rely on coherent processing of a train of pulses within a so-called coherent processing interval (CPI) without any frequency change. This is especially the case in the Doppler processing mode for moving target detection and location in the presence of clutter. The operator of the jammer may then learn our frequency quickly, and concentrate his noise power on our operating frequency band.

The jammers may be carried by special aircraft flying on holding patterns in a stand-off mode or may be installed on the ground. In these cases they are generally received by the sidelobes of our antenna. In special worst-case situations the jammers are received within the width of our mainbeam. This situation may occur especially in air-defence applications if attacking hostile aircraft are accompanied by escorting aircraft carrying the jammers. The complete jammer scenario may be encompassed by stand-off, escorting and ground-based jammers. Certain jammer scenarios have been generated by military planning authorities but the real possible future scenario is unknown. Jammers are inexpensive compared with radar systems and it may be worthwhile for our counterpart to apply a larger number of jammers to make our radar ineffective.

For our own radar it would be advisable to combine available ECCM techniques to achieve maximum effectiveness against an ECM threat. The most effective contribution to ECCM would be to form the antenna pattern to eliminate or at least to reduce the reception of the jamming noise signals by the formation of very low sidelobes in the jammer directions. One common technique is to form antenna patterns by applying a tapering function (Taylor for the sum beam, Bayliss for the difference beams) to achieve low sidelobes everywhere outside the mainbeam, as described in chapter 4. We then have to accept as the price a broadened main beam pattern and a loss in gain.

With multiple jammers this technique of sidelobe reduction will nevertheless be insufficient. As an example Figure 10.1 shows the computed reduction of radar sensitivity expressed as the normalised signal-to-noise-plus-interference ratio (*SNIR*) dependent on the beam direction $u = \sin(\varphi)$ for ten jammers present at the indicated directions (dashed lines) [2]. The decrease of SNIR is even more dramatic the more closely the jammers are spaced because then they are more frequently within the main beam part of the antenna pattern, for example on its flanks. In Figure 10.2 the course of the *SNIR* is converted to a reduction of the radar range coverage: only the area

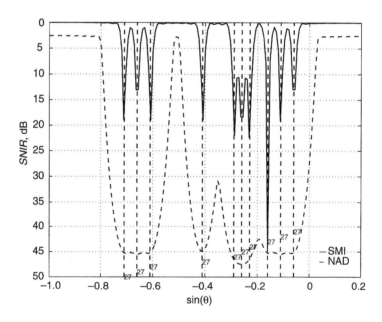

Figure 10.1   SNIR for ten jammers, low sidelobe antenna (dashed line), adaptive antenna (solid line) (courtesy U. Nickel)

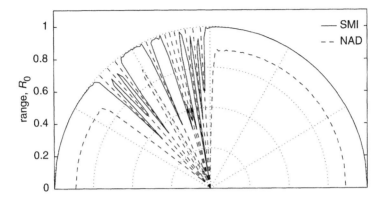

Figure 10.2   *Radar coverage according to Figure 10.1, low sidelobe (dashed line), adaptive antenna (solid line) (courtesy U. Nickel)*

below the dashed line is left for target detection. The range reduction is especially effective on the left-hand side with the neighbouring jammers.

An effective countermeasure against such jamming threats may be achieved by an adaptation of the antenna pattern to the actual jammer situation [2–8]. This is possible in the sidelobe region and even, to a certain extent, in the area of the main lobe. In the sidelobe region notches have to be created in the jammer directions. The mainlobes must be deformed to minimise jammer interference as far as possible. In Figures 10.1 and 10.2 the recovered *SNIR* and radar coverage, respectively, are indicated by the curves with solid lines. Between the jammer directions almost the full *SNIR* or radar range is recovered.

Because the jammer situation with respect to the beam direction changes with each new beam position, adaptation of the antenna pattern has to be performed after each beam switching. Because there are always certain errors and deviations in the antenna elements and receiving channels this adaptation cannot be achieved by deterministic beamforming but has to be performed by an adaptive procedure using the received jammer signals at the outputs of the array's receiving channels as the basis for the learning process.

10.1   Deterministic generation of pattern notches

As a first step for discussing the techniques for jammer suppression and to develop an improved insight we will derive the weighting coefficients for a receiving array to generate notches into given jammer directions.

Assume a linear array with N antenna elements at a regular spacing with $d = \lambda/2$. The direction of the main beam is given by $u_0 = (\sin \phi_0)$ and the directions of the K jammers by u_k with $k = 1, \ldots, K$. The received signals from these sources impinging on the array with normalised amplitudes equal to 1 are given by the columns

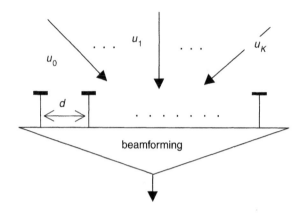

Figure 10.3   Jamming signals at a linear array

$\mathbf{a}_k$ in matrix $\mathbf{A}$ with elements $n = 1, \ldots, N$ (Figure 10.3):

$$a_{kn} = \exp\left(j\pi u_k n\right)$$

$$\mathbf{A} = \begin{bmatrix} a_{01} & a_{11} & \cdots & a_{K1} \\ \vdots & \vdots & \cdots & \vdots \\ a_{0N} & a_{1N} & \cdots & a_{KN} \end{bmatrix}$$

The antenna vectors given by the columns $\mathbf{a}_k$ are mutually orthogonal, that is the corresponding patterns have a zero in the respective other beam directions, only if the directions u_k are on a directional raster $u: -1 : 2/N : 1$, or $u_k = 2k/N$. Then, as is discussed in chapter 2 in conjunction with equation 2.22:

$$a_k^* a_l = \sum_{n=1}^{N} \exp\left(-\frac{2\pi j}{N}(k - l)n\right)$$

$$= 0 \quad \text{for } k \neq l$$

$$= N \quad \text{for } k = l$$

Each beam would have zeros in the pattern for all other beam directions. But for arbitrary directions u_k this is not the case.

10.1.1   LMS weighting

In a first approach we will now develop the computation of the weighting vector in order to form a receive beam with jammer suppression. For this task we apply the least-mean-square method already treated in chapter 3 section 3.5.2.

We assume a signal vector containing the unknown signal amplitudes of the sources (target and jammers) in the column matrix **s**:

$$\mathbf{s} = \begin{bmatrix} s_0 \\ s_1 \\ \vdots \\ s_K \end{bmatrix}$$

The receiver noise is given by the noise vector:

$$\mathbf{n} = \begin{bmatrix} n_1 \\ \vdots \\ n_N \end{bmatrix}$$

The complete received signal vector **x** at the antenna elements is the superposition of all signal components $k = 0, \ldots, K$, including the main beam signal for $k = 0$, and the noise vector **n**:

$$\mathbf{x} = \mathbf{As} + \mathbf{n}$$

By a weighting matrix **W** we want to get from **x** an estimate **ŝ** of the source signal with a least-mean-square error between **s** and **ŝ**:

$$\hat{\mathbf{s}} = \mathbf{W}^* \mathbf{x}$$

From chapter 3 we can directly apply equation 3.79 for this task:

$$\mathbf{W} = (\mathbf{APA}^* + \mathbf{Q})^{-1} \mathbf{AP}$$

For **P**, the covariance of **s**, we assume the identity matrix $\mathbf{I}[K + 1, K + 1]$, because the jammer signals are mutually uncorrelated. **Q** is as before the covariance matrix of the noise **n**. Matrix **W***** has $(K + 1)$ rows and each row (with index k) is the beamforming vector in the direction u_k with minimum output signal from all other directions u_l and $k \neq l$. This means that each beam has notches in these other beam directions. So we have generated a set of mutually decoupled beams. These are of importance also for other applications, e.g. for the discussion of superresolution techniques (chapter 12).

For our main beam in direction u_0 we achieve notches in all jammer directions $u_1, \ldots, u_K$ as desired. The main beam is selected by the row matrix $\mathbf{e}^*[1, (K + 1)]$ given by:

$$\mathbf{e}^* = \begin{bmatrix} 1 & 0 & \cdots & 0 \end{bmatrix}$$

The weighting vector for the main beam then becomes:

$$\mathbf{w}_0^* = \mathbf{e}^* \mathbf{W}^*$$

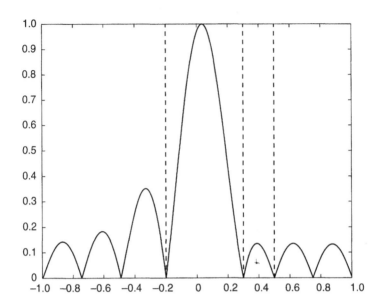

Figure 10.4   *Computed antenna pattern for a linear array, $N = 8$, 3 jammer directions indicated by dotted lines, horizontal axis: $u = sin\Phi$ from -1 to 1*

and finally:

$$\mathbf{w}_0^* = \begin{bmatrix} w_{01}^* & \cdots & w_{0N}^* \end{bmatrix} = \mathbf{e}^*\mathbf{A}^*(\mathbf{A}\mathbf{A}^* + \mathbf{Q})^{-1}$$

The term $\mathbf{e}^*\mathbf{A}^*$ is just the beamforming vector $\mathbf{a}_0^*$ for the direction u_0. The noise covariance matrix $\mathbf{Q}$ may be expressed as $\mathbf{Q} = q\mathbf{I}[K + 1, K + 1]$ giving:

$$\mathbf{w}_0^* = \mathbf{a}_0^*(\mathbf{A}\mathbf{A}^* + q\mathbf{I})^{-1} \tag{10.1}$$

The noise level q has to be selected according to the anticipated receiver noise to jammer power ratio, e.g. $q = 10^{-4}$. Without adding the noise matrix $\mathbf{Q}$ the matrix $(\mathbf{A}\mathbf{A}^* + \mathbf{Q})$ with dimension $[N, N]$ would be singular and inversion would be impossible. An illustration is given by the computed pattern for a linear array with $N = 8$ in Figure 10.4.

10.1.2   Multiple beam approach

We will now discuss a second approach to forming the main beam, using a weighting vector to achieve the required notches for jammer suppression.

We imagine that for all $(K + 1)$ directions conventional beams are formed by the rows of matrix $\mathbf{A}^*$ as a first step for beamforming with notches. The main beam is again formed for direction u_0. Each beam pointing to a jammer in direction u_k will measure the actual jammer signal and this may be subtracted, after multiplication with a suitable factor matched to the mainbeam sidelobe, from the output of the main beam for u_0. The factors v_k are combined in a row matrix $\mathbf{v}[1, K + 1]$:

$$\mathbf{v} = \begin{bmatrix} v_0 & \cdots & v_K \end{bmatrix}$$

The required weighting vector may now be a linear combination of the $(K + 1)$ beams:

$$\mathbf{w}^* = \mathbf{v}\mathbf{A}^*$$

If the signals received from the selected $K + 1$ directions are weighted and added by the beamforming vector $\mathbf{w}^*$ the resultant output signal shall be $\mathbf{e}^*$ as above, or in other words: only for the beam direction $k = 0$, our main beam for u_0, shall the output signal be 1. From all other beam directions the output signal shall be zero. This leads to:

$$\mathbf{w}^*\mathbf{A} = \mathbf{e}^* \quad \text{or} \quad \mathbf{v}\mathbf{A}^*\mathbf{A} = \mathbf{e}^*$$

This results, by multiplying with $(\mathbf{A}^*\mathbf{A})^{-1}$ from the right, in:

$$\mathbf{v} = \mathbf{e}^*(\mathbf{A}^*\mathbf{A})^{-1}$$

and:

$$\mathbf{w}^* = \mathbf{e}^*(\mathbf{A}^*\mathbf{A})^{-1}\mathbf{A}^* \tag{10.2}$$

Here, the matrix in ( ) has only the dimension $[K + 1, K + 1]$ and inversion is no problem. The idea behind this approach is again the principle: estimate and subtract. The computed result according to equation 10.2 is the same as with equation 10.1, shown in Figure 10.4.

10.1.3 Limit for number of notches

An important limitation exists for the number of possible notches K:

$$K \leq N - 1 \tag{10.3}$$

If we have only two antenna elements, only one notch is possible by subtracting the received signals. The direction of the notch may be determined by a complex factor or phase shift applied to one of the antenna outputs. By each additional antenna element an additional notch may be created. The limit is demonstrated in Figures 10.5 and 10.6. For $N = 8$ seven notches have been created, but if we try to produce eight notches this is not possible. Now the notches are no longer matched to the jammer directions. In other words: the number of degrees of freedom (DOF), given by the number of antenna elements, must be greater than the number of jammers or required notches.

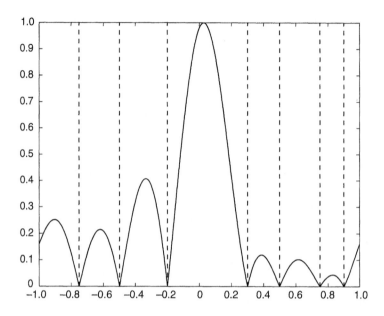

Figure 10.5   Linear array N = 8 and seven jammer directions

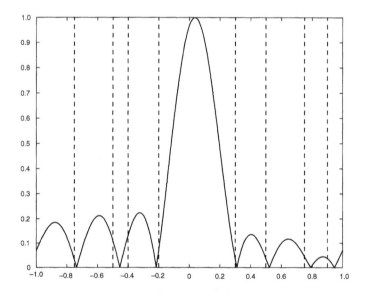

Figure 10.6   Linear array N = 8 and eight jammer directions

10.2   Adaptive jammer suppression

In real radar operation the directions of the jammers are unknown or at least not precisely known. There are some effects which influence the received jammer signals to be different from a model: mutual coupling between the antenna elements, unbalanced receiver channels and multipath effects. Therefore, only an adaptive solution can be efficient in practice.

10.2.1 Optimal processing

We may apply the basic equation 3.50 from chapter 3 for our task of beamforming with interference suppression: the beamforming vector $\mathbf{a}(u)$ for a selected direction u has to be multiplied by the inverse covariance matrix $\mathbf{Q}^{-1}$ and then by the received signal vector $\mathbf{z}$:

$$v = \mathbf{a}^*\mathbf{Q}^{-1}\mathbf{z} \tag{10.4}$$

v is the beamforming output and may be used for further signal processing such as pulse compression, Doppler filtering and sequential detection.

Two interpretations or realisations are possible for equation 10.4:

(i) The received signal vector $\mathbf{z}$ is first multiplied by the inverse of the covariance matrix $\mathbf{Q}$ resulting in a vector $\mathbf{y}$ with minimised jammer signals, then follows beamforming with $\mathbf{a}$:

$$\mathbf{y} = \mathbf{Q}^{-1}\mathbf{z} \quad \text{and} \quad v = \mathbf{a}^*\mathbf{y} \tag{10.5}$$

As explained in chapter 3 for each signal component z_n the most likely interference part $\hat{z}_n$ is estimated from all other signal components $z_m\,(m \neq n)$. By using $y_n = z_n - \hat{z}_n\ (n = 1, \ldots, N)$ the interference is optimally cancelled from the signal vector $\mathbf{y}$. $\mathbf{Q}$ has to be estimated from the received interference signal and noise from time segments without target echoes. For interference suppression by this matrix multiplication N^2 complex multiplications are necessary for each received data sample $\mathbf{z}$. Beamforming is performed afterwards by the weighting vector $\mathbf{a}^*$ applied to $\mathbf{y}$. Multiple beams may be formed from $\mathbf{y}$, that is after interference suppression, by applying a set of beamforming vectors.

(ii) An adapted beamforming weight vector $\mathbf{w}$ is first computed by:

$$\mathbf{w}^* = \mathbf{a}^*\mathbf{Q}^{-1}$$

and then follows:

$$v = \mathbf{w}^*\mathbf{z} \tag{10.6}$$

The N^2 multiplications with $\mathbf{Q}^{-1}$ have to be performed only for each new beamforming vector or beam position. For each data sample $\mathbf{z}$ the beam is formed by multiplication with $\mathbf{w}$ by N multiplications. Generally, this concept is much more economical.

The improvement by applying the adapted weight vector $\mathbf{w}$ may be described by the *SNIR* (signal-to-noise-plus-interference ratio). We assume a received signal $\mathbf{z} = b\mathbf{a}(u) + \mathbf{n}$. The target signal amplitude b multiplies with the direction-dependent antenna vector $\mathbf{a}(u)$. The interference vector $\mathbf{n}$ (from jammers and noise) adds to the signal. The *SNIR* at the beamformer output is:

$$\hat{\rho}(\mathbf{w}) = \frac{|b|^2|\mathbf{w}^*\mathbf{a}|^2}{\mathbf{w}^* E\{\mathbf{n}\mathbf{n}^*\}\mathbf{w}}$$

$$= \frac{|b|^2|\mathbf{w}^*\mathbf{a}|^2}{\mathbf{w}^*\mathbf{Q}\mathbf{w}} \qquad (10.7)$$

With the optimal weight vector from equation 10.6 in equation 10.7 follows:

$$\hat{\rho}_{opt} = |b|^2\mathbf{a}^*\mathbf{Q}^{-1}\mathbf{a}$$

In the case without jammers $\mathbf{Q} = \mathbf{I}$ follows:

$$\hat{\rho}_0 = |b|^2 N$$

and the optimal *SNIR* normalised to $\hat{\rho}_0$ is then:

$$\rho = \frac{1}{N}\mathbf{a}^*\mathbf{Q}^{-1}\mathbf{a} \qquad (10.8)$$

In the following we will discuss various suboptimal solutions instead of applying $\mathbf{Q}^{-1}$. For example, we may apply the sample covariance estimate $\hat{\mathbf{Q}}$. With:

$$\mathbf{w}^* = \mathbf{a}^*\hat{\mathbf{Q}}^{-1} \qquad (10.9)$$

for the *SNIR* then follows from equation 10.7:

$$\rho(u) = \frac{1}{N}\frac{\left|\mathbf{a}^*(u)\hat{\mathbf{Q}}^{-1}\mathbf{a}(u)\right|^2}{\mathbf{a}^*(u)\hat{\mathbf{Q}}^{-1}\mathbf{Q}\hat{\mathbf{Q}}^{-1}\mathbf{a}(u)} \qquad (10.10)$$

10.2.2   Illustration with a model

For an illustration and further insight into jammer suppression we use a model for $\mathbf{Q}$. The columns of matrix $\mathbf{A}[N, K]$ may now describe the directions of the jammers:

$$\mathbf{A} = \begin{bmatrix} \mathbf{a}_1 & \cdots & \mathbf{a}_K \end{bmatrix}$$

Jammer k emits a signal r_k. All jammer signals are combined in a column vector $\mathbf{r}[K, 1]$. The received signal vector thus will be:

$$\mathbf{z} = \mathbf{A}\mathbf{r} + \mathbf{n}$$

The covariance matrix for the jammer signal and noise is defined as:

$$\mathbf{Q} = E\{\mathbf{z}\mathbf{z}^*\}$$

or with $\mathbf{R} = E\{\mathbf{rr}^*\}$:

$$\mathbf{Q} = \mathbf{ARA}^* + \sigma^2\mathbf{I}$$

If we assume for all jammers the same power, that is $E\{|r_k|^2\} = 1$ for all k, we arrive at:

$$\mathbf{Q} = \mathbf{AA}^* + \sigma^2\mathbf{I} \qquad (10.11)$$

With the above assumption and a beamforming vector $\mathbf{a}_0$, for receiver noise alone the adapted beamforming vector $\mathbf{w}$ will result in:

$$\mathbf{w}^* = \mathbf{a}_0^*(\mathbf{AA}^* + \sigma^2\mathbf{I})^{-1} \qquad (10.12)$$

This equation is very similar to equation 10.1. $\mathbf{A}$ now contains only all $\mathbf{a}_k$ for the jammer directions and these are determining $\mathbf{Q}$. The computed adapted antenna pattern is the same as in Figure 10.4.

10.2.3 Orthogonalisation and eigenmatrix projection

One solution to the matrix inversion problem is offered by matrix orthogonalisation. The matrix $\mathbf{Q}$ may already be estimated. Then it is possible to develop the following representation of $\mathbf{Q}$ by the well-known vector iteration algorithm of equation 2.18:

$$\mathbf{Q} = \mathbf{X\Lambda X}^* \qquad (10.13)$$

The columns of matrix $\mathbf{X}[N, N]$ are the eigenvectors $\mathbf{x}_k$ with $k = 1, \ldots, N$. These are mutually orthogonal, which means that $\mathbf{x}_k^*\mathbf{x}_j = 0$ for $k \neq j$. The diagonal matrix $\mathbf{\Lambda}$ is given by the eigenvalues λ_l for $l = 1, \ldots, N$. The eigenvalues for $l = 1, \ldots, K$ correspond to the jammers and their values are comparable to the jammer powers. The eigenvalues for $l = K + 1, \ldots, N$ correspond to the receiver noise power σ^2. An example for the eigenvalues is given in Figure 10.7 for the same jammer scenario as used for Figure 10.1.

Using the power assumption leading to equation 10.11 we may write equation 10.13 in more detail:

$$\mathbf{Q} = \begin{bmatrix} \mathbf{x}_1 & \cdots & \mathbf{x}_K & \cdots & \mathbf{x}_N \end{bmatrix} \begin{bmatrix} 1+\sigma^2 & & & & & \\ & \ddots & & & & \\ & & 1+\sigma^2 & & & \\ & & & \sigma^2 & & \\ & & & & \ddots & \\ & & & & & \sigma^2 \end{bmatrix} \times \begin{bmatrix} \mathbf{x}_1^* \\ \vdots \\ \mathbf{x}_K^* \\ \vdots \\ \mathbf{x}_N^* \end{bmatrix}$$

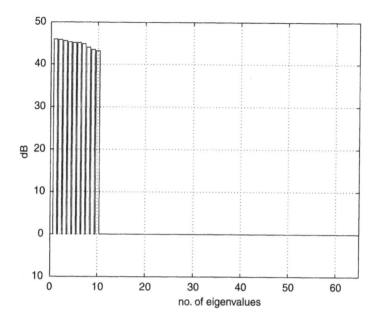

Figure 10.7   Eigenvalue distribution for ten jammers distributed as in Figure 10.1 (courtesy U. Nickel)

or by separating the jammer and receiver noise parts:

$$Q = X \left(\begin{bmatrix} 1 & & & & & \\ & \ddots & & & & \\ & & 1 & & & \\ & & & 0 & & \\ & & & & \ddots & \\ & & & & & 0 \end{bmatrix} + \begin{bmatrix} \sigma^2 & & & & & \\ & \ddots & & & & \\ & & \ddots & & & \\ & & & \ddots & & \\ & & & & \ddots & \\ & & & & & \sigma^2 \end{bmatrix} \right) X^*$$

(10.14)

The jammer part is described only by the eigenvectors $x_1, \ldots, x_K$, combined in the rectangular matrix X_K. So we get:

$$Q = X_K X_K^* + \sigma^2 I \qquad (10.15)$$

This equation corresponds to equation 10.11. The difference is that the eigenvectors x_k are a kind of beamforming vector which are mutually orthogonal. For the antenna vectors a_k this is the case only for the $N+1$ directions in a raster given by $u : -1 : 2/N : 1$. Moreover, the eigenvectors and their eigenvalues describe the jammer situation including multipath effects in contrast to the antenna vectors a_k.

In the general case we have to write for equation 10.13 with $\sigma^2 = 1$:

$$\mathbf{Q} = \mathbf{X} \begin{bmatrix} \lambda_1 & & & & & & \\ & \ddots & & & & & \\ & & \lambda_K & & & & \\ & & & 1 & & & \\ & & & & \ddots & & \\ & & & & & 1 \end{bmatrix} \mathbf{X}^*$$

The inversion then simply follows by:

$$\mathbf{Q}^{-1} = \mathbf{X} \begin{bmatrix} \lambda_1^{-1} & & & & & & \\ & \ddots & & & & & \\ & & \lambda_K^{-1} & & & & \\ & & & 1 & & & \\ & & & & \ddots & & \\ & & & & & 1 \end{bmatrix} \mathbf{X}^*$$

For very strong jammers the λ_k tend to infinity and we get:

$$\mathbf{Q}^{-1} \approx \mathbf{X} \begin{bmatrix} 0 & & & & & & \\ & \ddots & & & & & \\ & & 0 & & & & \\ & & & 1 & & & \\ & & & & \ddots & & \\ & & & & & 1 \end{bmatrix} \mathbf{X}^*$$

or denoting with $\mathbf{X}_n$ the matrix with the eigenvectors describing the noise space:

$$\mathbf{Q}^{-1} \approx \mathbf{X}_n \mathbf{X}_n^*$$

Because $\mathbf{X} = \mathbf{X}_K + \mathbf{X}_n$ we may express $\mathbf{Q}^{-1}$ also by:

$$\mathbf{Q}^{-1} \approx \mathbf{I} - \mathbf{X}_K \mathbf{X}_K^* = \mathbf{P} \qquad (10.16)$$

Remark: assume the matrices $\mathbf{X}_n, \mathbf{X}_K$ filled up with zeros to dimension $[N, N]$. Then we have:

$$\mathbf{X}\mathbf{X}^* = (\mathbf{X}_K + \mathbf{X}_n)(\mathbf{X}_K^* + \mathbf{X}_n^*) = \mathbf{X}_K \mathbf{X}_K^* + \mathbf{X}_n \mathbf{X}_n^* = \mathbf{I}$$

Equation 10.16 gives an approximation for $\mathbf{Q}^{-1}$ by the eigenvectors describing the jammer space. So the matrix inversion is replaced by the determination of K eigenvectors. Equation 10.16 gives us a so-called projection matrix $\mathbf{P}$: a jammer signal combination may be written as:

$$\mathbf{x} = \mathbf{X}_K \mathbf{r}$$

with **r** representing the jammer signals in the eigenvector space $\mathbf{X}_K$. Then:

$$\mathbf{x}^*\mathbf{P} = \mathbf{r}^*\mathbf{X}_K^*\left(\mathbf{I} - \mathbf{X}_K\mathbf{X}_K^*\right) = \mathbf{r}^*\left(\mathbf{X}_K^* - \mathbf{X}_K^*\right) = 0$$

The jammer signals are completely cancelled because they are projected into a signal space, given by **P**, which is orthogonal to the jammer space. It is worth noting that by this projection method the adaptation of the antenna pattern is independent of the individual jammer powers. This is of interest especially for jammer situations with varying jammer powers, e.g. blinking jammers. On the other hand we will see some difficulties for an application in practice if channel errors are considered. Then the determination of the dimension of the jammer space becomes difficult. There is also a high computational effort of $O(N^3)$ complex multiplications resulting in difficulties for real-time applications.

In the discussion so far we have had an idea of the fundamental advantage of an adaptive receiving array against interference by jammers. We have discussed a limit for the number of generated notches dependent on the element number (DOF) and how the necessary beamforming weighting vector may be computed. We have gained an insight into how the covariance matrix of the interference and noise depends on the jammer directions and powers by assuming a model for the jammer scenario. The interference suppression may be explained by the estimate and subtract principle discussed in chapter 3 or by orthogonal projection. In the reality of radar operation we also have to take into account environmental effects like multipath and technical effects such as errors in the receiving channels, which are unknown and varying. Therefore only adaptive operation will be useful.

10.3   Antenna architecture for adaptation

With equation 10.4 the basic processing instruction for two possible versions of realisation is given.

10.3.1   Adaptation before beamforming

In a multifunction radar system the beam direction is changed by phase steering much more rapidly than a possible change of the jammer scenario. So it would be advantageous to perform jammer suppression before beamforming. Then multiple beams, e.g. sum and difference beams, or even a beam cluster, could be formed without the interference. On the other hand, operational multifunction radars will have $N = 1000, \ldots, 10\,000$ antenna elements and a direct application of equation 10.4 for an optimal procedure is not worth discussion. For jammer suppression after equation 10.3 we only need a number of elements N corresponding to the number of jammers K. Therefore, a suboptimal concept was proposed [9]: to each receiving channel are assigned only K instead of $N - 1$ auxiliary elements, randomly selected out of the whole array. Despite the suboptimal concept the processing structure would be complex and worth discussion only for smaller arrays and rapid beam switching.

10.3.2 Adaptive beamforming with subarrays

The standard solution for multifunction active array radars would be the use of sub-arrays [10]. The active receiving array is divided into a number L of subarrays. Within each subarray all antenna elements within a connected area are combined as a part of the array. The value L must be at least equal to the number K of jammers. The block diagram for this concept is given in Figure 10.8.

The subarrays may be formed at the outputs of the receiving channels, which mainly consist of a low-noise amplifier and a phase shifter, by microwave combining networks. After the combining network further amplification, mixing to an intermediate frequency, bandpass filtering (block R) and finally analogue-to-digital conversion will be performed. The subarray output signals are then available for further processing as digital samples at a sampling rate corresponding to the signal bandwidth.

For each new selected beam direction, with the corresponding phase shifter settings, the jammer signals are combined in the subarray in a new manner, which means that each new beam direction affords a new adaptation. This has to be accomplished in a time interval which is short compared to the time segment for the radar task. Only under this condition can real-time radar operation be maintained. With the subarray outputs the final beams are formed by a weighting vector and summing (e.g. sum and difference beams).

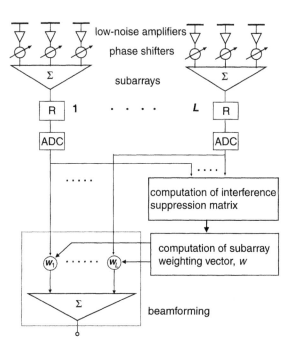

Figure 10.8 Subarray concept for adaptive beamforming

The subarray outputs are used to compute the adapted weights for final beam-forming with jammer suppression. Two steps have to be performed: gathering of interference data snapshots (learning or secondary data) and afterwards the computation of the weight vector. The subarray outputs are treated as single antenna elements as discussed so far. To avoid the grating-lobe effects discussed in chapter 4, that is repetition of jammer notches (grating notches) in the main beam direction and thus decreasing the gain unnecessarily, the subarrays should be irregular, especially concerning their phase centres or centres of gravity. With circular planar arrays even with a checkerboard division there is enough irregularity by the subarrays at the margin of the array plane to avoid undesired notches. Extensive studies have been made by Nickel [11] under various aspects to find an efficient subarray architecture (see chapter 5, section 5.1.4).

10.4   Adaptation algorithms

The computational problems for adaptive jammer suppression are the covariance matrix estimation and its inversion. After steering the beam into the desired direction we have to adapt the receiving pattern against the jammers with a minimum time delay. But first we have to gather jammer signals for estimation of the covariance matrix and afterwards we can start the weight computation. The required time intervals for these operations must be minimised for real-time radar operation.

10.4.1   Sample matrix estimation

The covariance matrix may be estimated from S data snapshots $\mathbf{z}_s$ (learning or secondary data set) at the array output by:

$$\hat{\mathbf{Q}} = \frac{1}{S} \sum_{s=1}^{S} \mathbf{z}_s \mathbf{z}_s^*$$

or with the data matrix $Z[N, S]$ combining all the S snapshots:

$$\hat{\mathbf{Q}} = \frac{1}{S} \mathbf{Z}\mathbf{Z}^* \tag{10.17}$$

The estimated $\hat{\mathbf{Q}}$ is after inversion applied in equation 10.6 to derive $\mathbf{w}$. This method is therefore known as the sample matrix inversion (SMI). The limited number of snapshots gives a fluctuating estimate $\hat{\mathbf{Q}}$ and a fluctuating weight vector. According to equation 10.13 the estimated $\hat{\mathbf{Q}}$ may be represented by orthogonalisation with its eigenvalues and eigenvectors. In particular, the small eigenvalues are estimated only with relatively great uncertainty because of the noise, and their inverse values fluctuate accordingly. This causes a fluctuation of the *SNIR* and an increased sidelobe level for the adapted antenna pattern.

The statistical distribution of the *SNIR* has been derived and analysed by Gierull [14,16]. The result is: to avoid a singularity of matrix $\hat{\mathbf{Q}}$ the snapshot number has to be $S \geq N$. For a mean *SNIR* loss of 3 dB the snapshot number must be even $S \geq 2N$.

A simple but effective solution to counter these fluctuation effects is the loaded sample matrix inversion method (LSMI). To limit the influence of the fluctuating estimate of $\hat{\mathbf{Q}}$ a noise matrix is added:

$$\hat{\mathbf{Q}} = \frac{1}{N}\mathbf{Z}\mathbf{Z}^* + \alpha\mathbf{I} \tag{10.18}$$

This especially reduces the relative variation of the small eigenvalues. The statistical property of the resulting *SNIR* has also been derived by Gierull in References 14 and 16 and the result is a *SNIR* loss of less than 3 dB for $S \geq 2K$. The 3 dB limit is achieved with a subarray number $L = 2K$ and a snapshot number $S = 2K$. The necessary number of snapshots now only depends on the number of jammers K and is much less than for the SMI method. According to simulations results, a recommended rule of thumb is the selection of $\alpha = 2, \ldots, 4\sigma^2$, with σ^2 for the receiver noise power.

For the computation of *SNIR* equation 10.10 is used for both methods.

10.4.2 Projection methods

In section 10.2.3 the favourable performance of an eigenvector projection was discussed. The disadvantage of this approach is the high computational effort with about $10N^3$ multiplications. Some proposals and studies have been given by Hung and Turner (HTP method) [12] to derive a suitable projection matrix directly from the data vectors **z**, combining all S snapshots in $\mathbf{Z}[N, S]$. Here again the number S of snapshots must be $S \geq K$, the number of jammers. Hung and Turner assumed that for a high jammer-to-noise ratio the jammer space is already sufficiently described by these S data vectors. The adapted weighting vector **w** is computed by an orthonormalised projection matrix and the beamforming vector **a** by:

$$\mathbf{w} = \left(\mathbf{I} - \mathbf{Z}(\mathbf{Z}^*\mathbf{Z})^{-1}\mathbf{Z}^*\right)\mathbf{a} \tag{10.19}$$

Remark: for a confirmation of the projection, a block of data vectors **Z** applied to **w** results in:

$$\mathbf{Z}^*\mathbf{w} = \mathbf{Z}^*\mathbf{a} - \mathbf{Z}^*\mathbf{Z}(\mathbf{Z}^*\mathbf{Z})^{-1}\mathbf{Z}^*\mathbf{a} = 0$$

Simulation studies and a statistical analysis [15] resulted in a loss of more than 3 dB of *SNIR* and this 3 dB limit was accomplished only if for L subarrays, K jammers and S samples the following relations hold:

$$L \geq 9K \quad \text{and} \quad S = \sqrt{K(L+1)} - 1 \tag{10.20}$$

This relatively high value for L is much more expensive in terms of subarray channels compared to the LSMI method. A further difficulty is the determination of the dimension of the jammer space. A high number of basis vectors decreases the remaining signal space and therefore the gain for signal detection.

There have also been studies to derive the projection matrix from columns of the sample covariance matrix as an improvement compared to the application of data vectors [19,20]. The expectation was that more data snapshots could be used for averaging without increasing the dimension of the jammer space. These approaches are described by applying a selection matrix $\mathbf{T}$:

$$\tilde{\mathbf{Q}} = \mathbf{QT} \tag{10.21}$$

Starting from the data matrix $\mathbf{Z}[N, S]$, containing the S snapshot vectors as columns, we get:

$$\hat{\tilde{\mathbf{Q}}} = \mathbf{ZZ}^*\mathbf{T} \tag{10.22}$$

The transformation or selection matrix $\mathbf{T}[N, K]$ may select K columns of $\mathbf{Q}$. For example, in each column one element is 1, the others are 0. If $\mathbf{T}$ has the first K columns of the identity matrix and the others zero then the first K columns of $\mathbf{Q}$ are selected. A projection matrix is then computed using equation 10.22 by:

$$\mathbf{P} = \mathbf{I} - \hat{\tilde{\mathbf{Q}}}(\hat{\tilde{\mathbf{Q}}}^*\hat{\tilde{\mathbf{Q}}})^{-1}\hat{\tilde{\mathbf{Q}}}^* \tag{10.23}$$

This projection was used with different choices for $\mathbf{T}$ in equation 10.22 to compare the course of the *SNIR* dependent on direction u for a certain selected jammer scenario [2] according to Figure 10.1.

Favourable results applicable for a realisation have been developed by Gierull at FGAN-FFM as the so-called matrix transformation-based projection (MTP) [14,17,18]. He proposed $\mathbf{T} = \mathbf{E}_1 + \mathbf{E}_2$ with $\mathbf{E}_{1/2}$ a block selection matrix $[N, K]$ for N array elements and K jammers, $[L, K]$ for L subarrays, with columns which have all elements equal to zero except one element with value 1 on a random position. $\mathbf{E}_1$, $\mathbf{E}_2$ have to satisfy:

$$\mathbf{E}_1^*\mathbf{E}_1 = \mathbf{E}_2^*\mathbf{E}_2 = \mathbf{I}[K, K] \quad \text{and} \quad \mathbf{E}_1^*\mathbf{E}_2 = 0$$

Matrix $\mathbf{T}$ therefore has K columns with two elements equal to 1, all other elements are zero.

Applied to the data matrix with all snapshots this results in:

$$\hat{\tilde{\mathbf{Q}}} = \mathbf{Z}\left(\mathbf{Z}^*\mathbf{E}_1 + \mathbf{Z}^*\mathbf{E}_2\right) \tag{10.24}$$

$(\mathbf{Z}^*\mathbf{E}_1 + \mathbf{Z}^*\mathbf{E}_2)$ forms a $[S, K]$ matrix. Using this term, only two are selected and added from each of the S data snapshots with N or L signals z_n from the array outputs. Figure 10.9 gives an example of the resulting *SNIR*. The loss compared to the optimum is only about 3 dB for $S \approx 2K$. This value is the same as with the eigenvector projection [16]. The advantage of this procedure is that the number of complex multiplications is only $0.5\ N^3$. By using averaging operations with this proposal the influence of receiver noise is reduced and therefore the noise influence on the weighting vector is also minimised. By selecting the columns randomly out of the whole covariance matrix the maximum aperture is used and the angular resolution for jammers is increased. This is comparable to the angular resolution given by a randomly-thinned array with which grating lobes are avoided.

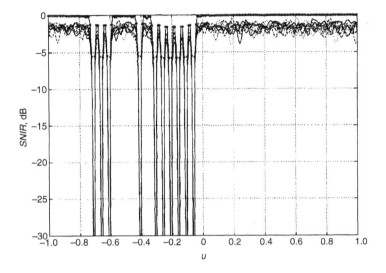

Figure 10.9   Ten realisations of SNIR for MTP, K = 32 (courtesy C. Gierulll)

10.4.3   Channel errors

Errors of the receiving channels decrease the achievable *SNIR*, especially if the interference signals at the array output are decorrelated [21]. This may be caused by *IQ* demodulation errors (orthogonality and amplification errors), by different or mismatched bandpass filter responses and different time delays.

These effects have been studied analytically with models by Nickel [22,23]. An analogue down conversion to the baseband followed by a lowpass filter, matched to the signal bandwidth, is assumed. For the channels $n = 1, \ldots, N$ are assumed offsets $C_n + j F_n$ and delays δ_n and for their I and Q components the amplifications R_n and H_n and phase errors φ_n and ψ_n, respectively.

The received signal may be expressed in complex form including these errors. For a deterministic source signal with complex amplitude b and with constant frequency offset f_m one obtains for the complex output of channel n:

$$S_n(t) = b \exp\{j2\pi f_m(t - \tau_n - \delta_n)\} a_n p_n$$
$$+ b^* \exp\{-j2\pi f_m(t - \tau_n - \delta_n)\} a_n^* q_n \qquad (10.25)$$

with $a_n = \exp((j2\pi/\lambda_0)(x_n u + y_n v))$ as the direction-dependent signal phase term at element n and:

$$p_n = R_n \exp(j\varphi_n) + H_n \exp(j\psi_n)$$

and:

$$q_n = \exp(-j4\pi f_0 \delta_n)[R_n \exp(-j\varphi_n) - H_n \exp(-j\psi_n)]$$

Equation 10.25 now represents two different signals instead of the one original signal for zero errors. With existing *IQ* orthogonality errors a single sinusoidal source with a certain frequency shift would describe an ellipse in the complex signal plane. This may be decomposed into two circles with the opposite frequency and direction-dependent initial phase.

Remark: in an extreme failure case, e.g. the y or Q component is zero, the x component would be represented by two complex signals with equal amplitude but opposite phase. One source is now represented by two sources with opposite frequencies and directions. The number of jammers seems to have doubled. Therefore the DOF, in this case the minimum number of subarrays, also has to be doubled.

The covariance matrix may also be derived from data $\mathbf{z}$ containing the channel errors. These data depend on the jammer signals $\mathbf{s}_k$, the noise vector $\mathbf{n}$ and the complex offset vector $\mathbf{G}$ [18]:

$$\mathbf{Q} = E\{\mathbf{z}\mathbf{z}^*\} = \sum_{k=1}^{K} E\{\mathbf{s}_k\mathbf{s}_k^*\} + E\{\mathbf{n}\mathbf{n}^*\} + \mathbf{G}\mathbf{G}^* \qquad (10.26)$$

For the noise part follows:

$$\mathbf{Q}_n = (\sigma^2/2)\,\mathrm{diag}\!\left(R_i^2 + H_i^2\right)$$

The diagonal noise covariance has now unequal elements with index i. For the jammer part follows [22] with a, p, q as above (with abbreviation si $(x) = \sin(x)/x$) for the matrix elements:

$$[\mathbf{Q}_k]_{mn} = \frac{b_k^2}{4}\,\mathrm{si}\,[\pi B(\tau_{mk} - \tau_{nk} + \delta_m - \delta_n)]\!\left(a_{mk}a_{nk}^* p_m p_n^* + a_{mk}^* a_{nk} q_m q_n^*\right)$$

$$(10.27)$$

This equation shows: the term in ( ) is the sum of two dyadic products representing a rank of two for a single source. We learn, therefore, that orthogonality errors will double the rank of the covariance matrix or double the number of eigenvalues. This corresponds to the remarks above. The offset results in an additional eigenvalue. So in effect we get $M = 2K + 1$ for the number of eigenvalues. This is the number of necessary DOF or the number of subarrays to suppress jammers and also compensate for channel errors.

The bandwidth also has a certain degrading effect. In equation 10.27 the si function determines the decrease of the correlation which is dependent on the difference of the direction-dependent time delays $\tau_{mk} - \tau_{nk}$. These are zero for broadside jammers but increase with off broadside directions. This decorrelation results in a shifting of jammer power into the noise space and a corresponding loss of *SNIR*. Surprisingly, this effect is partly compensated for by random delay differences $\delta_m - \delta_n$. If they compensate the directional delay difference for some receiving channels the respective correlation recovers and an improved jammer cancellation results. An example is shown in Figure 10.10. The improvement is observed between the dashed curve without random delays, the curves with solid lines demonstrating the effect of random

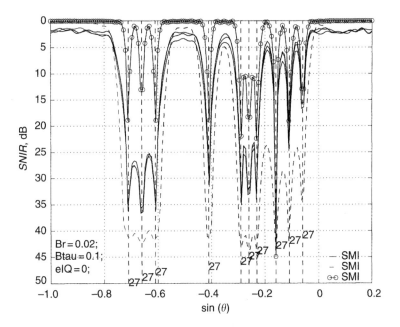

Figure 10.10   SNIR for 2% bandwidth with and without delay errors, noise without bandwidth (circled line), noise with bandwidth (dashed line), noise with bandwidth and random delays (solid line) (courtesy U. Nickel)

delays. As a reference the *SNIR* for a noise without bandwidth is also given, marked by circles. So it is not necessary to adjust delay differences below the values given by D/c (D = antenna diameter, c = velocity of light), the maximum travelling time across the array [22].

Another reason for a decorrelation of jammer signals is a different filter response of the receiving channels. The passband of the RF part, with a low-noise amplifier, phase shifter, mixer to intermediate frequency (IF), bandpass filter at IF and a mixer to the I and Q components at the baseband, is fairly broadband and has a flat response within the signal bandwidth. On the other hand, the baseband filters, which are matched to the signal bandwidth, may be critical. Nickel has studied these effects also with a model [22]. For the filter response is assumed:

$$a(f) = 1 + \varepsilon_n \cos\left(\frac{2\pi f}{B}\left(j_n - \frac{1}{2}\right)\right) \quad \text{for} \quad |f| < B/2$$

$$= 0 \quad \text{else}$$

with j_n an integer which denotes the number of ripples within the passband B. The result is a further decorrelation of the jammer signals and a loss in *SNIR*. An example is given by Figure 10.11. The ripple number is selected as 5 with a varying phase of the ripple function.

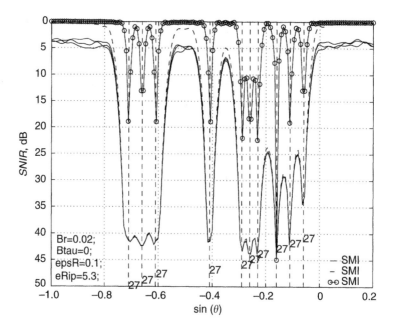

Figure 10.11   SNIR with filter errors: bandwidth 2%, five amplitude ripples 0.8 dB, reference curve for zero bandwidth (circled line), bandwidth but without filter ripple (dashed line), (courtesy U. Nickel)

By the application of digital lowpass filters after sampling and analogue-to-digital conversion at the intermediate frequency as described in chapter 6 an improved match between receiving channels will be achieved. Therefore, this technique is recommended for future designs.

In practice all the discussed errors cause a leakage of jammer power from the jammer space into the remaining noise or signal space and the clear separation of jammer and noise space is lost. This is demonstrated in Figure 10.12 [22], according to the jammer scenario in Figure 10.1. The eigenvalues are now decreasing more slowly compared to the distribution shown in Figure 10.7. The necessary DOF for jammer suppression are now increased depending on a threshold in relation to the noise level selected for the eigenvalues. If the number of eigenvalues is selected to be too large, an unnecessary reduction of the signal space and therefore a reduction of *SNIR* results. Methods to estimate effectively the dimension of the jammer space for small sample size have been studied by Nickel [23] and will be discussed in chapter 12.

10.4.4   Weighted projection: lean matrix inversion method

As discussed above, with channel errors the eigenvalues are distributed more widely, increasing the dimension of the jammer space. The jammer power seems to creep into

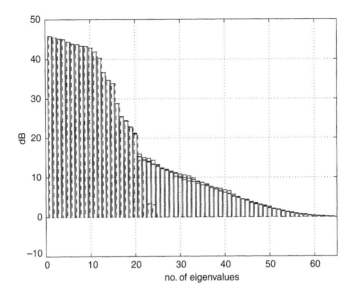

Figure 10.12 Eigenvalues for jammer scenario as in Figure 10.1 and IQ errors (courtesy U. Nickel)

the noise space and a clear separation between both spaces disappears, especially for strong jammers. As a consequence the value of K, the assumed number of jammers or dominant eigenvalues, would have to be increased. But this would reduce the signal space with a loss in gain or a loss in *SNIR*. The main consequence for the projection methods discussed in section 10.4.2 is the difficulty of determining the jammer space dimension for a jammer suppression as efficiently as with the LSMI method. The reason for this difficulty is that projection methods completely cancel all jammers independent of their power. This is not necessary in practice, especially for those low-power (low-eigenvalue) components introduced by channel errors. The LSMI method automatically achieves the optimal compromise.

The inverse covariance matrix estimate with a diagonal load may be approximated as in the following [18] to find the way to a weighted projection matrix:

$$\hat{Q} + \alpha I = X_K(\Lambda_K + \alpha I)X_K^* + X_n(\Lambda_n + \alpha I)X_n^*$$
$$\approx X_K(\Lambda_K + \alpha I)X_K^* + \alpha X_n X_n^*$$
$$\approx X_K(\Lambda_K + \alpha I)X_K^* + \alpha(I - X_K X_K^*) = X_K \Lambda_K X_K^* + \alpha I$$

With the matrix inversion identity given in section 2.1 with equation 2.17 we may derive the desired approximation for the inverse of the estimate $\hat{Q}$ loaded with a diagonal noise matrix (LSMI):

$$(\hat{Q} + \alpha I)^{-1} \approx (X_K \Lambda_K X_K^* + \alpha I)^{-1}$$
$$= \frac{1}{\alpha}(I - X_K G X_K^*) \quad \text{with } G = \text{diag}\left\{\frac{\lambda_l}{\lambda_l + \alpha}\right\} \quad (10.28)$$

G causes a weighting descending with the order $l = 1, \ldots, K$ of the eigenvalues. This weighting may be applied correspondingly to the projection matrix computed with the MTP procedure according to equation 10.24.

The weighting coefficients follow from a QR decomposition of $\hat{\mathbf{Q}}\mathbf{T}$ as the diagonal elements of the upper triangular matrix. Gierull [18] suggested starting from the sample matrix $\hat{\mathbf{Q}}$ and proceeding in the following way:

$$\hat{\mathbf{Q}}\mathbf{T} = \mathbf{Z}\big(\mathbf{Z}^*\mathbf{E}_1 + \mathbf{Z}^*\mathbf{E}_2\big) = \mathbf{U}\mathbf{L} = [\mathbf{U}_K\mathbf{U}_n]\mathbf{L}$$

UL is the result of a standard triangular decomposition, therefore **L** is an upper triangle matrix. The unitary matrix **U** (separated into $\mathbf{U}_K$ and $\mathbf{U}_n$) gives with its orthogonal columns the basis of the jammer space $\mathbf{U}_K$ and the noise space $\mathbf{U}_n$. The separation between both is decided by the descending values of the diagonal values L_{ll} of **L**, which correspond to the eigenvalues above in equation 10.28. They build a diagonal weighting matrix **G** with L_{ll} taken for the λ_l as in equation 10.28 above. A threshold for the jammer space is selected, e.g. $3\sigma^2$ to $5\sigma^2$. This defines the subspace dimension K (number of columns from U) and therefore $\mathbf{U}_K$ whereby the final projection matrix follows:

$$\mathbf{P} = \mathbf{I} - \mathbf{U}_K\mathbf{G}\mathbf{U}_K^* \tag{10.29}$$

This procedure is called lean matrix inversion (LMI), because only those $N \times K$ respective $L \times K$ elements of the sample covariance matrix have to be computed which are selected by matrix $\mathbf{E}_{1/2}$.

10.5   Realisation aspects

The number of complex operations to achieve a loss of less than 3 dB for the *SNIR* is given by Table 10.1 [17,23]. For K jammers the necessary number L of subarrays

Table 10.1   *Parameters for adaptive algorithms*
EVP: eigenvector projection
LSMI: loaded sample matrix inversion
HTP: Hung-Turner projection
LMI: lean matrix inversion
MTP: matrix transformation projection

Method	Subarrays (L)	Snapshots (S)	Complex operations
EVP	$2K$	$2K$	$10 \times L^3$
LSMI	$2K$	$2K$	$0.66 \times L^3$
HTP	$9K$	$3K$	$0.5 \times L^3$
LMI, MTP	$2K$	$2K$	$2K^2L \leq 0.5 \times L^3$

and S snapshots is given. An estimate of the number of complex operations is given in the last column.

For all adaptation methods the gain loss may be improved below the value of 3 dB by a higher number of snapshots. The number of operations for the LMI method depends on the number of jammers present. Therefore, the computation time depends on the actual number of jammers. For all other methods K represents the maximum number of jammers which may be coped with.

One problem to be considered is the sampling of data for the computation of the adaptive weights (secondary data). These secondary data should not contain the target signal to avoid signal cancellation, they should not contain clutter echoes to avoid adaptation against the clutter areas and, finally, the time period should be short in favour of a fast adaptation. The problem of signal cancellation has been studied by U. Nickel [24]. Before final beamforming, pulse compression and signal integration by Doppler filtering or sequential detection the signal-to-noise-plus-interference ratio ($SNIR$) is quite low for reasonable detectable targets. Therefore, the covariance matrix estimation is not much influenced by target signals and the signal cancellation effect can be tolerated. For stronger echo signals a loss can be tolerated because target detection is evident.

For adaptive processing two basic structures are applicable [24]. For a continuous data stream which is typical for a rotating radar with a necessary continuous adaptation of the receiving beam pattern a systolic array processor is well suited [25]. In contrast, with a multifunction phased-array radar and its stepwise and random beam switching to independent beam positions a blockwise operation seems to be more suitable. Results for adaptive weights for certain beam positions may be stored for a limited time interval and applied again because the beam positions are reproducible. This may be sensible for example within the multiplex target tracking process for multiple targets or for search operations within the same area with a high repetition rate.

For the implementation of the adaptive processor a compromise solution has to be found for combining parallel and distributed processing. This depends, of course, on the available digital signal processors. A major amount of processing is necessary for the beamforming, that is application of the adapted weights to the received data stream. In favour of high flexibility for the adaptive weight computation a high-level language should be applied for programming.

10.6   Experiments with the ELRA system

The experimental active phased-array radar system ELRA, described in chapter 17, was used for some adaptive beamforming experiments by I. Gröger [26], presented in the following. This system has 48 subarrays. The architecture for the assignment of antenna elements to the subarrays is shown in Figure 17.4. With randomly-thinned antenna elements and therefore also randomly-arranged subarrays the ELRA antenna is well suited for the purpose of adaptive jammer suppression according to the concept shown in Figure 10.8.

The aim of the experiments was to demonstrate the feasibility of sidelobe suppression or generating of notches for multiple jammer directions with the available hardware and to draw conclusions for future developments and improvements.

For larger array antennas it is not practical to measure the antenna pattern of the beam steered into certain directions. This would require a turntable for the array and a fixed probe antenna at a certain distance, or in reverse an accurately controlled movement of the probe antenna around the array. This kind of measurement would be very expensive and would have to be repeated for all beam positions relative to broadside. Instead, it is very easy to measure the antenna scan pattern by electronically steering the beam across a fixed probe antenna, as described in chapter 17. In the case of the receiving array antenna a probe or auxiliary antenna radiates a constant RF signal towards the array. The beam output signal is observed while scanning the receiving beam by the phase shifter steering across the probe antenna. If the probe is in the nearfield of the array the phases have to be focused correspondingly into the nearfield by adding appropriate phase values to those for forming a plane wave. From this equivalent antenna measurement the parameters of the antenna pattern (main beamwidth, sidelobe level) can also be estimated.

Because all the subarrays have a different antenna-element distribution they produce different patterns. As an example we show in Figure 10.13 the scan pattern of three out of the available 48 subarrays.

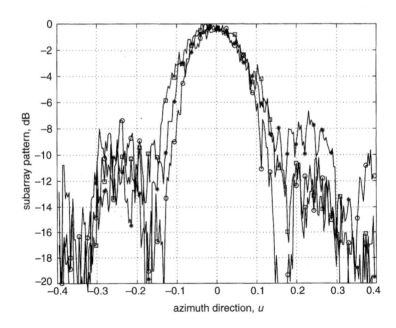

Figure 10.13   Scan pattern of three out of 48 subarrays of the ELRA receiving array
(courtesy I. Gröger)

For adaptive beamforming experiments the aim was to show the following: while the beam is scanned the scan pattern determines the beam output from the probe antenna. Now we require for selected multiple beam directions a suppression of the probe signal by creating notches in the scan pattern. By scanning across the probe we then expect notches for these selected directions. This is equivalent to the creation of notches in the antenna pattern for multiple directions. This behaviour has to be achieved by one weighting vector for final beamforming at the subarray output, as indicated in Figure 10.8. The basic equation for computing the weight vector is given by equation 10.6. Because the phases are steered for the desired direction the beamsteering vector for the subarray output is given by $\mathbf{a}^* = (1, \ldots, 1)$. The covariance matrix $\mathbf{Q}$ is estimated by gathering at the subarray outputs signal samples for all the desired look directions where notches will be generated. With all these samples together the interference distribution is described by a sample matrix $\hat{\mathbf{Q}}$ according to equation 10.17. For weight computation all methods may then be applied which have been described in section 10.4. After applying the computed weight vector the receiving beam is scanned across the probe antenna and the resulting scan pattern is measured. Many such measurements have been performed. Some examples are given here.

In Figure 10.14 the antenna scan pattern is shown against u (azimuth) for $-0.4 < u < +0.4$. Ten notches are created. The usual beam pattern without adaptation is

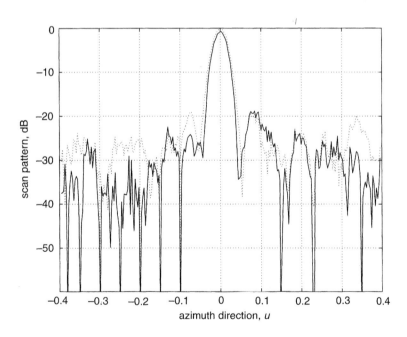

Figure 10.14   *Adapted scan pattern (solid line) with ten notches, together with unadapted pattern (dotted line), both produced from 48 subarrays (ELRA) (courtesy I. Gröger)*

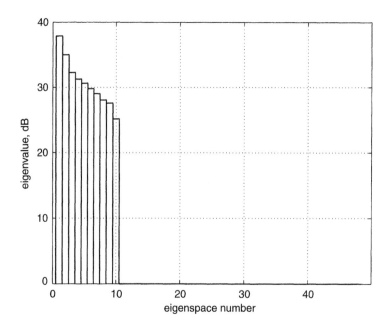

Figure 10.15   Eigenvalues corresponding to adaptation of Figure 10.14 for ten jammers (courtesy I. Gröger)

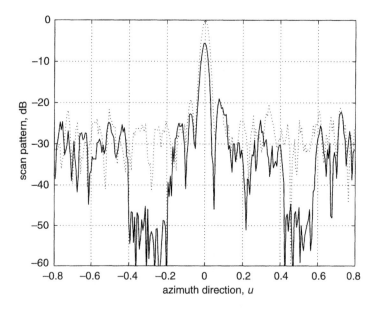

Figure 10.16   Two regions with clustered jammer directions, adapted scan pattern (solid line) and unadapted pattern (dotted line) (courtesy I. Gröger)

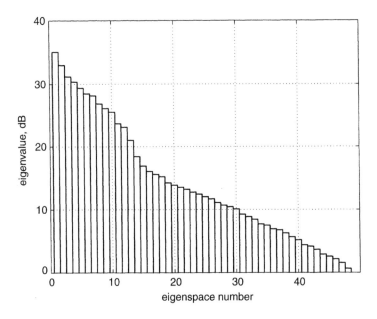

Figure 10.17   Eigenvalues corresponding to adaptation of Figure 10.16 for two sets of clustered jammers (courtesy I. Gröger)

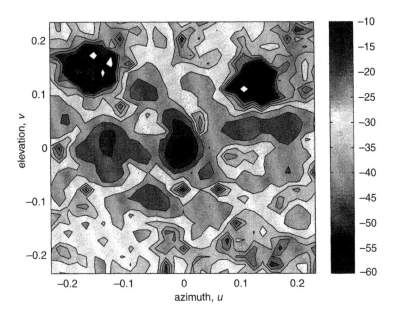

Figure 10.18   Adapted two-dimensional sidelobe suppression areas and the main beam in the centre (measured with ELRA) (courtesy I. Gröger)

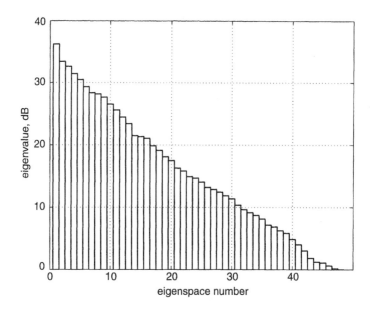

Figure 10.19   Eigenvalues corresponding to adaptation of Figure 10.18 for two-dimensional sidelobe suppression areas (courtesy I. Gröger)

given by the dotted line, the adapted pattern with notches at the required directions is represented by the solid line. The notch depth is more than 20 dB below the sidelobe level. The price for this adaptation is a reduction of the main beam gain by about 1 dB.

In Figure 10.15 the corresponding eigenvalues are shown. We recognise that there are only ten dominating eigenvalues corresponding to the ten required notch directions. There are enough remaining degrees of freedom for forming the main beam in the signal direction. In Figure 10.16 the notch directions have been chosen in adjacent positions to form two broader angular sectors with sidelobe reduction. Figure 10.17 shows the corresponding eigenvalues. These are distributed over the whole available signal space which is now reduced. This results in a corresponding gain reduction of the main beam of about 4 dB. Even a two-dimensional region with lowered sidelobes may be created as shown by a contour plot in Figure 10.18 with the corresponding eigenvalues shown in Figure 10.19.

10.7   Conclusions

An adaptive array, based on an active receiving array with subarrays, results in an efficient countermeasure against the threat of multiple jammers. This possibility represents a major advantage and a motivation for implementing an active phased-array radar. The adaptive technique is superior to a low sidelobe antenna in the case of multiple jammers.

If there are K jammers then from theory at least $K + 1$ subarrays with receiving channels are necessary to create notches in the antenna pattern for the suppression of the jammer signals. In practice, to counter some inaccuracies of the receiving antenna, it is recommended to have at least $2K$ subarrays.

Some algorithms for computing the adapted weighting vector, to form the beam from the subarray outputs, are discussed. For this adaptation algorithm the loaded sample matrix inversion is efficient and recommended. Projection methods based on eigenvector matrices describing the jammer and signal space offer a jammer suppression independent of their power. Their difficulty is the separation between these spaces, especially for high jammer power and channel inaccuracies. The lean matrix-inversion algorithm needs the lowest number of computations and is especially recommended for a higher number of jammers. For the learning phase only about $2K$ snapshots of the jammer signals are then necessary.

By some first experimental results the adaptive principles are demonstrated and theoretical expectations are confirmed.

10.8   References

1 JOHNSTON, S. L.: 'Radar electronic counter-countermeasures', *IEEE Trans. Aerosp. Electron. Syst.*, 1978, **14**, pp. 109–117

2 NICKEL, U.: 'Comparison of a non-adaptive low-sidelobe array antenna with an adaptive phased array antenna'. Proceedings of IEEE international conference on *Radar*, Arlington, Va, USA, 1990, pp. 486–490

3 MARR, J. D.: 'A selected bibliography on adaptive antenna arrays', *IEEE Trans. Aerosp. Electron. Syst.*, 1986, **22**, pp. 781–798

4 APPLEBAUM, S. P.: 'Adaptive arrays', *IEEE Trans. Antennas Propag.*, 1976, **24**, pp. 585–598

5 BUCCIARELLI, T., *et al.*: 'The Gram-Schmidt sidelobe cancellor'. Proceedings of IEE conference on *Radar*, 1982, pp. 486–490

6 FIELDING, J. G., *et al.*: 'Adaptive interference cancellation in radar systems'. Proceedings of IEE conference on *Radar*, 1977, pp. 212–217

7 FARINA, A., and STUDER, F. A.: 'Evaluation of sidelobe canceller performance', *IEE Proc. F, Signal Radar Process.*, 1982, **129**, pp. 52–58

8 WHITE, W. D.: 'Adaptive cascade networks for deep nullling', *IEEE Trans. Antennas Propag.*, 1978, **26**, pp. 396–402

9 WIRTH, W. D.: 'Suboptimal suppression of directional noise by a sensor array before beam forming', *IEEE Trans. Antennas Propag.*, September 1976, pp. 741–744

10 NICKEL, U., and WIRTH, W. D.: 'Beamforming and array processing with active arrays'. Proceedings of IEEE *Antennas and propagation* symposium, Ann Arbor, USA, 1993, pp. 1540–1543

11 NICKEL, U.: 'Subarray configurations for digital beam forming with low sidelobes and adaptive interference suppression'. Proceedings of IEEE international conference on *Radar*, Alexandria, Va, USA, 1995, pp. 714–719

12 HUNG, E. K., and TURNER, R. M.: 'A fast beamforming algorithm for large arrays', *IEEE Trans. Aerosp. Electron. Syst.*, 1983, **19** (4), pp. 589–607

13 BUEHRING, W.: 'Adaptive orthogonal projection for rapid converging interference suppression', *Electron. Lett.*, 1978, **14** (16), pp. 515–516

14 GIERULL, C. H.: 'Schnelle Signalraumschätzung in Radaranwendungen'. Dissertation, Aachen 1995, Verlag Shaker

15 GIERULL, C. H.: 'Performance analysis of fast projections of the Hung-Turner type for adaptive beamforming', *Signal Process.*, 1996, **50**, pp. 17–28

16 GIERULL, C. H.: 'Statistical analysis of the eigenvector projection method for adaptive spatial filtering of interference', *IEE Proc. Radar, Sonar Navig.*, 1997, **144** (2), pp. 57–63

17 GIERULL, C. H.: 'A fast subspace estimation method for adaptive beamforming based on covariance matrix transformation', *AEÜ International Journal of Electronics and Communications*, 1997, (4), pp. 196–205

18 GIERULL, C. H.: 'Fast and efficient method for low rank interference suppression in presence of channel noise', *Electron. Lett.*, 1998, **34** (6), pp. 518–520

19 BRANDWOOD, D. H.: 'Noise-space projection: MUSIC without eigenvectors', *IEE Proc. H, Microw. Antennas Propag.*, 1987, **134**, pp. 303–309

20 YEH, C. C.: 'Projection approach to bearing estimation', *IEEE Trans. Acoust. Speech Signal Process.*, 1986, **34**, pp. 1347–1349

21 STEYSKAL, H.: 'Array error effects in adaptive beamforming', *Microwave J.*, September 1991, pp. 101–112

22 NICKEL, U.: 'On the influence of channel errors on array signal processing methods', *AEÜ International Journal of Electronics and Communications*, 1993, **47** (4), pp. 209–219

23 NICKEL, U.: 'On the application of subspace methods for small sample size', *AEÜ International Journal of Electronics and Communications*, 1997, **51** (6), pp. 279–289

24 NICKEL, U.: 'Adaptive beam forming for phased array radars'. DGON international symposium on *Radar*, IRS, München, Germany, 1998, pp. 897–906

25 PROUDLER, I. K., and McWHIRTER, J. G.: 'Algorithmic engineering in adaptive signal processing: worked examples', *IEE Proc., Vis. Image Signal Process.*, 1994, **141** (1), pp. 19–26

26 GRÖGER, I.: 'Antenna pattern shaping'. International conference on *Radar*, Paris, France, 1989, pp. 283–287

Chapter 11

Monopulse direction estimation

After the detection of targets in the search mode there follows target location with range and direction estimation which is as precise as possible. This procedure may in principle be performed with the same receiving data as that already used for detection. Multifunction radar systems with phased arrays may apply an additional acquisition mode after target detection to confirm the first target detection and thereby cancel to a large extent false alarms produced by noise and interference. This radar task may be performed with increased transmit energy compared to that used for the search mode, because there are relatively fewer acquisition orders. Because of the higher SNR for the acquisition mode the location accuracy will be improved. The waveform will be chosen for improved range resolution, as discussed in chapter 7. After target acquisition the location function will also be applied to the target tracking process.

11.1 Likelihood direction estimation

The parameter estimation may be implemented according to chapter 3 as a maximum-likelihood estimation. The assumed probability density distribution for the received signal depends on the unknown parameter direction, which has to be estimated.

A planar array with antenna element positions $r_n = (x_n, y_n)$ with $n = 1, \ldots, N$ receives from a target with the unknown direction $\omega = (u, v)$ and complex amplitude b the signal vector and the unavoidable noise vector $\mathbf{n}$, the vector $\mathbf{z}$ given by:

$$\mathbf{z} = \mathbf{a}b + \mathbf{n}$$

with:

$$
\mathbf{a} =
\begin{bmatrix}
\exp\left[\dfrac{2\pi j}{\lambda} r_1 \omega\right] \\
\vdots \\
\exp\left[\dfrac{2\pi j}{\lambda} r_n \omega\right] \\
\vdots \\
\exp\left[\dfrac{2\pi j}{\lambda} r_N \omega\right]
\end{bmatrix}
=
\begin{bmatrix}
\exp\left[\dfrac{2\pi j}{\lambda}(x_1 u + y_1 v)\right] \\
\vdots \\
\exp\left[\dfrac{2\pi j}{\lambda}(x_n u + y_n v)\right] \\
\vdots \\
\exp\left[\dfrac{2\pi j}{\lambda}(x_N u + y_N v)\right]
\end{bmatrix}
\tag{11.1}
$$

Vector **a** describes the direction of arrival. Because of the Gaussian distributed noise we have following equation 3.41 for the probability density for $\mathbf{z} = \mathbf{x} + j\mathbf{y}$ dependent on parameters ω and b and for uncorrelated noise with $\sigma = 1$ for the single components x and y of $\mathbf{z}$:

$$p(\mathbf{z}/\omega, b) = \frac{1}{(2\pi)^N} \exp\left[-\frac{1}{2}(\mathbf{z} - \mathbf{a}b)^*(\mathbf{z} - \mathbf{a}b)\right] \qquad (11.2)$$

To determine the estimate $\hat{\omega}$ for the target direction and amplitude $\hat{b}$ the value of p has to be maximised dependent on ω and b:

$$p(\mathbf{z}/\omega, b)|_{\omega=\hat{\omega}, b=\hat{b}} = \max \qquad (11.3)$$

Therefore in equation 11.2 the quadratic expression $QE = (\mathbf{z} - \mathbf{a}b)^*(\mathbf{z} - \mathbf{a}b)$ has to be minimised.

As a first step, following Nickel [1,4], an estimate for $\hat{b}$ is derived by the partial derivative to b:

$$\hat{b} = (\mathbf{a}^*\mathbf{a})^{-1}\mathbf{a}^*\mathbf{z} = \frac{1}{N}\mathbf{a}^*\mathbf{z}$$

because $(\mathbf{a}^*\mathbf{a})^{-1}$ is a factor $1/N$. With this expression the unknown variable b may be eliminated:

$$QE = -\mathbf{z}^*\mathbf{z} + \frac{1}{N}\mathbf{z}^*\mathbf{a}\,\mathbf{a}^*\mathbf{z}$$

Only **a** depends on the unknown variable ω, therefore we may drop $\mathbf{z}^*\mathbf{z}$. So we have to maximise the expression:

$$P(\omega) = |\mathbf{a}^*\mathbf{z}|^2 = |S(\omega)|^2 = S\,S^* \quad \text{with } S = \mathbf{a}^*(\omega)\mathbf{z} \qquad (11.4)$$

P is the power of the sum beam output S, which has to be maximised by variation of ω to find the estimate $\hat{\omega}$. We could vary ω by steering the beam and observe ω to find a maximum value of P. This would be very time and energy consuming.

On the other hand, if we know the function $P(\omega)$ the maximum and the related direction may be found by computation. Following Reference 1 the maximum of $F = ln(P)$ is determined instead of P, to achieve an estimation independent of the target SNR. This is permitted, because ln is a monotonically increasing function.

The beam is steered to $\omega_0 = (u_0, v_0)$, given by a search result, near to the target direction $\omega = (u, v)$ within a beamwidth separation. The derivative of F with respect to u and v is now approximated by a Taylor series around $\hat{\omega}$, with F_ω, $F_{\omega\omega}$ the first and second derivatives of F with respect to ω:

$$F_\omega(\omega) = F_\omega(\hat{\omega}) + F_{\omega\omega}(\hat{\omega})(\omega - \hat{\omega}) + \cdots \qquad (11.5)$$

$F_\omega(\hat{\omega}) = 0$, because at the maximum the first derivative is zero. Writing equation 11.5 in detail we then have:

$$\begin{pmatrix} F_u(u, v) \\ F_v(u, v) \end{pmatrix} = \begin{pmatrix} F_{uu}(\hat{u}, \hat{v}) & F_{uv}(\hat{u}, \hat{v}) \\ F_{uv}(\hat{u}, \hat{v}) & F_{vv}(\hat{u}, \hat{v}) \end{pmatrix} \begin{pmatrix} u - \hat{u} \\ v - \hat{v} \end{pmatrix} \tag{11.6}$$

F_u, F_v, F_{uu}, F_{uv} and F_{vv} are the first and second partial derivatives with respect to u and v at the point $(\hat{u}, \hat{v})$. We insert into equation 11.6 for (u, v) the known beam direction (u_0, v_0) and resolve for the desired $(\hat{u}, \hat{v})$:

$$\begin{pmatrix} \hat{u} \\ \hat{v} \end{pmatrix} = \begin{pmatrix} u_0 \\ v_0 \end{pmatrix} - \begin{pmatrix} F_{uu}(\hat{u}, \hat{v}) & F_{uv}(\hat{u}, \hat{v}) \\ F_{uv}(\hat{u}, \hat{v}) & F_{vv}(\hat{u}, \hat{v}) \end{pmatrix}^{-1} \begin{pmatrix} F_u(u_0, v_0) \\ F_v(u_0, v_0) \end{pmatrix} \tag{11.7}$$

This is the general monopulse equation for direction estimation. We may compute with one data sample z the values F_u, F_v for the actual beam direction (u_0, v_0) and from this the target direction estimate $(\hat{u}, \hat{v})$. This can also be called the off-boresight estimation, because it is not necessary to scan the beam across the target.

Next we have to determine the derivatives of $F = \ln P$ from equation 11.4. With $\partial \mathbf{a}/\partial u = \mathbf{a}_u$ we get:

$$F_u = \frac{\partial F}{\partial u} = \frac{1}{P}\frac{\partial P}{\partial u} = \frac{\mathbf{a}_u^* \mathbf{z} \mathbf{z}^* \mathbf{a} + \mathbf{a}^* \mathbf{z} \mathbf{z}^* \mathbf{a}_u}{\mathbf{a}^* \mathbf{z} \mathbf{z}^* \mathbf{a}}$$

$$= 2\,\text{Re}\left\{\frac{\mathbf{a}_u^* \mathbf{z}}{\mathbf{a}^* \mathbf{z}}\right\}$$

$$= 2\,\text{Re}\left\{\frac{S_u}{S}\right\} \tag{11.8}$$

The weighting vector $\mathbf{d}_x^* = \mathbf{a}_u^*$ with the elements:

$$d_{x_i} = \frac{\partial a_i^*}{\partial u} = -\frac{2\pi}{\lambda}jx_i a_i^* \quad i = 1, \ldots, N \tag{11.9}$$

may be written with a diagonal matrix $\mathbf{D}_x = \text{diag}(-jx_i)$:

$$\mathbf{d}_x^* = \frac{2\pi}{\lambda}\mathbf{a}^* \mathbf{D}_x \tag{11.10}$$

For the variable v the derivation is the same, resulting in a weight vector $\mathbf{d}_y$ with $\mathbf{D}_y = \text{diag}(jy_i)$. So we have modified the beamforming vectors for direction estimation in both dimensions: the original beamforming vector $\mathbf{a}$ is multiplied by the element coordinates in the respective dimension x and y. The centre of the antenna is the origin of the coordinates (x, y), therefore the weighting vectors are linear and unsymmetrical. The contribution of the edge elements is the most important.

Especially for higher signal-to-noise ratios, the signal amplitude cancels in equation 11.8. That was the reason for choosing $\ln P$ for the maximising operation [1].

Because the antenna halves have opposite signs, the beam output may be seen as the difference of the outputs of these halves and therefore these beams are also called difference beams. The beam shape corresponds to the classical monopulse pattern with an approximately linear function through the origin within the 3 dB beamwidth of the sum beam. This pattern is generated with reflector antennas by a special monopulse feed horn. The name monopulse resulted from the possibility of estimating a target direction with only one pulse, in contrast to older estimation procedures which scan across a target or use conical scanning. Nevertheless, the monopulse procedure will usually be applied for pulse series, which are necessary for Doppler filtering or signal integration. Then the result of this signal integration has to be applied as the signal vector $\mathbf{z}$ or preferably the output of the sum and difference beams S, S_u, S_v are integrated and then used in the monopulse ratio equation 11.8.

It remains for us to determine the matrix with the second derivatives in equation 11.7. We start from $F_u = P_u/P$ given in equation 11.8. The second derivative is then with the basic rule for deriving a quotient $((f/g)_x = f_x g - f g_x/g^2)$:

$$F_{uu} = \frac{(P_{uu}P - P_u^2)}{P^2} = \frac{P_{uu}}{P} \tag{11.11}$$

because P_u is zero at $(\hat{u}, \hat{v})$. With equation 11.4 follows for $P_u = S_u S^* + S S_u^*$, and:

$$P_{uu} = S_{uu}S^* + S_u S_u^* + S_u S_u^* + S S_{uu}^*$$

Because S_u is zero at $(\hat{u}, \hat{v})$ it follows with equation 11.11:

$$F_{uu} = \frac{S_{uu}S^* + S S_{uu}^*}{SS^*}$$

$$= \frac{S_{uu}}{S} + \frac{S_{uu}^*}{S^*}$$

$$= 2\,\mathrm{Re}\left\{\frac{S_{uu}}{S}\right\}$$

For a high SNR we may assume at $(\hat{u}, \hat{v})$: $\mathbf{z} = \mathbf{a}b$ and $S = \mathbf{a}^*\mathbf{a}b = Nb$. For S_{uu} we get:

$$S_{uu} = \mathbf{a}_{uu}^* \mathbf{z} = \mathbf{a}_{uu}^* \mathbf{a}b = -b\left(\frac{2\pi}{\lambda}\right)^2 \sum_{i=1}^{N} x_i^2$$

and finally:

$$F_{uu}(\hat{u}, \hat{v}) = -\frac{2}{N}\left(\frac{2\pi}{\lambda}\right)^2 \sum_{i=1}^{N} x_i^2 \tag{11.12}$$

For the other derivatives follows accordingly:

$$F_{vv}(\hat{u}, \hat{v}) = -\frac{2}{N}\left(\frac{2\pi}{\lambda}\right)^2 \sum_{i=1}^{N} y_i^2 \tag{11.13}$$

$$F_{uv}(\hat{u}, \hat{v}) = -\frac{2}{N}\left(\frac{2\pi}{\lambda}\right)^2 \sum_{i=1}^{N} x_i y_i \tag{11.14}$$

For a planar array the element coordinates are uniformly distributed and for the origin at the centre of the array we have the relation $\sum_{i=1}^{N} x_i y_i \approx 0$ and therefore $F_{uv} \approx 0$. The slope correction matrix:

$$-\begin{pmatrix} F_{uu}(\hat{u}, \hat{v}) & F_{uv}(\hat{u}, \hat{v}) \\ F_{uv}(\hat{u}, \hat{v}) & F_{vv}(\hat{u}, \hat{v}) \end{pmatrix}^{-1}$$

in equation 11.7 has therefore only elements on the main diagonal and the equations for estimates u and v are therefore decoupled. Equation 11.7 can be written as follows:

$$\hat{u} = u_0 + \left(\frac{2}{N}\left(\frac{2\pi}{\lambda}\right)^2 \sum_{i=1}^{N} x_i^2\right)^{-1} 2\,\mathrm{Re}\left\{\frac{\mathbf{d}_x^*\mathbf{z}}{\mathbf{a}^*\mathbf{z}}\right\}$$

$$= u_0 + k_x\,\mathrm{Re}\left\{\frac{\mathbf{d}_x^*\mathbf{z}}{\mathbf{a}^*\mathbf{z}}\right\}$$

with:

$$k_x = \left(\frac{1}{N}\left(\frac{2\pi}{\lambda}\right)^2 \sum_{i=1}^{N} x_i^2\right)^{-1}$$

and for the elevation dimension:

$$\hat{v} = v_0 + k_y\,\mathrm{Re}\left\{\frac{\mathbf{d}_y^*\mathbf{z}}{\mathbf{a}^*\mathbf{z}}\right\}$$

with:

$$k_y = \left(\frac{1}{N}\left(\frac{2\pi}{\lambda}\right)^2 \sum_{i=1}^{N} y_i^2\right)^{-1} \tag{11.15}$$

In these equations the beamforming vector $\mathbf{a}^*$ is looking to the direction (u_0, v_0). Both equations show an almost linear function between the real target and the estimated direction difference to the beam-steering direction, that is between $(u - u_0, v - v_0)$ and $(\hat{u} - u_0, \hat{v} - v_0)$ according to the assumption used for equation 11.5 and then equation 11.7. The slope at the origin is given by the factors k_x, k_y. An example is shown in Figure 11.1 for the dimension u and $N = 64$. The abscissa indicates the target direction u and the ordinate the estimated $\hat{u}$. The half-power beamwidth follows with equation 4.19: $\Delta u = 0.027$. The ideal monopulse characteristic would be $\hat{u} = u$ (straight line).

If an amplitude tapering for sidelobe reduction is applied for beamforming this has to be used as a modification of $\mathbf{a}$ and $\mathbf{d}$. The direction estimation is derived for only one target present. If there are more targets within the beamwidth, e.g. from multipath effects above sea, we have to expect large errors, which are called the glint

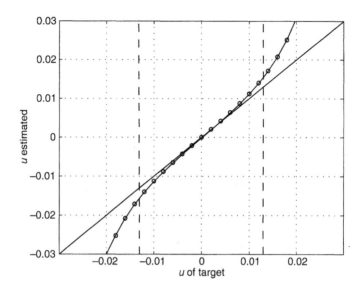

Figure 11.1   Monopulse characteristic without jamming (circled line), half-power beamwidth (dashed line), N = 64

effect. In these cases the monopulse technique has to be extended to superresolution, discussed in chapter 12.

The monopulse equation 11.15 depends on the array geometry with the distribution of the antenna elements. Subarrays could also be treated as elements with the coordinates of their phase centres. Different amplitudes of elements or subarrays (by tapering of receiving gain) could also be included into these formulas.

11.2   Experimental monopulse correction for failing elements

The receiving array of the experimental array radar system ELRA (chapter 17) is divided into 48 subarrays. The output signals of these subarrays are used to form the sum and difference beams with the appropriate weighting vectors. To demonstrate the possibility of recalibration for the case of subarray failure some subarray output signals have been turned off. In Figure 11.2 these subarrays are indicated.

For comparison, the contour plots of the sum and difference beams and the monopulse errors within the main beam area are shown in the Figures 11.3–11.6 for the original array without switched-off subarrays. The half-power beamwidth of the ELRA antenna is about $\Delta u = 0.03°$ or $1.8°$. The monopulse errors are indicated in Figure 11.6 by the lines drawn from the correct direction (*) to the estimated direction (+).

The turning off of the indicated subarrays (reduced array) results in a distortion of the sum and monopulse patterns, shown in Figures 11.7–11.10. The increased errors are seen in Figure 11.10.

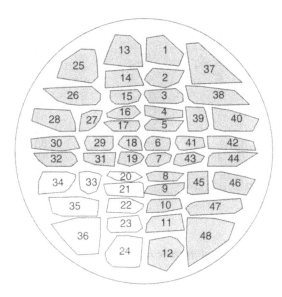

Figure 11.2 *Subarray configuration of ELRA receiving antenna, switched off subarrays indicated by white filling*

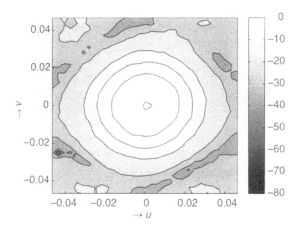

Figure 11.3 *Original array, sum beam*

After recomputation of the monopulse equation 11.7 with new coefficients using equations 11.12–11.14, considering the missing subarrays, a correction was achieved, shown in Figures 11.11–11.14.

Now the slope correction matrix in equation 11.7 has nonzero off-diagonal elements F_{uv}.

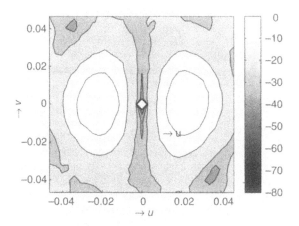

Figure 11.4   Original array, difference beam u

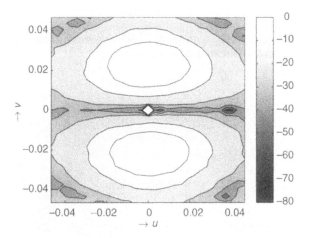

Figure 11.5   Original array, difference beam v

The monopulse errors have been reduced almost to the original values. A software repair of an array with defective elements or subarrays is demonstrated to be possible.

The suggestion and realisation of these experiments is due to I. Gröger (FGAN-FFM).

11.3   Variance of monopulse estimate

The Cramér–Rao inequality (see chapter 3 section 3.4.1) may be applied to determine the achievable accuracy expressed by the variance of the estimate. Because we applied a maximum-likelihood estimator we can expect the variance σ^2 to reach the lower

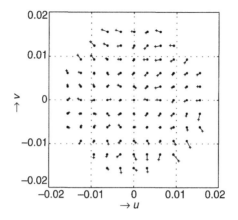

Figure 11.6   Original array, monopulse errors

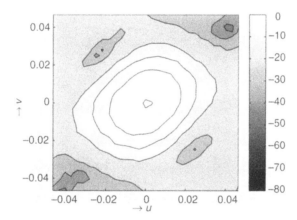

Figure 11.7   Reduced array, sum beam

limit. We may assume the signal amplitude error independent from the direction error, corresponding to the Doppler estimation case discussed in chapter 8 section 8.5.1. The basic equation 3.67 applied for the variance σ_ω^2 for the estimated parameter ω is:

$$\sigma^2 \geq \frac{1}{E\{(\partial \ln p(z, \omega)/\partial\omega)^2\}} \tag{11.16}$$

Here we confine ourselves on the variance for direction u. The estimation problem is similar to that in section 8.5.1 for the Doppler frequency. With equation 11.2 follows for log p:

$$\ln p(\mathbf{z}/\omega) = const - \left[\tfrac{1}{2}(\mathbf{z} - \mathbf{a}b)^*(\mathbf{z} - \mathbf{a}b)\right]$$

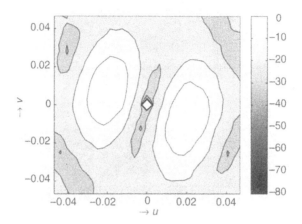

Figure 11.8   Reduced array, difference beam u

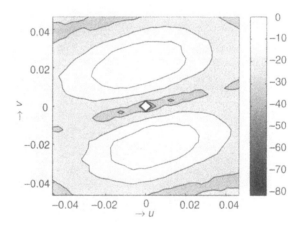

Figure 11.9   Reduced array, difference beam v

and:

$$\frac{\partial \ln p}{\partial u} = -\tfrac{1}{2}[-\mathbf{a}_u^* b^* (\mathbf{z} - \mathbf{ab}) + (\mathbf{z} - \mathbf{ab})^* (-\mathbf{a}_u b)]$$

$$\left(\frac{\partial \ln p}{\partial u}\right)^2$$

$$= \frac{1}{4}\left[\begin{array}{l} b^* \mathbf{a}_u^* (\mathbf{z} - \mathbf{ab}) b^* \mathbf{a}_u^* (\mathbf{z} - \mathbf{ab}) + \\ (\mathbf{z} - \mathbf{ab})^* \mathbf{a}_u b (\mathbf{z} - \mathbf{ab})^* \mathbf{a}_u b + 2 b^* \mathbf{a}_u^* (\mathbf{z} - \mathbf{ab}) (\mathbf{z} - \mathbf{ab})^* \mathbf{a}_u b \end{array} \right]$$

The first two terms contain noise, expressed by $(\mathbf{z} - \mathbf{ab})$, in a product without the conjugate sign, that means these terms have a random phase and by forming the expectation for both terms the result is zero.

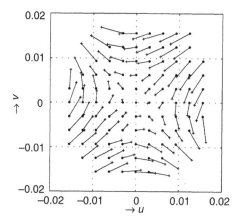

Figure 11.10 Reduced array, monopulse errors

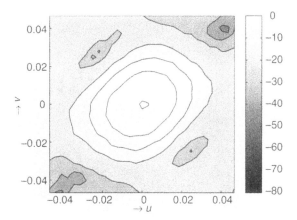

Figure 11.11 Reduced array, sum beam

The expectation of the third term leads to $E\{(\mathbf{z} - \mathbf{a}b)(\mathbf{z} - \mathbf{a}b)^*\} = diag_N(2)$. This is explained as follows: the quadratic matrix has on the main diagonal only products with noise signals which are both from antenna element n. By the conjugate operation the phases are cancelled and we have the noise power from both orthogonal components, each $\sigma^2 = 1$ and together giving a power equal to 2. So we get:

$$E\left\{\left[\frac{\partial \ln p(z/u)}{\partial u}\right]^2\right\} = bb^*\mathbf{a}_u^*\mathbf{a}_u$$

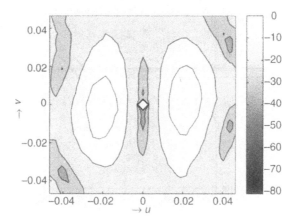

Figure 11.12   Reduced array, corrected difference beam u

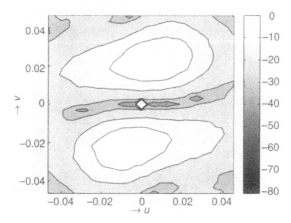

Figure 11.13   Reduced array, corrected difference beam v

and with equation 11.9 and x_i now again the x coordinates of the array elements:

$$= \left(\frac{2\pi}{\lambda}\right)^2 |b|^2 \sum_{i=1}^{N} x_i^2$$

As the final result the variance is:

$$\sigma_u^2 \geq \frac{1}{(2\pi/\lambda)^2 |b|^2 \sum_{i=1}^{N} x_i^2} \tag{11.17}$$

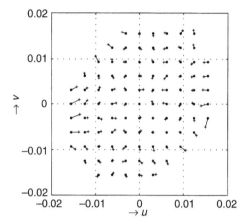

Figure 11.14   Monopulse errors after correction

The formula for σ_v^2 is of course correspondingly:

$$\sigma_v^2 \geq \frac{1}{(2\pi/\lambda)^2 |b|^2 \sum_{i=1}^{N} y_i^2} \tag{11.18}$$

As expected, the variance is inversely proportional to the SNR $= |b|^2/2$ at each single receiving channel output.

For illustration we give an example: for a linear array with $N + 1 = 65$ antenna elements at a regular distance $d = \lambda/2$ and therefore $x_i = i\lambda/2$ and $i = -N/2, \ldots, 0, \ldots, N/2$ the denominator of equation 11.17 is:

$$\text{denominator} = 4\pi^2 \frac{|b|^2}{2} \sum_{i=1}^{i=N/2} i^2$$

With the formula $\sum_{i=1}^{K} i^2 = \frac{1}{6}K(K+1)(2K+1)$, the half-power beamwidth according to chapter 4 equation 4.19 is given by:

$$\Delta u = \lambda/L = 1.768/N$$

and the SNR at the sum beam output given by:

$$(SNR)|_{out} = N(SNR)|_{in} = N|b|^2/2$$

the relative standard deviation results to:

$$\frac{\sigma_u}{\Delta u} = \frac{0.44}{\sqrt{SNR|_{out}}} \tag{11.19}$$

For the usual value $SNR = 13\,\mathrm{dB}$ or $SNR = 20$ then we have:

$$\frac{\sigma_u}{\Delta u} = 0.0986$$

This gives, as a rule of thumb: the standard deviation by monopulse estimation is 1/10 of the beamwidth. This result corresponds to that in section 8.5.1 for the Doppler frequency estimation.

11.4   Monopulse correction against jamming

The direction estimate $(\hat{u}, \hat{v})$ of a target based on equation 11.7 depends completely on the sum and difference patterns of the array, given by the weighting vectors $\mathbf{a}$, $\mathbf{d}_x$ and $\mathbf{d}_y$. In the case of a jammer scenario these patterns have to be adapted against the jammer threat by creating notches in the jammer directions, as is discussed in chapter 10. This also implies a change in the antenna patterns between the notches. The monopulse estimation is especially affected by a jammer close to the main beam. Then the difference pattern distortion causes a large estimation error [2]. The monopulse characteristic given by equation 11.15 for the noise-only case will now have a bias and a change in the slope. An example is given in Figure 11.15. If bias and correction factor for the slope are known, the correct estimate may be computed.

Starting from suggestions first given by Davies, Brennan and Reed for a linear array [3], Nickel developed estimation methods for planar and volume arrays for

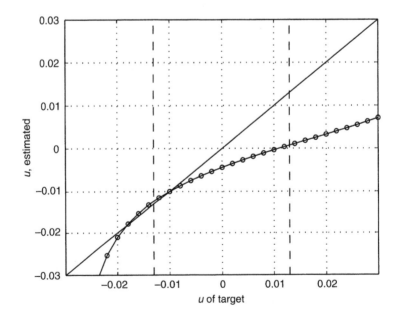

Figure 11.15   Monopulse characteristic (circled line) with jammer at $u = 0.9$ beamwidth, without correction, $N = 64$

processing the received signals from all single antenna elements [4] and recently also for larger array antennas divided into subarrays. He also solved the problem of maintaining low sidelobes for the sum and difference beams [5].

11.4.1 Correction with likelihood estimation

First we discuss the approach following section 11.1 for the likelihood estimation of the direction ω. For the treatment of the jammer case we simply have to extend the probability density function given by equation 11.2 by introducing the covariance matrix $\mathbf{Q}$ of noise and jammer interference signals:

$$p(\mathbf{z}/\omega, b) = \frac{1}{\pi^N |\mathbf{Q}|^{-1}} \exp\left[-(\mathbf{z} - \mathbf{a}b)^* \mathbf{Q}^{-1}(\mathbf{z} - \mathbf{a}b)\right] \tag{11.20}$$

With a derivation produced in the same way as in section 11.1 the following expression for P has to be maximised:

$$P(\omega) = |\mathbf{w}^*\mathbf{z}|^2 \quad \text{with } \mathbf{w} = \left(\mathbf{a}^*\mathbf{Q}^{-1}\mathbf{a}\right)^{-1/2}\mathbf{Q}^{-1}\mathbf{a} \tag{11.21}$$

$\mathbf{w}$ is now the sum beamforming vector adapted against the jammers and replaces vector $\mathbf{a}$ in equation 11.4. The term $(\mathbf{a}^*\mathbf{Q}^{-1}\mathbf{a})$ is a scalar value for normalisation. For the application of the general monopulse equation given by equation 11.7 again we have to derive the first derivatives F_u, F_v at (u_0, v_0) and the second derivatives F_{uu}, F_{uv}, F_{vv} at $(\hat{u}, \hat{v})$. Using again, as in section 11.1, $F = ln(P)$ we get:

$$F_u = \frac{P_u}{P} = \frac{\mathbf{w}_u^*\mathbf{z}\mathbf{z}^*\mathbf{w} + \mathbf{w}^*\mathbf{z}\mathbf{z}^*\mathbf{w}_u}{\mathbf{w}^*\mathbf{z}\mathbf{z}^*\mathbf{w}}$$

$$= 2\,\mathrm{Re}\left\{\frac{\mathbf{w}_u^*\mathbf{z}}{\mathbf{w}^*\mathbf{z}}\right\} \tag{11.22}$$

For the derivative $\mathbf{w}_u$ follows from equation 11.21 and considering the basic rule for derivatives of a product of functions $f(x)g(x)$:

$$(fg)_x = f_x g + f g_x$$

$$\mathbf{w}_u = \left(\mathbf{a}^*\mathbf{Q}^{-1}\mathbf{a}\right)^{-1/2}\mathbf{Q}^{-1}\mathbf{a}_u$$

$$\quad - \frac{1}{2}\left(\mathbf{a}^*\mathbf{Q}^{-1}\mathbf{a}\right)^{-3/2}\left(\mathbf{a}_u^*\mathbf{Q}^{-1}\mathbf{a} + \mathbf{a}^*\mathbf{Q}^{-1}\mathbf{a}_u\right)\mathbf{Q}^{-1}\mathbf{a}$$

$$= \left(\mathbf{a}^*\mathbf{Q}^{-1}\mathbf{a}\right)^{-1/2}\mathbf{Q}^{-1}\mathbf{a}_u - \frac{\mathrm{Re}\{\mathbf{a}_u^*\mathbf{Q}^{-1}\mathbf{a}\}}{\mathbf{a}^*\mathbf{Q}^{-1}\mathbf{a}}\mathbf{w} \tag{11.23}$$

The expression for $\mathbf{w}_u$ is inserted into equation 11.22 and this gives the result for F_u:

$$F_u = 2\,\mathrm{Re}\left\{\frac{\hat{\mathbf{d}}_x^*\mathbf{z}}{\mathbf{w}^*\mathbf{z}}\right\} - 2\mu_x = 2r_x - 2\mu_x \tag{11.24}$$

with $\hat{\mathbf{d}}_x$ corresponding to equation 11.10:

$$\hat{\mathbf{d}}_x = (\mathbf{a}^*\mathbf{Q}^{-1}\mathbf{a})^{-1/2}\mathbf{Q}^{-1}\frac{2\pi}{\lambda}\mathbf{D}_x\mathbf{a} \tag{11.25}$$

and μ_x given by:

$$\mu_x = \frac{\mathrm{Re}\{\mathbf{d}_x^*\mathbf{Q}^{-1}\mathbf{a}\}}{\mathbf{a}^*\mathbf{Q}^{-1}\mathbf{a}} \tag{11.26}$$

$\hat{\mathbf{d}}_x$ is now the difference beamforming vector adapted against the jammers and normalised. The bias term μ_x may be computed from the unadapted difference and sum beamforming vectors $\mathbf{d}_x$ and $\mathbf{a}$. $\mathbf{Q}$ has to be estimated for the actual jammer situation. If there is only noise we have $\mathbf{Q} = \mathbf{I}$ and it follows $\mu_x = 0$, because $\mathbf{d}_x$ and $\mathbf{a}$ are orthogonal and the result is as with equation 11.15.

For the elevation v of course we have the same derivation for F_v:

$$F_v = 2\,\mathrm{Re}\left\{\frac{\hat{\mathbf{d}}_y^*\mathbf{z}}{\mathbf{w}^*\mathbf{z}}\right\} - 2\mu_y = 2r_y - 2\mu_y \tag{11.27}$$

with:

$$\hat{\mathbf{d}}_y = (\mathbf{a}^*\mathbf{Q}^{-1}\mathbf{a})^{-1/2}\mathbf{Q}^{-1}\frac{2\pi}{\lambda}\mathbf{D}_y\mathbf{a} \tag{11.28}$$

and μ_y given by:

$$\mu_y = \frac{\mathrm{Re}\{\mathbf{d}_y^*\mathbf{Q}^{-1}\mathbf{a}\}}{\mathbf{a}^*\mathbf{Q}^{-1}\mathbf{a}} \tag{11.29}$$

It remains for us to calculate the expressions for the second derivatives for the slope-correcting matrix. We repeat here the result given by Nickel [4]:

$$F_{uu} = 2\mu_x^2 - 2\hat{\mathbf{d}}_x^*\mathbf{d}_x/\mathbf{w}^*\mathbf{a} \tag{11.30}$$

$$F_{uv} = 2\mu_x\mu_y - 2\mathrm{Re}\{\hat{\mathbf{d}}_x^*\mathbf{d}_y\}/\mathbf{w}^*\mathbf{a} \tag{11.31}$$

$$F_{vv} = 2\mu_y^2 - 2\hat{\mathbf{d}}_y^*\mathbf{d}_y/\mathbf{w}^*\mathbf{a} \tag{11.32}$$

So, finally, we have a procedure given by:

$$\begin{pmatrix}\hat{u}\\\hat{v}\end{pmatrix} = \begin{pmatrix}u_0\\v_0\end{pmatrix} + \begin{pmatrix}c_{xx} & c_{xy}\\c_{yx} & c_{yy}\end{pmatrix}\begin{pmatrix}r_x - \mu_x\\r_y - \mu_y\end{pmatrix} \tag{11.33}$$

The monopulse ratios r_x and r_y are given by equations 11.24 and 11.27:

$$r_x = \mathrm{Re}\left\{\frac{\hat{\mathbf{d}}_x^*\mathbf{z}}{\mathbf{w}^*\mathbf{z}}\right\}, \qquad r_y = \mathrm{Re}\left\{\frac{\hat{\mathbf{d}}_y^*\mathbf{z}}{\mathbf{w}^*\mathbf{z}}\right\}$$

μ_x and μ_y are given by equations 11.26 and 11.29 and the matrix $\mathbf{C}$ is determined according to equation 11.7 by the second derivatives of F:

$$\mathbf{C} = \begin{pmatrix} c_{xx} & c_{xy} \\ c_{yx} & c_{yy} \end{pmatrix} = -2 \begin{pmatrix} F_{uu}(\hat{u}, \hat{v}) & F_{uv}(\hat{u}, \hat{v}) \\ F_{uv}(\hat{u}, \hat{v}) & F_{vv}(\hat{u}, \hat{v}) \end{pmatrix}^{-1} \tag{11.34}$$

We apply the processing rule first to the one-dimensional case according to Figure 11.15, using equations 11.7, 11.24 and 11.30:

$$\hat{u} = u_0 - \left(\mu_x^2 - \frac{\hat{\mathbf{d}}_x^* \mathbf{d}}{\mathbf{w}^* \mathbf{a}} \right)^{-1} \left(\mathrm{Re} \left\{ \frac{\hat{\mathbf{d}}_x^* \mathbf{z}}{\mathbf{w}^* \mathbf{z}} \right\} - \mu_x \right) \quad \text{with } \mu_x = \frac{\mathrm{Re}\{\mathbf{d}_x^* \mathbf{Q}^{-1} \mathbf{a}\}}{\mathbf{a}^* \mathbf{Q}^{-1} \mathbf{a}} \tag{11.35}$$

A result is shown in Figure 11.16 for a jammer direction assumed $0.9 \times$ beamwidth apart from the look direction as an example. This has to be compared with Figure 11.15. Now the monopulse characteristic is approximating the ideal curve around $u = 0$ without bias and with the same slope.

The estimation procedure was also tested in two dimensions (u, v) for a randomly-thinned array antenna with 192 elements, grouped into 24 subarrays. The element distribution of an intermediate development state of the ELRA experimental radar, described in chapter 17, is shown in Figure 11.17. It was used as an example for some simulation trials.

In Figure 11.18 the bias values are shown by arrows pointing from the true to the estimated direction for a grid of assumed target directions. The main beam is

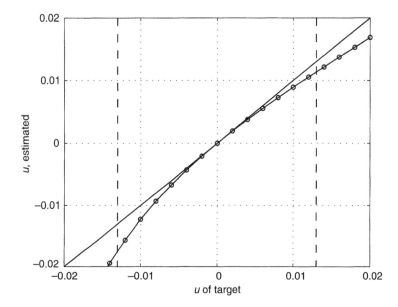

Figure 11.16 Monopulse characteristic (circled line) with jammer at $u_k = 0.9 \times$ beamwidth $= 0.023$ from beam axis, with correction, $N = 64$

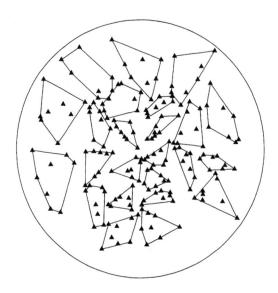

Figure 11.17   Antenna element distribution of ELRA 192 (courtesy U. Nickel)
            © *IEE 1993 [4]*

indicated by a dashed circle for the 3 dB width. A jammer, indicated by a small square, is assumed near the main beam. In this figure no correction is applied.

In Figure 11.19 correction is applied with an efficient reduction of errors. In the centre the bias errors tend to zero, because the Taylor expansion is developed around this point.

Nickel [4] suggested repeating the procedure to reduce the errors also in the vicinity of the jammer. The result in Figure 11.20 demonstrates the improvement. This second step could be computed with the same received signal vector **z** but with a new beamforming vector **a**, now computed for the estimated direction of the first step. So the monopulse operation, measurement of only one data vector **z**, would be maintained.

11.4.2   Correction in expectation

The estimation procedure discussed so far was further developed by Nickel [5] after the following observations. For operational systems the weighting vector for the difference beam has to be chosen according to the Bayliss function to achieve low sidelobes. Therefore, the weighting vector is not the derivative of the sum beam vector. This is especially true for the processing of subarrays. For future multifunction radar systems it is suggested applying a Taylor weighting at the elements to achieve low sidelobes for the sum beam. At the subarray level for the sum beamforming simple summing is applied, but for the difference beams special weighting vectors have to be applied to approximate a Bayliss weighting by the combination of element weighting (Taylor) and weighting at the subarray output. This has been discussed in chapter 5

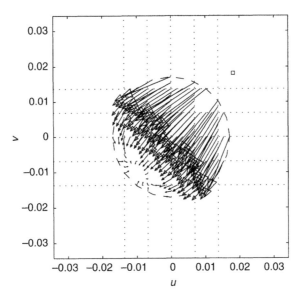

Figure 11.18 Monopulse under jamming, bias without correction, $s = 0\,dB$, $j = 27\,dB$ (courtesy U. Nickel)

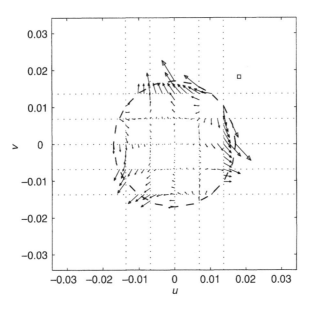

Figure 11.19 Monopulse under jamming, bias with correction, $s = 0\,dB$, $j = 27\,dB$ (courtesy U. Nickel)

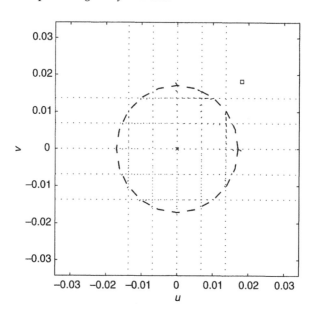

Figure 11.20   *Two-step correction monopulse under jamming, s = 0 dB, j = 27 dB, mstep = 2 (courtesy U. Nickel)*

section 5.1.4. The approach given by the general monopulse linear equation 11.33 is used:

$$\begin{pmatrix} \hat{u} \\ \hat{v} \end{pmatrix} = \begin{pmatrix} u_0 \\ v_0 \end{pmatrix} + \begin{pmatrix} c_{xx} & c_{xy} \\ c_{yx} & c_{yy} \end{pmatrix} \begin{pmatrix} r_x - \mu_x \\ r_y - \mu_y \end{pmatrix}$$

with r_x, r_y again given by equations 11.4 and 11.27. Nickel computed the values c and μ by defining a function $\mathbf{M}$:

$$\mathbf{M} = \begin{pmatrix} M_x(u, v) \\ M_y(u, v) \end{pmatrix} = \begin{pmatrix} c_{xx} & c_{xy} \\ c_{yx} & c_{yy} \end{pmatrix} \begin{pmatrix} R_x(u, v) - \mu_x \\ R_y(u, v) - \mu_y \end{pmatrix} \quad (11.36)$$

The R_x, R_y are the expectations of r_x, r_y, therefore the expectation of the estimated direction should be correct. This monopulse characteristic $\mathbf{M}$ has to fulfil two conditions:

$$\mathbf{M}(u_0, v_0) = 0 \quad \text{(zero bias)}$$

$$\left. \frac{\partial \mathbf{M}}{\partial u} \right|_{(u_0, v_0)} = \begin{pmatrix} 1 \\ 0 \end{pmatrix} \quad \text{(slope = 1)} \quad (11.37)$$

$$\left. \frac{\partial \mathbf{M}}{\partial v} \right|_{(u_0, v_0)} = \begin{pmatrix} 0 \\ 1 \end{pmatrix}$$

Using $\mathbf{M}$ as defined by equation 11.37 in equation 11.35 forces:

$$E\left\{\begin{pmatrix} \hat{u} - u_0 \\ \hat{v} - v_0 \end{pmatrix}\right\} = \mathbf{M}(u, v) \approx \begin{pmatrix} u - u_0 \\ v - v_0 \end{pmatrix}$$

This is the linear approximation around the beam-steering direction. The expectations $R_{x,y}$ of the monopulse ratios may be computed using the monopulse ratios given in equations 11.24 and 11.27. The expectations of numerator and denominator can be computed independently for Gaussian variables. Then the bias values are:

$$\mu_x = R_x(u_0, v_0) \quad \text{and} \quad \mu_y = R_y(u_0, v_0)$$

Both values are computed as an approximation for a large SNR which is useful also for all reasonable SNRs. Then the result is:

$$\mu_x = \mathrm{Re}\left\{\frac{\mathbf{d}_x^* \mathbf{Q}^{-1} \mathbf{a}_0}{\mathbf{m}^* \mathbf{Q}^{-1} \mathbf{a}_0}\right\} \quad \text{and} \quad \mu_y = \mathrm{Re}\left\{\frac{\mathbf{d}_y^* \mathbf{Q}^{-1} \mathbf{a}_0}{\mathbf{m}^* \mathbf{Q}^{-1} \mathbf{a}_0}\right\} \tag{11.38}$$

The vectors $\mathbf{d}_{x,y}$ and $\mathbf{m}$ are the beamforming vectors for the difference and sum beams respectively. They may be modified compared to section 11.4.1 to achieve low sidelobes. $\mathbf{a}_0$ is the normalised signal vector for the beam-steering direction (u_0, v_0). All beamforming vectors may be applied for the complete array or after subarrays. After subarrays these vectors and also $\mathbf{Q}$ have to be transformed by a subarray transform matrix $\mathbf{T}$:

$$\mathbf{T} = \begin{pmatrix} g_{11} & 0 & 0 \\ \vdots & \vdots & \vdots \\ g_{1K_1} & 0 & \vdots \\ 0 & g_{21} & \vdots \\ \vdots & \vdots & \vdots \\ \vdots & g_{2K_2} & \vdots \\ \vdots & 0 & \ddots & 0 \\ \vdots & \vdots & g_{L1} \\ \vdots & \vdots & \vdots \\ 0 & 0 & g_{LK_L} \end{pmatrix}$$

$$\tilde{\mathbf{a}} = \mathbf{T}^* \mathbf{a}, \quad \tilde{\mathbf{d}} = \mathbf{T}^* \mathbf{d}, \quad \tilde{\mathbf{m}} = \mathbf{T}^* \mathbf{m}, \quad \tilde{\mathbf{Q}} = \mathbf{T}^* \mathbf{Q} \mathbf{T} \tag{11.39}$$

The elements g_{lk} are the weighting factors for achieving low sidelobes for sum and difference beams. The matrix $\mathbf{C}$ in equation 11.36 is computed by combining the second part of equation 11.37:

$$\mathbf{C} \begin{pmatrix} \dfrac{\partial R_x}{\partial u} & \dfrac{\partial R_x}{\partial v} \\[2ex] \dfrac{\partial R_y}{\partial u} & \dfrac{\partial R_y}{\partial v} \end{pmatrix}_{(u_0, v_0)} = \mathbf{I}$$

The result for high SNR is given in Reference 4:

$$\left(\frac{\partial R_x}{\partial u} \right)_{(u_0, v_0)} = \frac{\mathrm{Re}\{\hat{\mathbf{d}}_x^* \mathbf{a}_{u,0} \mathbf{a}_0^* \mathbf{w} + \hat{\mathbf{d}}_x^* \mathbf{a}_0 \mathbf{a}_{u,0}^* \mathbf{w}\}}{|\mathbf{a}_0^* \mathbf{w}|^2} - \mu_x 2\mathrm{Re}\left\{ \frac{\mathbf{w}^* \mathbf{a}_{u,0}}{\mathbf{w}^* \mathbf{a}_0} \right\} \qquad (11.40)$$

with $\mathbf{a}_0$ now the expected normalised signal from the look direction (u_0, v_0) at the elements including weighting or at the outputs of the subarrays. $\mathbf{a}_{u,0}$ is the derivative with respect to u of this signal vector, $\hat{\mathbf{d}}_x$ and $\mathbf{w}$ the adapted difference and sum beamforming vectors, respectively, applied at antenna elements or subarrays. For the other elements of the matrix the results are found accordingly. All quantities may be computed after estimation of $\mathbf{Q}$ to realise the estimation with correction.

For an illustration this more generalised procedure may be compared for the u dimension with the corresponding result of section 11.4.1 given by equation 11.35. Now $\mathbf{m} = \mathbf{a}$ and $\hat{\mathbf{d}}_x = \mathbf{d}_x \mathbf{Q}^{-1}$. With this we have for μ_x the same value as given by equation 11.35. For c_{xx} we get [5]:

$$c_{xx} = \left(\frac{\partial R_x}{\partial u} \right)^{-1}$$

with:

$$\begin{aligned} \frac{\partial R_x}{\partial u} &= \mathrm{Re}\left\{ \frac{\mathbf{d}_x^* \mathbf{Q}^{-1} \mathbf{d}_x}{\mathbf{a}^* \mathbf{Q}^{-1} \mathbf{a}} + \left(\frac{\mathbf{d}_x^* \mathbf{Q}^{-1} \mathbf{a}}{\mathbf{a}^* \mathbf{Q}^{-1} \mathbf{a}} \right)^2 \right\} - 2\mu_x^2 \\[1.5ex] &= \mathrm{Re}\left\{ \frac{\mathbf{d}_x^* \mathbf{Q}^{-1} \mathbf{d}_x}{\mathbf{a}^* \mathbf{Q}^{-1} \mathbf{a}} + (\mu_x + j\kappa)^2 \right\} - 2\mu_x^2 \\[1.5ex] &= \mathrm{Re}\left\{ \frac{\mathbf{d}_x^* \mathbf{Q}^{-1} \mathbf{d}_x}{\mathbf{a}^* \mathbf{Q}^{-1} \mathbf{a}} + \mu_x^2 - \kappa^2 \right\} - 2\mu_x^2 \end{aligned}$$

because the second term equals μ_x but may have also an imaginary part. The imaginary part cancels because we have to take only the real part. For symmetrical arrays κ^2 is approximately zero [5] and we arrive at:

$$\hat{u} = u_0 - \left(\mu_x^2 - \frac{\hat{\mathbf{d}}_x^* \mathbf{d}_x}{\mathbf{w}^* \mathbf{a}} \right)^{-1} (r_x - \mu_x)$$

already given by equation 11.35, as expected.

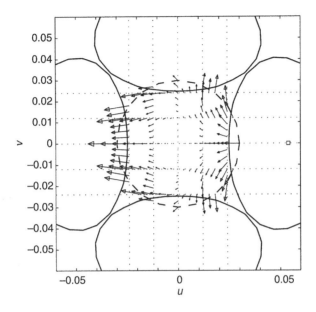

Figure 11.21   Monopulse corrected, one step, s = 0 dB, j = 27 dB (courtesy U. Nickel)

The procedure has been tested with simulations [5,6]. For an array divided into subarrays as shown in Figure 5.3 the bias values are presented in Figure 11.21 for one-step processing and in Figure 11.22 for two steps. The jammer direction is indicated by a small square.

As a result, we now have available a correcting estimation procedure for adapted weighting vectors based on low sidelobe weighting e.g. according to Taylor and Bayliss.

11.4.3   Statistical performance analysis of estimation

To characterise the performance of estimation procedures with correction against jammers Nickel derived the probability density function of the estimates for Rayleigh targets and jammers [5]. For nonfluctuating targets it was possible to derive mean and variance, so the dependence of bias and variance on the jammer scenario and the subarray configuration may be determined and compared. Regrettably, the results are given by some inconvenient expressions, which have to be evaluated numerically. Therefore only some demonstrative results are given in the following for the Rayleigh target case.

A circular planar array with 902 elements is divided into 32 irregular subarrays as shown in Figure 5.3 in chapter 5. For the sum beam, tapering coefficients

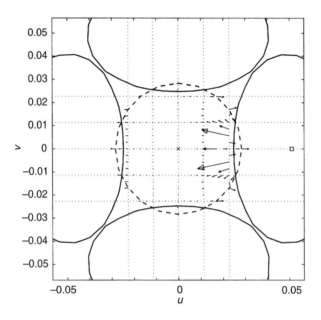

Figure 11.22   Monopulse corrected, two steps, s = 0 dB, j = 27 dB (courtesy U. Nickel)

according to a 40 dB Taylor weighting are applied at the elements. The difference beams are formed by an independent subarray weighting to approximate a Bayliss weighting.

In Figure 11.23 the bias and standard deviation ellipses without jammers, that is for the basic monopulse estimation are shown as a reference. The little arrows show the bias. The dashed circle indicates the sum beam half-power beamwidth (−3 dB contour) and the solid lines indicate the −3 dB contour of both difference beams. The SNR is assumed to be 0 dB at the elements or 28 dB at the sum beam output (increased by the array gain). We recognise a bias less than the standard deviation within the −3 dB contour of the sum beam. The linear approximation with equation 11.5 of the monopulse characteristic is therefore confirmed.

Figure 11.24 shows the influence of a jammer, separated by 0.9 of a beamwidth from the look direction and with a jammer-to-noise ratio (JNR) of 27 dB at the elements. The jammer location is marked by a small square. No correction is applied.

Figure 11.25 shows the effect of the correction for the same situation. Around the look direction the bias is zero, only the standard deviation has increased compared to the case without jamming.

In Figure 11.26 the jammer direction is chosen on the flanks of both difference beams but the same distance from the look direction. The error ellipses are now smaller because the difference beam patterns are less distorted.

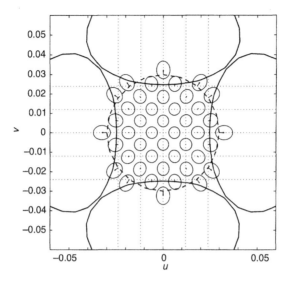

Figure 11.23 *Mean and standard deviation ellipses for monopulse estimates without jammers, s = 0 dB (courtesy U. Nickel) © IEE 1999 [6]*

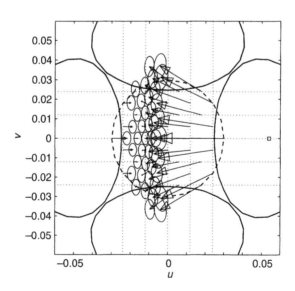

Figure 11.24 *Mean and standard deviation ellipses for monopulse estimates with a jammer and correction, s = 0 dB, j = 27 dB (courtesy U. Nickel)*

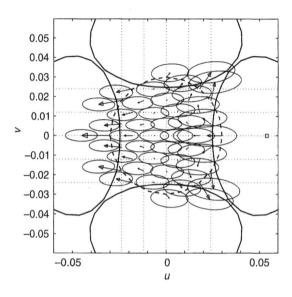

Figure 11.25   Mean and standard deviation ellipses for monopulse estimates with jammer and correction, s = 0 dB, j = 27 dB (courtesy U. Nickel)

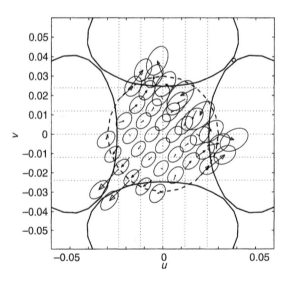

Figure 11.26   Mean and standard deviation ellipses for monopulse estimates with jammer and correction, s = 0 dB, j = 27 dB (courtesy U. Nickel), © IEE 1999 [6]

11.5   Polarisation independence

The conventional reflector antenna with a monopulse feed horn is matched to a design polarisation. If signals with an orthogonal polarisation are received this results in an error for direction estimation [8]. This applies for active radar operation for the location of targets and of course also for passive location of jammer sources.

With array antennas in contrast to reflector antennas all the single antenna elements have the same polarisation-dependent pattern. This is multiplied with the array factor (chapter 4); this also applies for the sum and difference patterns. The resultant form of the beam pattern will thus be independent of polarisation. Therefore, the monopulse operation will also not be disturbed.

With the experimental phased-array system ELRA of FGAN-FFM the monopulse scan pattern was measured with vertically-polarised transmission and horizontally-polarised reception by illuminating a target of opportunity. In Figure 11.27 the result is shown: the monopulse pattern has the notch as expected in the look direction of the beam, indicated by the maximum of the sum pattern.

11.6   Indication of multiple targets

For the monopulse direction estimation procedure discussed so far only a single target is assumed to be present. In the case of more targets within the beamwidth

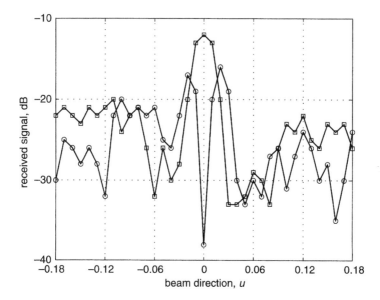

Figure 11.27   Vertical transmission and horizontal receiving polarisation: monopulse beam with notch in look direction (measured with ELRA)

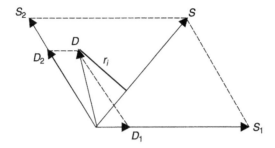

*Figure 11.28   Monopulse with two targets and resultant sum and difference beam
outputs*

an error occurs dependent on the relative phase of the target signals. In these cases a
superresolution in angle could be applied. This method will be discussed in chapter 12.
An indication for a multiple-target situation may be derived from the monopulse ratio
[9] given by equation 11.15. For radar operation by this indication superresolution
processing may be activated.

We denote the sum beam output by $S = \mathbf{a}^*\mathbf{z}$ and the difference beam output by
$D = \mathbf{d}_x^*\mathbf{z}$. Both outputs are complex numbers. The ratio $r = D/S$ may be written:

$$r = \frac{D}{S} = \frac{DS^*}{SS^*} = \frac{DS^*}{|S|^2} \tag{11.41}$$

The complex angle of D is shifted back by the angle of S. The real part of r results
from the part of D parallel to S and the imaginary part of r results from the part of D
orthogonal to S.

In the case of a single target without noise both values are in phase and the ratio
D/S is real valued with the imaginary part equal to zero. In the sketch in Figure 11.28
we recognise for two targets that although the individual difference beam outputs are in
phase with their respective sum beam outputs this is not the case for the superposition.
Because the targets are at different angles with respect to the beam direction the ratios
D/S are different and the resultant D is out of phase with S. Therefore the imaginary
part r_i will be unequal to zero.

The absolute value of r_i may be compared with a threshold to decide for a
multitarget situation. An example is given in Figure 11.29. For ten trials with a
random phase between the two target signals of equal amplitudes the course of the
absolute value of r_i is shown. It is always well above zero. Asseo [10] presented
statistical investigations for the detection of target multiplicity.

11.7   Conclusions

Direction estimation for target acquisition and tracking is performed by forming a
difference pattern with the receiving array. For an active array this will be realised with

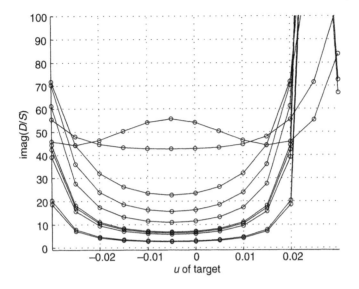

Figure 11.29   Imag(D/S) to indicate multiple targets; example: two targets with equal amplitude, $\Delta u = 0.01$

the outputs of the subarrays by applying a special weighting vector. The procedure has been derived by applying estimation theory.

If there are known channel failures, at element or subarray level, the weighting vector can be recomputed to reestablish the difference pattern.

The variance of the direction estimate has been derived from the Cramér-Rao limit. It depends on the signal-to-noise ratio of the received signal.

Adaptation against jammers near the main beam deteriorates the sum and especially the difference pattern. A correction can be applied to reestablish the performance even if a jammer is within the main beam. The limit is an angular difference of about only 0.3 of a beamwidth between target and jammer.

Because the difference pattern is determined by the array factor it is independent of polarisation, therefore no direction estimation errors are to be expected.

A situation with multiple targets within a beam leads to direction errors by the monopulse procedure. This situation may be indicated by the output of the monopulse ratio. Then superresolution methods have to be applied.

11.8   References

1 NICKEL, U.: 'Radar target parameter estimation with array antennas', in HAYKIN, S., LITVA, J., and SHEPHERD, T. J. (Eds.): 'Radar array processing' (Springer Verlag Berlin-Heidelberg-New York, 1993), chapter 3 pp. 47–98

2   LIN, F. C., and KRETSCHMER, F. F.: 'Angle measurement in the presence of mainbeam interference'. IEEE international conference on *Radar*, Washington, USA, 1990, pp. 444–450

3   DAVIES, R. C., BRENNAN, L. E., and REED, I. S.: 'Angle estimation with adaptive arrays in external noise fields', *IEEE Trans. Aerosp. Electron. Syst.*, 1976, **12**, pp. 179–186

4   NICKEL, U.: 'Monopulse estimation with adaptive arrays', *IEE Proc. F, Radar Signal Process.*, 1993, **140** (5), pp. 303–308

5   NICKEL, U.: 'Monopulse estimation with subarray-adaptive arrays and arbitrary sum and difference beams', *IEE Proc., Radar Sonar Navig.*, 1996, **143** (4), pp. 232–238

6   NICKEL, U.: 'Performance of corrected adaptive monopulse estimation', *IEE Proc., Radar Sonar Navig.*, 1999, **146** (1), pp. 17–24

7   PAINE, A. S.: 'Minimum variance monopulse technique for an adaptive phased array radar', *IEE Proc., Radar Sonar Navig.*, 1998, **145**, pp. 374–380

8   HOWARD, D. D.: 'Tracking radar', in SKOLNIK, M. I. (Ed.): 'Radar Handbook' (McGraw-Hill, New York, 1990) chapter 18

9   SHERMANN, S. M.: 'Complex indicated angles applied to unresolved radar targets and multipath', *IEEE Trans. Aerosp. Electron. Syst.*, 1971, **7** (1), pp. 160–170

10 ASSEO, S. J.: 'Detection of target multiplicity using monopulse quadrature angle', *IEEE Trans. Aerosp. Electron. Syst.*, 1981, **17** (2), pp. 271–280

Chapter 12

Superresolution in angle

12.1 Introduction

As discussed in chapter 11, direction estimation using the monopulse procedure assumes only one target being present within the beamwidth. Several targets within the main beam will cause mutual estimation errors. The resolution limit for an individual direction estimation for single targets is given by the beamwidth of the sum beam. This is the well-known Rayleigh limit for all conventional radar systems with reflector antennas and also for phased-array radars, which only form sum and difference beams. The limit is given according to chapter 4 by the fundamental relation for the beamwidth $\Theta = \lambda/D$, λ the wavelength and D the antenna or aperture diameter. Both parameters are determined by the general system requirements and cannot be changed arbitrarily.

A resolution beyond the limit given by the beamwidth can only be achieved by suitable array processing methods with an active receiving array with access to individual receiving antenna elements or at least to subarrays. This improved angular resolution is called superresolution.

For the target range R and beamwidth θ the lateral resolution is given simply by $b = R\theta$. For a beamwidth of $\theta = 1°$ or $\theta = 0.01745$ (radians) and $R = 30\,\text{km}$ follows $b = 523\,\text{m}$. Because the range resolution is usually about $100\,\text{m}$ there is a mismatch between these resolution values and an improvement would be very desirable. With superresolution a resolution improvement by a factor of four to five seems to be possible. Therefore, in our example, the resolutions may be approximately matched.

The angular resolution improvement for targets at the same range may be applied for the following purposes:

- Resolution of airplane formations. This is of special interest for target tracking and may be applied on special request by the radar system's control to improve multitarget tracking.

- Correction of elevation errors caused by multipath reflections of a target above sea by treating the mirror image as a second target and thus resolving both.
- Individual passive location of neighboured jammers.
- Reduction of glint errors caused by multiple targets seen by seeker heads approaching a formation of targets. Glint refers to the distortion of the phase front by superposition of waves from several radiation sources. The phase front of target echoes is used by the radar to estimate the target direction with the monopulse procedure. A distorted phase front results in direction errors.
- Countermeasure against crosseye deception jamming. With crosseye two jamming sources are steered in relative phase to produce an artificial glint effect.

In chapter 11 section 11.6 an indication of a multitarget situation by the imaginary part of the monopulse ratio has already been discussed. Using this indication, for example, the superresolution procedure may be activated as a second step or an additional option after the monopulse procedure. Superresolution may be activated for selected situations by the radar control system. It has to be performed automatically in the sense of producing results for the unknown target number and their amplitudes and directions without any inspection of scan patterns by a human operator.

For some weapon systems, especially when high resolution in elevation against multipath effects is required, an extra radar at a higher frequency and corresponding improved angular resolution has been installed in the past to counter the multipath effect for approaching targets. By the application of superresolution this extra radar is unnecessary and could be saved.

The idea of angular superresolution based on receiving array antennas was first discussed in publications by Ksienski and McGhee [1] and Young [2] in 1968. Some first experiments with a vertical array, added to a naval radar to counter the multipath effect, have been reported by White [3] and Howard [4]. The applicable algorithms result from likelihood parameter estimation based on the probability density function for the received signal dependent on the unknown parameters as discussed in a general form in chapter 3. For each single target a model with the questioned unknown parameters is assumed. A superposition of multiple model targets is the basis for the estimation procedure. It is then possible to resolve even targets with completely correlated echo signals as is typical for the radar situation with the targets illuminated by one transmitter.

Other approaches for superresolution have been developed from seismology and spectral analysis. These methods assume uncorrelated source signals, e.g. jammers.

12.2   Parametric estimation

Resolution of several targets within the beamwidth may be treated as a parameter-estimation problem [5,6]. If it is possible to estimate the correct parameters of all M targets then they are resolved. An array antenna, with N antenna elements, provides generally enough independent measurements for the determination of the unknown variables, that is $N \geq 2M$, although these measurements are superimposed by receiver and measurement noise. First some preliminary remarks and assumptions are to be considered:

- We want to discriminate flying objects from one another but we don't want to resolve cross section elements of targets. Airplanes may be seen as point targets, because they are small compared to the radar resolution cell, even with superresolution.
- For resolution we need the estimation of the target number and the estimation of angle parameters.
- The two-target situation is of predominant importance for the solution of the multipath problem above reflecting ground and sea.
- Only above a certain amplitude threshold will a target in a certain direction be accepted.
- A maximum number of targets to be resolved is accepted.
- We need as a prerequisite an increased SNR (signal-to-noise power ratio) to achieve superresolution.

Regarding the last point: as a price for successful superresolution a high measurement accuracy and a higher signal-to-noise power ratio (SNR) is required compared to simple target-detection techniques and monopulse direction estimation. Fortunately, an increasing SNR is available for approaching targets according to the radar equation with SNR $\approx 1/R^4$. For example if a target is first detected at maximum range R_0 then after approaching to $R = 0.3R_0$ the SNR would have increased by a factor $(1/0.3)^4 = 123$ or 21 dB.

For the solution of the parameter-estimation problem we start as discussed in chapter 3 section 3.4 with the probability density function for the received signal vector $\mathbf{z}$, depending on the target parameters: b_m for the target complex amplitudes and ω_m for the target directions, with $m = 1, \ldots, M$ for the target index and M the number of targets. The parameters for all M targets are summarised in Ω_M, and that means we have to maximise the expression:

$$p(\mathbf{z}/M, \Omega_M)|_{\hat{M}, \hat{\Omega}} = \max \tag{12.1}$$

The probability density can be assumed to be multidimensional Gaussian from the assumption of Gaussian distributed receiver noise, superposed to the mean values produced by the unknown target signals. As in chapter 10 we define an antenna matrix $\mathbf{A}$ with columns $\mathbf{a}_m$ for the antenna element outputs from a normalised point target with amplitude 1 and with index m at direction ω_m. Each matrix element a_{nm} is formed for each antenna element n with coordinates (x_n, y_n) and target m with direction (u_m, v_m) and is given by:

$$a_{nm} = \exp\left(j\frac{2\pi}{\lambda}(u_m x_n + v_m y_n)\right) \tag{12.2}$$

The matrix $\mathbf{A}$ with dimension $[N, M]$ is given by:

$$\mathbf{A} = \begin{bmatrix} a_{11} & a_{21} & \cdots & a_{M1} \\ \vdots & \vdots & \cdots & \vdots \\ a_{1N} & a_{2N} & \cdots & a_{MN} \end{bmatrix} \tag{12.3}$$

The complex target amplitudes are combined in a vector **b**:

$$\mathbf{b} = \begin{bmatrix} b_1 \\ \vdots \\ b_M \end{bmatrix} \tag{12.4}$$

The received signal vector **z**, a column matrix with dimension $[N, 1]$, can then be written with the superposed noise vector **n**:

$$\mathbf{z} = \mathbf{Ab} + \mathbf{n} \tag{12.5}$$

The orthogonal I and Q components of noise n_{In} and n_{Qn} $(n = 1, \ldots, N)$ are assumed to be uncorrelated and to have variances $\sigma^2 = 1$. As discussed in chapter 3 the multidimensional Gaussian distribution for complex signals is with equation 3.41 and the mean vector **Ab**:

$$p(\mathbf{z}/M, \Omega_M) = \frac{1}{(2\pi)^N} \exp\left[-\frac{1}{2}(\mathbf{z} - \mathbf{Ab})^*(\mathbf{z} - \mathbf{Ab}) \right] \tag{12.6}$$

In order to maximise this expression for p we have to minimise the quadratic expression:

$$Q = (\mathbf{z} - \mathbf{Ab})^*(\mathbf{z} - \mathbf{Ab}) \tag{12.7}$$

As in chapter 11 section 11.1, the first step is to determine the best suitable vector **b** for each matrix **A** by setting the gradient of Q with respect to **b** equal to zero:

$$\nabla_b Q = 0$$
$$\nabla_b Q = -\mathbf{A}^*(\mathbf{z} - \mathbf{Ab}) + (\mathbf{z} - \mathbf{Ab})^*(-\mathbf{A})$$
$$= -2\,\mathrm{Re}\{\mathbf{A}^*\mathbf{z} - \mathbf{A}^*\mathbf{Ab}\}$$

This expression is zero if:

$$\hat{\mathbf{b}} = (\mathbf{A}^*\mathbf{A})^{-1}\mathbf{A}^*\mathbf{z} \tag{12.8}$$

The meaning of $\mathbf{A}^*\mathbf{z}$ is, as discussed in chapter 10, the formation of M parallel beams in unknown directions ω_m $(m = 1, \ldots, M)$. Also, M is still unknown. The effect of the matrix $(\mathbf{A}^*\mathbf{A})^{-1}$ is to decouple the M beams, that is to create zeros in all other $(M - 1)$ beam directions, respectively. With these decoupled beams an independent measurement of the M target signals b_m is accomplished. The decoupling is effective independent of the angular distance of the beams, so it is in principle also effective for overlapping main beams.

We eliminate **b** in equation 12.7 and arrive at:

$$Q = \left(\mathbf{z} - \mathbf{A}(\mathbf{A}^*\mathbf{A})^{-1}\mathbf{A}^*\mathbf{z}\right)^*\left(\mathbf{z} - \mathbf{A}(\mathbf{A}^*\mathbf{A})^{-1})\mathbf{A}^*\mathbf{z}\right) \tag{12.9}$$

Or in another form:

$$Q = \mathbf{z}^* \boldsymbol{\Gamma} \mathbf{z} \quad \text{with} \quad \boldsymbol{\Gamma} = \mathbf{I} - \mathbf{A}(\mathbf{A}^*\mathbf{A})^{-1}\mathbf{A}^* \tag{12.10}$$

using $\boldsymbol{\Gamma}^2 = \boldsymbol{\Gamma}$ and $\boldsymbol{\Gamma} = \boldsymbol{\Gamma}^*\boldsymbol{\Gamma}$ is a projection matrix with respect to $\mathbf{A}$ because $\boldsymbol{\Gamma}\mathbf{A}$ is zero.

So we are left with the problem to determine the unknown directions. The expression for Q may be also written as:

$$Q = (\mathbf{z} - \hat{\mathbf{z}})^*(\mathbf{z} - \hat{\mathbf{z}}) = \|\mathbf{z} - \hat{\mathbf{z}}\|^2 \tag{12.11}$$

$\hat{\mathbf{z}}$ is the estimated receiving signal for an assumed antenna matrix $\mathbf{A}$. Therefore Q is the mean-squared error between $\mathbf{z}$ and the model signal $\hat{\mathbf{z}}$.

The task of solving the superresolution problem now becomes: find a target model with a matrix $\hat{\mathbf{A}}$ minimising Q. Additionally, we also have to find $\hat{M}$. For stepwise increasing M, starting from $M = 2$, we determine the respective optimal estimate $\hat{\mathbf{A}}$ for a minimum value of Q_M. So we achieve a minimum least-square error between the measured signal $\mathbf{z}$ and our model signal $\hat{\mathbf{z}}$ for each selected M.

For the estimation of the target number M an initial simple idea is the following: if there are several targets we can expect that by increasing M a better fit of the model to the real situation will be achieved and therefore we will get $Q_M < Q_{M-1}$. If M equals the actual target number no significant improvement will be achieved by increasing M. Therefore the value of Q_M, finally determined by residual noise, may be compared with a suitable threshold to reach a decision for M. Moreover, if M is selected too high, the estimated target amplitude for ghost targets would be, under certain conditions, near zero and these targets could be eliminated by a threshold comparison.

The problem to be solved now is the determination of the directions ω_m, which determine the matrix $\mathbf{A}$. Since matrix $\mathbf{A}$ contains, according to equation 12.2, non-linear transcendental functions ($\exp(jx)$ or $\cos(x)$, $\sin(x)$) there exists no direct solution from linear algebra techniques.

Different proposals have been made for the solution of this problem, e.g. a $2M$-dimensional raster search or a random search by varying the parameters ω_m. A promising solution is a standard gradient search.

So with the next step we derive the gradient of Q with respect to the directions $\omega_m = (u_m, v_m)$ [7].

Using the abbreviation as before:

$$\frac{\partial Q}{\partial u_m} = Q_{u_m}$$

we can write:

$$Q_{u_m} = \mathbf{z}^* \boldsymbol{\Gamma}_{u_m} \mathbf{z} = \mathbf{z}^* \left(-\frac{\partial}{\partial u_m} \mathbf{A}(\mathbf{A}^*\mathbf{A})^{-1}\mathbf{A}^* \right) \mathbf{z}$$

with $\mathbf{\Gamma}$ from equation 12.10. Now we have to use the relation with the abbreviation $\mathbf{X}_u = \partial\mathbf{X}/\partial u$:

$$\mathbf{X}_u^{-1} = -\mathbf{X}^{-1}\mathbf{X}_u\mathbf{X}^{-1}$$

because with $\mathbf{X}\mathbf{X}^{-1} = \mathbf{I}$ follows:

$$\mathbf{X}_u\mathbf{X}^{-1} + \mathbf{X}\mathbf{X}_u^{-1} = 0$$

$$
\begin{aligned}
Q_{u_m} &= -\mathbf{z}^*\big[\mathbf{A}_{u_m}(\mathbf{A}^*\mathbf{A})^{-1}\mathbf{A}^* - \mathbf{A}(\mathbf{A}^*\mathbf{A})^{-1} \\
&\quad \times (\mathbf{A}_{u_m}^*\mathbf{A} + \mathbf{A}^*\mathbf{A}_{u_m})(\mathbf{A}^*\mathbf{A})^{-1}\mathbf{A}^* + \mathbf{A}(\mathbf{A}^*\mathbf{A})^{-1}\mathbf{A}_{u_m}^*\big]\mathbf{z} \\
&= -\mathbf{z}^*\big[\mathbf{\Gamma}\mathbf{A}_{u_m}(\mathbf{A}^*\mathbf{A})^{-1}\mathbf{A}^* + \mathbf{A}(\mathbf{A}^*\mathbf{A})^{-1}\mathbf{A}_{u_m}^*\mathbf{\Gamma}\big]\mathbf{z} \\
&= -2\operatorname{Re}\big\{\mathbf{z}^*\mathbf{A}(\mathbf{A}^*\mathbf{A})^{-1}\mathbf{A}_{u_m}^*\mathbf{\Gamma}\mathbf{z}\big\} \\
&= -2\operatorname{Re}\big\{\hat{\mathbf{b}}^*\mathbf{A}_{u_m}^*(\mathbf{z} - \mathbf{A}\hat{\mathbf{b}})\big\}
\end{aligned}
$$

Matrix $\mathbf{A}_{u_m}$ has only column m with elements $d_{n_{x_m}} = \partial a_{m_n}/\partial u_m$. All other elements of $\mathbf{A}_{u_m}$ are zero because they are independent of u_m. So with equation 12.2 we can write:

$$\hat{\mathbf{b}}^*\mathbf{A}_{u_m}^* = j\hat{b}_m^*\mathbf{a}_m^*\mathbf{D}_x$$

with

$$\mathbf{D}_x = diag\left(-\frac{2\pi}{\lambda}x_n\right) \quad \text{for } n = 1, \ldots, N$$

With this relation we get for the gradient Q_{u_m}:

$$Q_{u_m} = -2\operatorname{Re}\big\{j\hat{b}_m^*\big(\mathbf{a}_m^*\mathbf{D}_x\mathbf{z} - \mathbf{a}_m^*\mathbf{D}_x\mathbf{A}\hat{\mathbf{b}}\big)\big\} \tag{12.12}$$

with $\hat{\mathbf{b}} = (\mathbf{A}^*\mathbf{A})^{-1}\mathbf{A}^*\mathbf{z}$ according to equation 12.8. Equation 12.12 can be summarised for $\mathbf{u}$ for all u_m by:

$$
\begin{aligned}
\mathbf{Q}_u &= 2\operatorname{Im}\big\{diag(\hat{b}_m^*)(\mathbf{A}^*\mathbf{D}_x\mathbf{z} - \mathbf{A}^*\mathbf{D}_x\mathbf{A}\hat{\mathbf{b}})\big\} \\
&= 2\operatorname{Im}\big\{diag(\hat{b}_m^*)(\mathbf{A}^*\mathbf{D}_x(\mathbf{z} - \hat{\mathbf{z}}))\big\}
\end{aligned} \tag{12.13}
$$

For $\mathbf{Q}_v$ the corresponding equation results. With $\mathbf{Q}_\omega$ we denote the $2M$-dimensional gradient of Q with respect to all parameters contained in ω_M. The gradient $\mathbf{Q}_\omega$ is now a vector with $2M$ components.

The expression $\mathbf{A}^*\mathbf{D}_x\mathbf{z}$ in equation 12.13 is the parallel formation of M difference beams in directions u_m. The vector $\hat{\mathbf{b}}$ is produced by M parallel sum beams with mutual decoupling. The main computational task may thus be performed by a beamforming subsystem. The additional processing only deals with matrices of dimension M.

The gradient search for the solution shall find the minimum of Q by following the direction of steepest descent, given by the gradient $\mathbf{Q}_\omega$. For a single step it is given by:

$$\varpi_{k+1} = \varpi_k - \mu \mathbf{Q}_{\varpi k} \tag{12.14}$$

The recursion constant μ determines the step size and has to be chosen suitably. If μ is chosen to be too small, it results in a long search and with a too high value one will jump across the minimum. This procedure may be applied for only one measured data vector $\mathbf{z}$. Monopulse processing is therefore also possible in principle for superresolution, but the SNR must be high enough (> 30 dB at the beamforming output).

Alternatively, we could apply a new data sample $\mathbf{z}_k$ for each iteration step k for the determination of the gradient. This leads to a gradient search in the form of a stochastic approximation, proposed by Nickel [6]:

$$\varpi_{k+1} = \varpi_k - \mu_k \mathbf{Q}_\varpi (\mathbf{z}_k, \varpi_k) \tag{12.15}$$

This procedure was named parametric target model fitting (PTMF). The series $\mu_k = \mu/(\beta + k)$ may be selected for convergence [8] with μ, β being suitable chosen constants. Some additional means are used to stabilise the convergence, e.g. a limitation of the gradient length. The convergence is accomplished usually with about 30 iterations. For an adaptive operation to follow moving sources the factor μ_k would be selected as a constant.

Now we will discuss the estimation of the target number M in more detail [8]. We want to decide between the hypothesis H_M, that is the target number to be $\leq M$, or the alternative hypothesis H_A, that is the target number is $>M$. This test is performed for stepwise increasing $M = 1, 2, 3 \ldots$. The decision will result in H_A until for the first time we reach a decision for H_M. Then we stop and take that specific M as the estimated target number.

The likelihood ratio test may be applied to decide between H_A and H_M. The test statistic is given by:

$$T(M) = \frac{\max p(\mathbf{z}/H_A)}{\max p(\mathbf{z}/H_M)} \tag{12.16}$$

The meaning of $\max p(\mathbf{z}/.)$ is to take the maximum possible value after estimating the best fitting parameter set for the target signals. In case of H_M we get with equation 12.6:

$$\max p = \frac{1}{(2\pi)^N} \exp\left(-\frac{1}{2} Q_M\right)$$

Q_M is minimised under the constraint of equation 12.10, that means using the antenna matrix $\mathbf{A}$ after equation 12.3 with M columns describing M possible target directions.

In case of H_A the number of assumed targets is arbitrary. To maximise $p(\mathbf{z}/H_A)$ we can take $M = N$, so we can make the model signal $\hat{\mathbf{z}}$ equal to $\mathbf{z}$, that is, even the noise may be represented by a signal model with N signal components, resulting in $Q_N = 0$. Thus we get:

$$\max p(\mathbf{z}/H_A) = \frac{1}{(2\pi)^N} \exp{(0)} = \frac{1}{(2\pi)^N}$$

We use the ln of $T(M)$ for the decision function and from equation 12.16 follows:

$$\ln T(M) = \tfrac{1}{2} Q_M \qquad (12.17)$$

The decision is made for the increasing series of $M = 1, 2, 3 \ldots$ and decreasing $T(M)$ by a comparison with threshold η:

$$\text{if } \ln T(M) > \eta : \quad \text{decision for } H_A$$

$$\text{if } \ln T(M) < \eta : \quad \text{decision for } H_M$$

With K data snapshots we have according to equation 12.10:

$$Q_M = \frac{1}{K} \sum_{k=1}^{K} \mathbf{z}_k^* \boldsymbol{\Gamma} \mathbf{z}_k$$

Q_M is χ^2 distributed because it is the sum of quadratic Gaussian variables. Projection by $\boldsymbol{\Gamma}$ in the expression for Q_M subtracts just M target vectors or eigenvectors. The number of degrees of freedom is therefore in the case of K snapshots $2K(N - M)$. The threshold η may thus be determined by this χ^2 ditribution for a certain selected probability α for overestimating the target number. One has to choose for η the α percentage point of the χ^2 distribution:

$$\eta = \chi^2_{\alpha;2K(N-M)}$$

A higher threshold decreases this error probability.

The PTMF procedure may be applied to arbitrary arrays and subarrays and in fact is able to resolve completely correlated targets. For superresolution in elevation v a vertical array would be sufficient. For a two-dimensional resolution in azimuth u and elevation v we need of course a planar or volume array. The antenna element or sub-array positions should be chosen to be irregular to avoid problems with grating-lobe effects if the distance between elements or subarray centres is greater than $\lambda/2$. The elements or subarrays should use the whole aperture to achieve the highest possible resolution by narrow single sum beams. The limited beamwidth of the subarray patterns is no problem because the targets of interest to be resolved by superresolution are within the beamwidth of the antenna's main beam, which is much narrower than the subarray patterns.

The number of antenna elements or subarrays must be above the number of parameters to be estimated. The more channels that are available the more relaxed are the requirements to minimise the channel errors, because by averaging effects the influence of errors decreases. These errors are produced by channel offset, amplification and orthogonality errors, inter-channel mismatch, analogue-to-digital conversion, differences in antenna patterns and mutual coupling effects. For the application of corrections by using calibration results the channels must be stable enough in time.

The PTMF algorithm given by equation 12.15 has a signal integration effect by using a series of data vectors $\mathbf{z}_k$, so the necessary SNR is decreased correspondingly [9,10]. The data vectors enter the algorithm by the beamforming functions and do not have to be stored. The starting values for the PTMF procedure should be chosen within the main beamwidth. Together with a test procedure for M the algorithm is an adaptive multitarget estimator. For only one target the algorithm reduces to nulling one difference beam equivalent to the monopulse procedure.

The achievable resolution enhancement depends on the SNR and the errors of the measurements by the antenna channels. For target separations below the half-power beamwidth the required SNR increases rapidly.

In Figure 12.1 the basic functions for the superresolution estimation procedure are shown. Above we have an estimation section for computing the Q function. The gradient of Q may be computed using a beamforming subsystem according

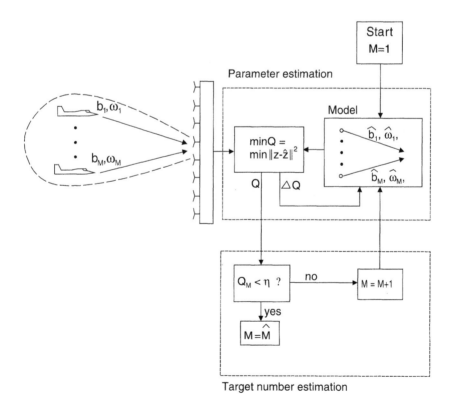

Figure 12.1 Superresolution processing scheme

to equation 12.13. The test for the number of targets M is performed below. The procedure is started with $M = 1$. This state corresponds to the conventional detection and monopulse procedure. Then M is stepwise increased until the test value Q_M decreases below the decision threshold η.

12.2.1   Resolution limit from Cramér–Rao inequality

The limit for resolution improvement compared with conventional resolution will now be discussed on the basis of the standard-deviation values of the direction estimates. These depend on the signal-to-noise power ratios (SNR) and the angular distribution of the targets. By noise we may understand the receiver noise and all errors in connection with the measurement and analogue-to-digital conversion of the received signals.

As in chapter 11 section 11.3 (monopulse direction estimation accuracy) we may apply the Cramér-Rao inequality to compute a lower limit for the variances of the direction estimates. For M targets there are $L = 3M$ or $4M$ single parameters for linear or planar arrays, respectively. For the set of parameters $\boldsymbol{\omega} = \omega_1, \ldots \omega_i, \ldots, \omega_L$ the variances are given by the matrix inequality, as discussed in chapter 3 section 3.4:

$$\mathbf{G} = E\{(\hat{\boldsymbol{\omega}} - \boldsymbol{\omega})(\hat{\boldsymbol{\omega}} - \boldsymbol{\omega})^*\} \geq \left[E\left\{ \frac{\partial \log p^*}{\partial \boldsymbol{\omega}} \frac{\partial \log p}{\partial \boldsymbol{\omega}} \right\} \right]^{-1} = \mathbf{F}^{-1}$$

The meaning of $\mathbf{G} \geq \mathbf{F}^{-1}$ is that the difference of the matrices is positive semi-definite, i.e. all elements on the main diagonal of matrix $\mathbf{G}$, representing the variances of the estimated parameters, are equal to or greater than the elements on the main diagonal of $\mathbf{F}^{-1}$. The variances are asymptotically minimal for maximum-likelihood estimation of the parameters.

We apply equation 3.74 from chapter 3:

$$F_{\omega_i \omega_k} = -\int \frac{\partial^2 \log p}{\partial \omega_i \, \partial \omega_k} p \, d\mathbf{z}$$

and with equations 12.6 and 12.7:

$$\log p(\mathbf{z}/\boldsymbol{\omega}) = const - \tfrac{1}{2} Q$$

$$F_{\omega_i \omega_k} = \frac{1}{2} \int \frac{\partial^2 Q}{\partial \omega_i \, \partial \omega_k} p \, d\mathbf{z} \qquad (12.18)$$

where:

$$Q = (\mathbf{z} - \mathbf{s}(\boldsymbol{\omega}))^*(\mathbf{z} - \mathbf{s}(\boldsymbol{\omega})) \quad \text{with } \mathbf{s} = \mathbf{Ab}$$

$$Q_{\omega_i} = -2\,\mathrm{Re}\,\mathbf{s}^*_{\omega_i}(\mathbf{z} - \mathbf{s})$$

and

$$Q_{\omega_i \omega_k} = -2\,\mathrm{Re}[\mathbf{s}^*_{\omega_i \omega_k}(\mathbf{z} - \mathbf{s}) - \mathbf{s}^*_{\omega_i}\mathbf{s}\omega_k]$$

Because the expectation of the first term results in zero for $F_{\omega_i \omega_k}$ we get:

$$F_{\omega_i \omega_k} = \mathrm{Re}\, \mathbf{s}^*_{\omega_i} \mathbf{s}_{\omega_k} \tag{12.19}$$

We now apply this equation to a regular linear array, as an example, and assume two targets with the parameter set ω given by the azimuth directions u_1 and u_2 and the complex amplitudes $b_1 = g_1 + jh_1$, $b_2 = g_2 + jh_2$. The parameter set may be denoted by $(u_1, u_2, g_1, g_2, h_1, h_2)$.

The matrix $\mathbf{F}$ is symmetric to its main diagonal. To develop this matrix we start with the derivatives with respect to u_i:

$$\mathbf{s}_{u_i} = \mathbf{A}_{u_i} \mathbf{b} = \frac{\partial}{\partial u_i} \mathbf{a}_i b_i \quad i = 1, 2$$

The columns of $\mathbf{A}$ with index $k \neq i$ are independent of u_i and the derivatives therefore are zero. For a regular linear array with $d = \lambda/2$ the element coordinates are $x_n = n\lambda/2$ ($n = -N/2 \ldots N/2$). According to equation 12.2 we get:

$$a_{i_n} = \exp(j\pi u_i n) \quad \text{for } n = -N/2, \ldots, N/2$$

$$\frac{\partial a_{i_n}}{\partial u_i} = j\pi n \exp(j\pi u_i n)$$

$$\frac{\partial \mathbf{a}_i}{\partial u_i} = j\pi \, diag(n)\mathbf{a}_i$$

and:

$$\frac{\partial \mathbf{a}_i^*}{\partial u_i} = -j\pi \, diag(n)\mathbf{a}_i^*$$

and with equation 12.19:

$$F_{u_i u_k} = \mathrm{Re}\, b_i^* \mathbf{a}^*_{i_{u_i}} \mathbf{a}_{k_{u_k}} b_k$$

$$= \mathrm{Re}\, \pi^2 b_i^* \mathbf{a}_i^* diag(n^2) \mathbf{a}_k b_k$$

For $F_{u_i g_k}$ with $i, k = 1, 2$ we get again from equation 12.19:

$$F_{u_i g_k} = \mathrm{Re}\, \mathbf{s}^*_{u_i} \mathbf{s}^*_{g_k} = \mathrm{Re}\{-j\pi b_i^* \mathbf{a}_i^* diag(n) \mathbf{a}_k\}$$

and correspondingly:

$$F_{u_i h_k} = \mathrm{Re}\, \pi b_i^* \mathbf{a}_i^* diag(n) \mathbf{a}_k$$

$$F_{g_i g_k} = \mathrm{Re}\, \mathbf{a}_i^* \mathbf{a}_k = F_{h_i h_k}$$

$$F_{g_i h_k} = \mathrm{Re}\, \mathbf{a}_i^* (j\mathbf{a}_k)$$

All elements $F_{\omega_i \omega_k}$ are used to build up the matrix $\mathbf{F}$ according to the following structure:

$$\mathbf{F} = \begin{bmatrix} F_{u_1 u_1} & F_{u_1 u_2} & F_{u_1 g_1} & F_{u_1 g_2} & F_{u_1 h_1} & F_{u_1 h_2} \\ F_{u_2 u_1} & F_{u_2 u_2} & F_{u_2 g_1} & F_{u_2 g_2} & F_{u_2 h_1} & F_{u_2 h_2} \\ F_{g_1 u_1} & F_{g_1 u_2} & F_{g_1 g_1} & F_{g_1 g_2} & F_{g_1 h_1} & F_{g_1 h_2} \\ F_{g_2 u_1} & F_{g_2 u_2} & F_{g_2 g_1} & F_{g_2 g_2} & F_{g_2 h_1} & F_{g_2 h_2} \\ F_{h_1 u_1} & F_{h_1 u_2} & F_{h_1 g_1} & F_{h_1 g_2} & F_{h_1 h_1} & F_{h_1 h_2} \\ F_{h_2 u_1} & F_{h_2 u_2} & F_{h_2 g_1} & F_{h_2 g_2} & F_{h_2 h_1} & F_{h_2 h_2} \end{bmatrix}$$

The desired bounds on $\sigma_{u_1}^2, \sigma_{u_2}^2$ we will find in matrix $\mathbf{G}$, the inverse of $\mathbf{F}$, in place of $F_{u_1 u_1}$, $F_{u_2 u_2}$ or $G(1,1)$ and $G(2,2)$, respectively. With a MATLAB program the example is numerically investigated for the following parameters:

$N = 16$, 2 equal target signals with a phase difference of $45°$

This is between the most favourable case $90°$ and the most unfavourable case $0°$ or $180°$. The SNR is varied in 3 dB steps from 24 to 45 dB at the beamforming output.

In Figure 12.2 the bounds on the standard deviation σ_u dependent on the angular distance $|u_i - u_k|$ are given by the curves. Both variables are normalised on the half-power beamwidth BW. The intuitive resolution limit $\sigma_{u_i} + \sigma_{u_k} \le |u_i - u_k|$ is indicated by the dashed line.

The target's angular separation must be greater than that given by the intersection point for the respective SNR. The usual SNR required for monopulse is 13 dB. So

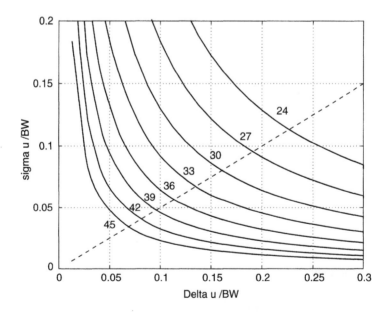

Figure 12.2   Cramér–Rao standard deviation and limit for resolution, two targets with equal amplitudes, phase difference 45°, parameter SNR (dB) at beamforming output

we have to increase the SNR by 10 to 20 dB to improve the resolution by a factor of four to six.

12.3 Experimental verification of superresolution

The effectiveness of superresolution has been confirmed not only with simulation studies but also by encouraging experiments. Measured and recorded data from a vertical array with eight antenna elements were processed offline to demonstrate its effectiveness against the multipath effect above sea [14]. An approaching low-flying aircraft above sea, illuminated by an X-band radar, produced direct and multipath echoes which superposed at the receiving array. By using the PTMF algorithm it was possible to distinguish between the direct and multipath echoes as shown as an example in Figure 12.3. The true elevation from an optical tracker coincides closely

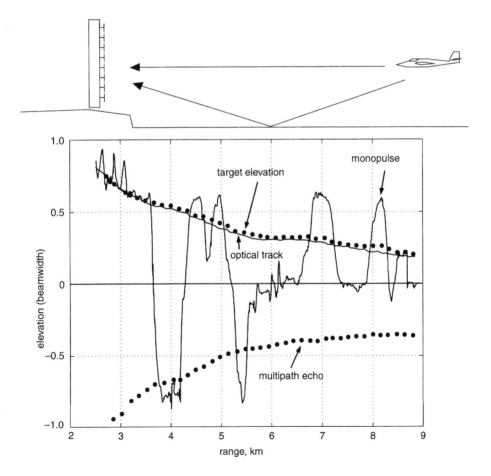

Figure 12.3 Multipath resolution for aircraft above sea, dots indicate elevation estimates by PTMF

with the upper elevation estimate for the aircraft target. In contrast, the monopulse estimation applied to the same data showed large fluctuations with useless results. Only for the very near-range measurements with an increased elevation angle above half of the beamwidth did the monopulse procedure finally result in correct estimates.

A special experimental system DESAS (demonstration of superresolution array signal processing) for investigating superresolution procedures in practice was developed at FGAN-FFM [6]. An eight-element active planar-array antenna at S band was equipped with receiving channels with 4.5 MHz bandwidth for each antenna element up to analogue-to-digital conversion with nine bits for the I and Q channels. Three sources were installed about 4 m in front of this array to simulate targets. The signals, transmitted from the sources, could be selected as sinusoids at the carrier frequency f_0 or Doppler sinusoids shifted by a selectable f_D, or noise. One source can be moved to vary the lateral distance to the other sources.

The block diagram of the whole experimental system is given by Figure 12.4.

Figure 12.5 shows the antennas installed on the roof.

Signal processing for the PTMF algorithm and other algorithms derived from spectral estimation, discussed in the next section, is performed on a personal computer

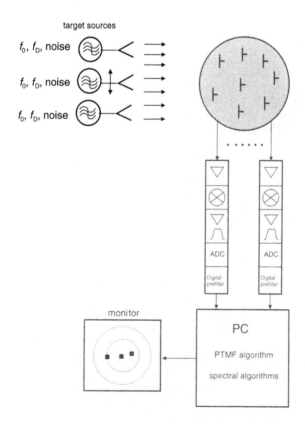

Figure 12.4   DESAS block diagram

Figure 12.5   DESAS array antenna and signal sources © Springer-Verlag, 1993 [15]

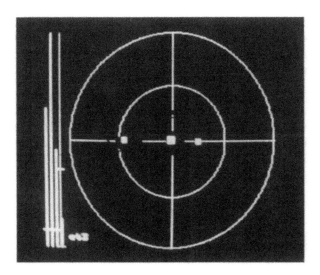

Figure 12.6   DESAS monitor with estimation results © Springer-Verlag, 1993 [15]

with programs written in MATLAB. The result is displayed as in Figure 12.6. The
three dots within the inner circle, corresponding to the 3 dB beamwidth, indicate
the estimated directions. On the left-hand side the estimated amplitudes are shown
by the three vertical bars. A threshold for target acceptance is indicated by a short

horizontal line. The outermost right bar is a test statistic for increasing the target number according to the white noise test.

Switching the sources on and off allows real-time demonstration of the closed-loop procedure. A number of iterations of the order of 30 for the gradient algorithm proved adequate.

Simulation and experimental results from currently available real data demonstrated the attractive performance of this PTMF test, so we recommend its application for radar superresolution.

12.4   Resolution by angular spectral estimation algorithms

Superresolution methods have first been developed for the investigation of signal-frequency spectra from signal samples in the time domain [15–20]. They are also applicable for the angular resolution of sources by processing of spatial sampled signals from an array. In chapter 4 we discussed the Fourier relationship between the signals distributed on the antenna aperture and the corresponding pattern in space.

The spectral superresolution algorithms are applicable only for the resolution of sources with mutual uncorrelated signals, for example for jammer mapping.

If there are coherent sources it is possible to achieve a decorrelation by spatial smoothing, that is averaging the array outputs within a window sliding across the array. Then the following methods may also be applied. The aperture, which determines the resolution, is reduced in this case by the window size.

All methods result in a scan pattern which has to be interpreted by inspection. Therefore, practical application seems to be limited especially if it should be applied to a multifunction radar with automated functions. For completeness they are mentioned in the following.

Linear prediction methods are used in the time domain for spectral analysis. They are also known under the names maximum entropy and autoregression method. By applying the computationally attractive algorithm of Burg [17,21] the time function of the prediction error filter is extended compatible with the measured or known correlation function. The prolonged error filter **h** is then used to compute a spectrum with improved frequency resolution:

$$S_{AR} = \frac{1}{|\mathbf{a}^*(\theta)\mathbf{h}|^2} \tag{12.20}$$

$\mathbf{a}(\theta)$ is the beamforming vector which has to be scanned computationally through the directions θ of interest. Applying this method to array antennas to compute the angular spectrum means virtually extending the available aperture and therefore improving the angular resolution. The algorithm is only well suited to regular equally-spaced linear arrays and is not applicable to planar arrays and not for irregular subarrays at all.

The *maximum likelihood estimator* for the signal power as a function of the direction θ after Capon [15, p. 62] can be derived from equation 3.77 by setting $S = E\{\hat{s}\hat{s}^*\}$

and $\mathbf{C} = \mathbf{a}$:

$$S_{ML}(\theta) = \frac{1}{\mathbf{a}^*(\theta)\mathbf{Q}^{-1}\mathbf{a}(\theta)} \qquad (12.21)$$

The inverse of the estimated covariance matrix $\mathbf{Q}^{-1}$ has to be computed for the received signal $\mathbf{z}$ at the array output from the existing sources, for example jammers. The antenna elements may be randomly distributed. This power estimator results in an accurate value especially for the powers of the resolved sources. The amount of computation is high because of the matrix inversion and multiplication. The resolution performance is moderate.

Signal subspace methods determine the lower dimensional space $\mathbf{X}$ for representing, as an orthonormalised system, the signal sources, as is discussed in chapter 10 for the jammer space. This method is known under the name MUSIC for multiple signal classification [19,20,28].

$\mathbf{X}$ determines a projection $\mathbf{P}$ orthogonal to the signal space:

$$\mathbf{P} = \mathbf{I} - \mathbf{X}\mathbf{X}^*$$

The signal spectral estimator:

$$S_P(\theta) = \frac{1}{\mathbf{a}^*(\theta)\mathbf{P}\mathbf{a}(\theta)} \qquad (12.22)$$

is similar to the S_{ML} above, but the peaks in the scan pattern are now independent of the source powers. Therefore no estimates of the signal powers can be expected.

The signal space may be derived from the dominant eigenvectors of the covariance matrix estimated from the array outputs as discussed in chapter 10 for jammer cancellation. This method may also be applied to random arrays, shows good resolution results but is computationally expensive.

An example of a scan pattern is given in Figure 12.7. It shows two uncorrelated sources with a distance of half the beamwidth and the SNR at the elements of 9.5 dB. The linear array has 64 elements with $\lambda/2$ spacing, but only 43 data snapshots are used. All ten trials show unbiased estimation and resolution.

The projection matrix may alternatively be derived directly from the array output data vectors as proposed by Hung and Turner (equation 10.19) [29]. Because only a few data are used in the algorithm, the number of data vectors corresponds to the signal subspace dimension, the noise influence is not averaged out and the method is suitable only for a large SNR. The method can also be used for random arrays but not for completely correlated target signals. Further development resulted in the Hung-Turner-Yeh method: instead of data vectors one could apply some selected columns of the estimated covariance matrix, thereby using the averaging process of the matrix estimation, as suggested by Yeh and Brandwood [30,31]. The noise contribution in the covariance matrix can be eliminated by computation.

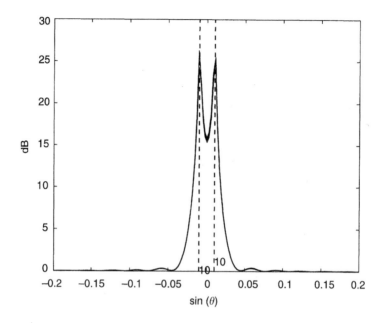

Figure 12.7   MUSIC scan pattern for two uncorrelated sources (courtesy U. Nickel)

A discussion on the relationships of the various spectral estimators has been given by Nickel in Reference 11.

It is left to mention the ESPRIT algorithm [33]: this may be applied only to arrays with the shift invariant property. Its performance with radar parameters is yet unknown.

12.4.1   Resolution and the dimension of the signal subspace

For all subspace methods the dimension of the signal subspace according to the target number has to be determined by a suitable test procedure. This detection procedure must be processed completely automatically. For practical applications the problem of estimating the dimension of the signal subspace with a small number of data snapshots arises. The reasons are the limited time budget for a special task within the multifunction operation of a phased-array radar or nonstationary jammer scenarios, varying with time.

We may take the MUSIC algorithm, given by equation 12.22, as being typical for subspace methods. It has been shown that for a large sample size this algorithm approaches the Cramér-Rao resolution bound. But its good resolution property can be retained for a small sample size only if the dimension of the signal subspace is known, as shown by Nickel [24,32]. For the resolution of targets or sources we first have to estimate their correct number with a high probability and afterwards their directions.

In chapter 10 section 10.4.3 we have already discussed the impact of channel errors. Decorrelation by dispersive errors deteriorates a clear separation between signal and noise subspaces. This leakage is effective especially for a high source signal power. The consequences for resolution properties is illustrated by the following examples.

For a regular linear array with 64 antenna elements and ten jammers resolution using Capon's method (equation 12.21) and the MUSIC method (equation 12.22) have been compared by Nickel [24]. For the *I* and *Q* components random amplitude errors of 0.9 dB and orthogonality errors of 7° are assumed. The bandwidth is assumed to be fairly large at one per cent of the carrier frequency.

Two scenarios are applied. A high-power jammer scenario (strong scenario) with an SNR of 27 dB for each signal at the antenna elements and a low-power scenario (weak scenario) with an SNR of only 0 dB. For the strong scenario the eigenvector distribution is shown in Figure 12.8. The longer series belongs to *IQ* errors, the short series is produced by bandwidth effects only.

Figure 12.9 shows the Capon and MUSIC angular spectrum.

Although there are only ten jammers assumed the number of principal eigenvalues and vectors is 40. The resolution by MUSIC is poor. The resolution by the Capon algorithm (equation 12.21) is much more clear, because by the covariance matrix the relative power of the eigenvalues is in principle taken into account. In contrast, the MUSIC method treats all eigenvectors as with an equal eigenvalue or power. For comparison in Figure 12.9 the MUSIC result without errors and with good resolution is also shown.

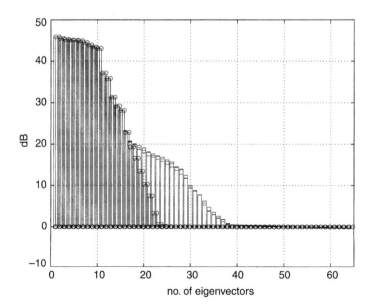

Figure 12.8   Eigenvector distribution for strong scenario (courtesy U. Nickel)

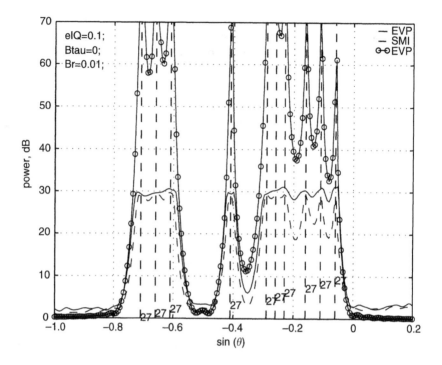

Figure 12.9   MUSIC and Capon angular spectrum, 40 eigenvectors, MUSIC (solid line), Capon (dashed line), MUSIC without errors (circled line) (courtesy U. Nickel)

For further comparison the respective results for the weak scenario are shown in Figures 12.10 and 12.11.

Now the number of dominant eigenvalues is reduced to 22 and the resolution result is improved considerably.

If the true dimension ten is applied the result for the strong scenario, given by Figure 12.12, is excellent although the channel errors are effective. The conclusion therefore is: the main problem with real data subject to errors is the determination of the dimension of the signal space. By a correct choice even error effects can be removed.

12.4.2   Estimation of the signal subspace dimension

Two test methods to determine the dimension of the signal subspace proved to be most suitable for the spectral resolution methods, according to comparative investigations by Nickel [26,32].

The diagonally-loaded minimum description length (LMDL) criterion is derived from the estimated covariance matrix $\hat{\mathbf{R}} + \alpha\mathbf{I}$. The load factor α can be chosen to be only $0.2, \ldots, 2$ times the receiver noise power level. Again we have N antenna

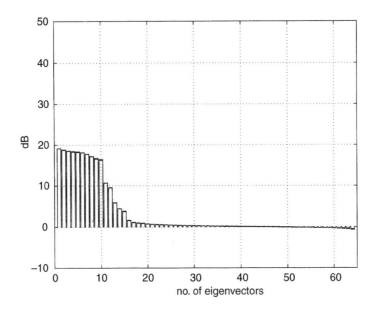

Figure 12.10   Eigenvector distribution for weak scenario (courtesy U. Nickel)

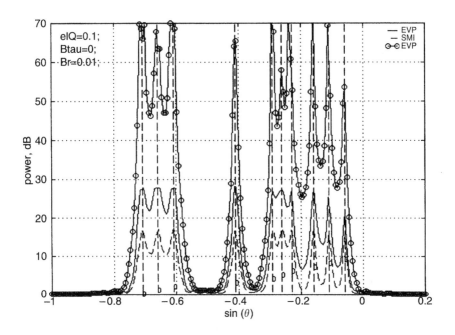

Figure 12.11   MUSIC and Capon angular spectrum for weak scenario, 22 eigen-vectors, MUSIC (solid line), Capon (dashed line), MUSIC without errors (circled line) (courtesy U. Nickel)

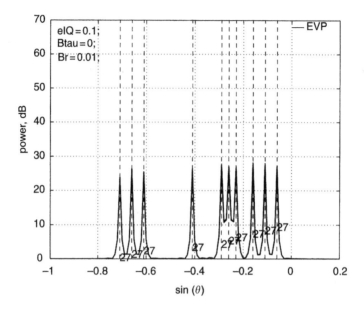

Figure 12.12   MUSIC angular spectrum for strong scenario, ten eigenvectors, with errors (courtesy U. Nickel)

or subarray channels. The number of used data snapshots is K. Then the following expression is formed by using the series of eigenvalues λ_i in descending order:

$$T(m) = \frac{(1/N - m) \sum_{i=m+1}^{N} \lambda_i}{\left(\prod_{i=m+i}^{N} \lambda_i\right)^{(1/N-m)}}$$

$T(m)$ is the ratio of arithmetic to geometrical mean for the remaining low eigenvalues λ_i with $i > m$. $T(m)$ is also a measure of the equality of the smallest eigenvalues. If the eigenvalues belong to the noise space they are all equal. With $T(m)$ the following expression is formed:

$$LMDL(m) = K(N - m) \log T(m) + \frac{m}{2}(2N - m) \log K \qquad (12.23)$$

The minimum value of *LMDL* for $m = 1, 2, 3, \ldots, N$ determines $m = M$.

The second method is a sequence of white noise threshold tests (WNT) [26], similar to the test in section 12.2 for the PTMF method. We assume that the noise power σ^2 is known by a measurement. Then M is estimated by a sequence of hypothesis tests for H(M) against the alternative H$_A$ with the following question at each step: is the target number $m \leq M$, that is hypothesis H$_M$, or is $m > M$, that is hypothesis H$_A$? The log-likelihood ratio of equation 12.16 for both hypotheses will be decreasing with increasing M because above $M = m$ (M for the estimated target number) the probability for H(M) increases.

The expression for Q_M by equation 12.10 is treated as a projection into space orthogonal to the space by the m signals which belong to the m greatest eigenvalues. $\hat{\lambda}_i$ are the descending eigenvalues from the estimated covariance matrix. The power of the remaining signal is then just the sum of eigenvalues with index $m + 1, \ldots, N$. The log-likelihood ratio may be expressed similar to equations 12.16 and 12.17 by:

$$L(m) = \frac{2K}{\sigma^2} \sum_{i=m+1}^{N} \hat{\lambda}_i \qquad (12.24)$$

For the decision as usual a comparison with a threshold is performed:

$$L(m) \leq \eta \quad \text{for the decision } m = M$$

By the threshold the probability α of overestimating the target number may be determined, as already discussed for the PTMF procedure. The expression $L(m)$ results from a quadratic expression of Gaussian-distributed signals and L is χ^2 distributed with $2K(N - M)$ remaining degrees of freedom for the space without the source signals and the threshold is selected as for PTMF.

Overestimating M is not a great problem, because by the following estimation procedure, e.g. by the Capon method, the power of the resolved sources is estimated and may also be compared for each suspected target with a threshold.

12.5 Conclusions

Superresolution is offered by an active receiving array and special signal-processing procedures as an additional and new capability. Targets at the same range cell and within the beamwidth can be resolved to a limit of about $0.2\, \Theta_B$. The necessary SNR will be available by approaching target formations. The procedure for superresolution is derived from likelihood-estimation theory. The resulting parametric target model fitting algorithm, combined with stochastic approximation, is recommended. It is applicable for coherent target signals. One prominent application is resolving the multipath problem, especially for locating low-flying targets above the sea.

A test for the target number has to be combined with the resolution procedure. Theoretical resolution limits are derived with the Cramér–Rao limit.

Experimental verification has been achieved already.

Resolution algorithms from the spectral domain have been summarised. They are suited for location with superresolution of uncorrelated sources, e.g. jammers. A special problem is the determination of the dimension of the signal subspace equivalent to the number of sources. Relevant test procedures have been discussed.

12.6 References

1 KSIENSKI, A. A., and McGHEE, R. B.: 'A decision theoretic approach to the angular resolution and parameter estimation problem for multiple targets', *IEEE Trans. Antennas Electron. Syst.*, 1968, **4** (3), pp. 443–455

2 YOUNG, G. O.: 'Optimum space-time signal processing and parameter estimation', *IEEE Trans. Antennas Electron. Syst.*, 1968, **4** (3), pp. 334–341
3 WHITE, W. D.: 'Low angle radar tracking in the presence of multipath', *IEEE Trans. Antennas Electron. Syst.*, 1974, **10** (6), pp. 835–852
4 HOWARD, J. E.: 'A low angle tracking system for fire control radars'. Proceedings IEEE international conference on *Radar*, Arlington, Va, USA, 1975, pp. 412–417
5 WIRTH, W. D.: 'Radar signal processing with an active receiving array'. IEE international conference on *Radar*, London, UK, 25–28 October 1977, pp. 218–221
6 NICKEL, U.: 'Superresolution using an active antenna array'. IEE international conference on *Radar*, London, UK, 18–20 October 1982, pp. 87–91
7 NICKEL, U.: 'Winkelauflösung eng benachbarter Ziele mit Gruppenantennen' (angular resolution of closely spaced targets with array antennas). Doctoral dissertation, Rheinisch-Westfälische Hochschule Aachen, December 1982
8 NICKEL, U.: 'Superresolution by spectral line fitting', in SCHUESSLER, H. W. (Ed.): 'EUSIPCO signal processing II: theories and applications' (North Holland 1983) pp. 645–648
9 NICKEL, U.: 'Angular superresolution with phased array radar: a review of algorithms and operational constraints', *IEE Proc. F, Commun. Radar Signal Process.*, 1987, **134** (1), pp. 53–59
10 NICKEL, U.: 'Angle estimation with adaptive arrays and relation to superresolution', *IEE Proc. F, Commun. Radar Signal Process.*, 1987, **134** (1), pp. 77–82
11 NICKEL, U.: 'Algebraic formulation of Kumaresan-Tufts superresolution method, showing relation to ME and MUSIC methods', *IEE Proc. F, Commun. Radar Signal Process.*, 1988, **135** (1), pp. 7–10
12 NICKEL, U.: 'Application of array signal processing to phased array radar', in 'EUSIPCO signal processing IV: theories and applications' (North Holland, 1988) pp. 467–474
13 NICKEL, U.: 'Angular superresolution by antenna array processing'. International conference on *Radar*, Paris, France, 1989, pp. 48–58
14 NICKEL, U., and WIRTH, W. D.: 'Beam forming and array processing with active arrays'. IEEE *Antennas and propagation* symposium, Ann Arbor, USA, 1993, pp. 1540–1543
15 NICKEL, U.: 'Radar target parameter estimation with array antennas', in HAYKIN, S., LITVA, J., and SHEPHE, T. J. (Eds.): Radar array processing' (Springer Verlag Berlin-Heidelberg-New York, 1993) chapter 3, pp. 53 ff
16 MULLIS, C. T., and SCHARF, L. L.: 'Quadratic estimators of the power spectrum', in HAYKIN, S. (Ed.): 'Advances in spectrum analysis and array processing, volume 1' (Prentice Hall, Englewood Cliffs, New Jersey, 1991) chapter 1
17 BURG, J. P.: 'A new analysis technique for time series data'. Proceedings of NATO Advanced Study Institute on *Signal Processing*, Enschede, Netherlands, 1968, paper 15

18 GABRIEL, W. F.: 'Spectral analysis and adaptive array superresolution techniques', *Proc. IEEE*, 1980, **68**, pp. 654–666

19 SCHMIDT, R.: 'Multiple emitter location and signal parameter estimation', *IEEE Trans. Antennas Propag.*, 1986, **34**, pp. 276–280

20 BIENVENUE, G., and KOPP, L.: 'Optimality of high resolution array processing using the eigensystem approach', *IEEE Trans. Acoust. Speech Signal Process.*, 1983, **31**, pp. 1235–1248

21 BURG, J. P.: 'A new analysis technique for time series data'. NATO Advanced Study Institute on Signal Processing with Emphasis on Underwater Acoustics, Enschede, NL, 1968

22 NICKEL, U.: 'Performance of adaptive arrays and superresolution methods subject to channel errors'. International conference on *Radar*, Paris, France, 1994, pp. 54–59

23 MAIWALD, D., and NICKEL, U.: 'Multiple signal detection and parameter estimation using sensor arrays with phase uncertainties', Signal Processing VIII: Theories and Applications (Proc. EUSIPCO-96, Trieste, Italy, 1996). Ramponi, G.; Sicuranza, G.; Carrato, S.; Marsi, S. (Eds.). Edizione Lint/EURASIP, **1**, pp. 551–554

24 NICKEL, U.: 'On the application of subspace methods for small sample size', *AEÜ International Journal of Electronics and Communications*, 1997, **51** (6), pp. 279–289

25 NICKEL, U.: 'Determination of the dimension of the signal subspace for small sample size'. Proceedings of the IASTED international conference on *Signal processing and communication*, 1998, Canary Islands, Spain, IASTED/Acta Press 1998, pp. 119–122

26 NICKEL, U.: 'Application of angular superresolution methods for jammer mapping'. RTO meeting proceedings 40 *High resolution radar techniques*, Granada, Spain, 22–24 March 1999, pp. 54.1–9

27 CHILDERS, D. G. (Ed.): 'Modern spectrum analysis' (IEEE Press, New York, 1978) p. 162

28 FARRIER, D. R., JEFFRIES, D. J., and MARDANI, R.: 'Theoretical performance prediction of the MUSIC algorithm', *IEE Proc. F, Commun. Radar Signal Process.*, 1988, **135** (3), pp. 216–224

29 HUNG, E., and TURNER, R.: 'A fast beamforming algorithm for large arrays', *IEEE Proc. Aerosp. Electron. Syst.*, 1983, **19** (4), pp. 598–607

30 YEH, C. C.: 'Projection approach to bearing estimation', *IEEE Trans. Acoust. Speech Signal Process.*, 1986, **34**, pp. 1347–1349

31 BRANDWOOD, D. H.: 'Noise space projection: MUSIC without eigenvectors', *IEE Proc. H, Microw. Antennas Propag.*, 1987, **134** (32), pp. 303–309

32 NICKEL, U.: 'Aspects of implementing super-resolution methods into phased array radar', *AEÜ International Journal of Electronics and Communications*, 1999, **53** (6), pp. 315–323

33 KAILATH, T., and ROY, R.: 'ESPRIT—estimation of signal parameters via rotational invariance techniques', *IEEE Trans. Acoust. Speech Signal Process.*, 1989, **37** (7), pp. 984–995

Chapter 13
Space-time adaptive processing

13.1 Introduction

For a radar onboard a flying platform (aircraft, drone, satellite) the direction-dependent relative velocities of the ground scatterers cause corresponding Doppler frequency shifts of the clutter echoes. Against this broadened clutter spectrum one can apply no common filter, because one would thereby suppress too broad a Doppler frequency range of possible targets and would thus limit target detection inadmissibly and unnecessarily. With an active receiving antenna array, however, a signal field with samples in the space and time dimension can be offered for optimal signal processing for the detection of moving targets [1–7], even if those targets are slowly moving. The antenna array elements may form a linear or a planar array. The antenna can be looking with its broadside orientation forward or sideways relative to the flight direction. For an onboard multifunction phased-array radar, the forward-looking case is particularly of interest [5–8].

13.2 Doppler-shifted clutter spectrum

The clutter Doppler frequency shift depends on the platform velocity v, wavelength λ and the angle α between the direction to the clutter scatterer and the flight velocity vector [8]:

$$f_D = \frac{2v}{\lambda} \cos \alpha \tag{13.1}$$

On a cone with its axis given by the flight velocity vector the Doppler frequency shift is constant, its value depends only on the angle α. The intersection of the cone with the flat ground plane is a hyperbola and on this curve we have a constant Doppler frequency. All curves with constant Doppler shifts are given by the set of hyperbolas as illustrated in Figure 13.1.

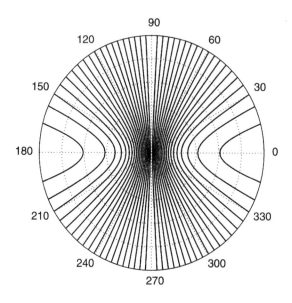

Figure 13.1   Isodops (courtesy R. Klemm) © IEE 1998 [8]

As a next step we are looking for the Doppler shift of clutter echoes at a certain range r and dependent on the azimuth angle φ. The geometry for a forward-looking array is explained with Figure 13.2 (a). The vector of the line of sight to a certain scatterer on the ground is given by azimuth angle φ and depression angle θ. The unit vector in the direction of a range element with a scatterer is given by:

$$\mathbf{r}^0 = (\cos\vartheta\ \cos\varphi,\ \cos\vartheta\ \sin\varphi,\ \sin\vartheta) \tag{13.2}$$

The antenna elements of the array are distributed in the y-z plane. The velocity vector of the platform is for the case of the forward-looking radar given by $\mathbf{v} = (v_x, 0, 0)$. The Doppler shift of echoes from the direction $\mathbf{r}^0$ is given by the projection of $\mathbf{v}$ on $\mathbf{r}^0$ (dot product):

$$v_r = v_x \cos\vartheta\ \cos\varphi \tag{13.3}$$

The echoes will be sampled in range according to the range resolution given by the signal bandwidth. Signal processing is then performed for signals belonging to individual range elements.

For a certain range $r = h/\sin\theta$ (h = platform height) the Doppler frequency f_d becomes:

$$f_d = \frac{2v_r}{\lambda} = \frac{2v_x}{\lambda}\sqrt{1 - \left(\frac{h}{r}\right)^2}\sqrt{1 - \sin^2\varphi} \tag{13.4}$$

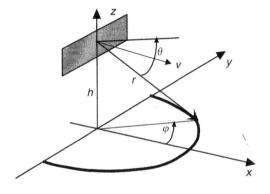

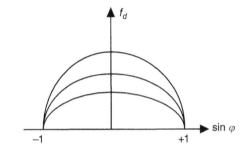

Figure 13.2   *Doppler distribution of clutter*
a geometry for airborne nose radar
b Doppler frequency distribution

or by writing $f_{d_{\max}} = 2v_x/\lambda\sqrt{1-(h/r)^2}$:

$$\left(\frac{f_d}{f_{d_{\max}}}\right)^2 + (\sin\varphi)^2 = 1$$

This is an ellipse equation for f_d over $\sin\varphi$ for a selected range element at range r. For each range element another ellipse develops with the maximum Doppler frequency $f_{d_{\max}}$ in each case equal to the vertical semiaxis (Figure 13.2 lower part). The Doppler spectrum of the clutter signals from range r is thus distributed along these curves in the $(\sin\varphi, f_d)$ plane. The clutter power is modulated additionally by the antenna patterns (transmit main beam and receiving antenna element pattern). This is demonstrated with a computed example in Figure 13.3.

The spectrum is ambiguous in frequency with the pulse recurrence frequency (PRF), therefore in the foreground we have a repetition for a clutter spectrum part.

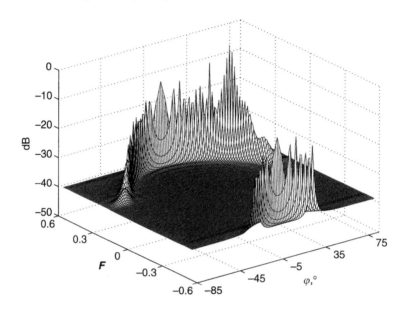

Figure 13.3   Clutter spectrum for forward-looking linear array (courtesy R. Klemm)
                © IEE 1998 [8]

For this forward-looking case (Figure 13.2) we recognise around $\sin \varphi = 0$ a small variation of f_d and near $\sin \varphi = 1$ a large variation of f_d for a certain interval of $\sin \varphi$, e.g. corresponding to the beamwidth. So we can expect a corresponding width for the clutter spectrum within a beam width.

13.3   Space-time processing

Let us now develop a first insight into the possibility of space-time processing.

In Figure 13.4 we find a sketch for the situation and conditions for a forward-looking radar. We want to detect a target with a certain Doppler frequency in a certain selected direction, given by the direction of the main beam. The target's Doppler frequency is unknown in advance and we have to cover the Doppler frequency range of interest with a bank of Doppler filters. We may form for each Doppler frequency an antenna beam in the desired direction. This beam should have low sidelobes in those directions where clutter echoes with the same Doppler frequency as the target are to be expected. In addition, each of these Doppler filters should have low sidelobes or even notches where the main beam ground clutter enters the Doppler filter.

This combination of Doppler filters and antenna patterns with suitable combined individual notches is possible obviously only with an active antenna array with its available spatial samples and resulting degrees of freedom, followed by the usual sampling in time. With this first basic idea for processing we can suppress the clutter echoes from the most harmful mutual sidelobes. This objective cannot be achieved

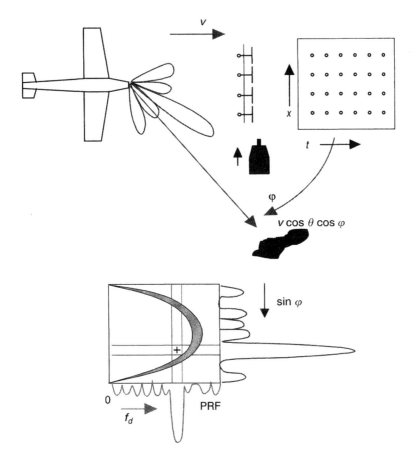

Figure 13.4   Individual beam pattern and Doppler filter from space-time sampled signals

by a single channel antenna beam followed by a Doppler filter bank. In this case only a multiplication of the filter and beam-pattern responses applies.

A further objective is to suppress as far as possible all clutter contributions along the ellipse in the ($\sin \varphi$, f_d) plane. This is possible with common processing of the signal samples from space and time. Because the distribution of the clutter spectrum in the ($\sin \varphi$, f_d) plane depends on the platform velocity and internal motion of the clutter scatterers, the space-time filter should be formed adaptively. For this concept of signal processing the abbreviation STAP (space-time adaptive processing) has been introduced and widely adopted. By application of STAP the achieved gain in signal-to-clutter-and-noise ratio in comparison to conventional processing may be relatively high, especially for antenna arrays with few antenna elements. Those arrays have naturally high sidelobes. So the disadvantages resulting from applying only a simple antenna array with few elements, especially from high sidelobes, may be compensated for by STAP processing.

After this preliminary discussion leading to the expectation that by suitable signal processing the problem of the suppression of clutter with a Doppler frequency spread can be solved in principle with antenna arrays, the processing rule has to be found by an appropriate approach [8].

As a first step only N spatial samples are assumed:

$$\mathbf{s} = \begin{bmatrix} s_1 \\ \vdots \\ s_N \end{bmatrix}, \qquad \mathbf{n} = \begin{bmatrix} n_1 \\ \vdots \\ n_N \end{bmatrix}$$

$\mathbf{s}$ is the vector of the complex target signal components, the components of $\mathbf{n}$ are samples of the interference from clutter and noise. $\mathbf{z} = \mathbf{s} + \mathbf{n}$ is thereby the vector of the discrete received signals, sampled all at the same time. The interference is described by its associated covariance matrix:

$$\mathbf{Q} = E\lfloor \mathbf{nn}^* \rfloor$$

$\mathbf{Q}$ may be computed on the basis of a clutter model [8]. For target detection the optimum processing rule follows according to chapter 3 with equation 3.50:

$$y = \mathbf{z}^* \mathbf{Q}^{-1} \mathbf{s} \tag{13.5}$$

y then has to be compared with a detection threshold. When the threshold is exceeded one decides 'target present', otherwise 'no target'.

$\mathbf{Q}^{-1}$ results in a suppression of the clutter signals contained in $\mathbf{z}$, and $\mathbf{s}$ represents a filter matched to the desired target Doppler signal.

The effectiveness of the processor after equation 13.5 is expressed by the gain of signal-to-clutter-and-noise power ratio from the input to the output of the processor according to equation 13.5:

$$G = E\left[\frac{\mathbf{s}^* \mathbf{Q}^{-1} \mathbf{s} \mathbf{s}^* \mathbf{Q}^{-1} \mathbf{s}}{\mathbf{s}^* \mathbf{Q}^{-1} \mathbf{nn}^* \mathbf{Q}^{-1} \mathbf{s}}\right] / \frac{trace(\mathbf{ss}^*)}{trace(\mathbf{Q})} = \mathbf{s}^* \mathbf{Q}^{-1} \mathbf{s} \cdot \frac{trace(\mathbf{Q})}{trace(\mathbf{ss}^*)} \tag{13.6}$$

This set-up must now be extended to the two dimensions space and time.

At each element output of the antenna array there appears a sequence with M signals. The entire signal is thus a two-dimensional data array with a temporal and a spatial dimension. Thus the vectors of signal and noise become:

$$\mathbf{z} = \begin{bmatrix} \mathbf{z}_1 \\ \mathbf{z}_2 \\ \vdots \\ \mathbf{z}_M \end{bmatrix}, \qquad \mathbf{s} = \begin{bmatrix} \mathbf{s}_1 \\ \mathbf{s}_2 \\ \vdots \\ \mathbf{s}_M \end{bmatrix}, \qquad \mathbf{n} = \begin{bmatrix} \mathbf{n}_1 \\ \mathbf{n}_2 \\ \vdots \\ \mathbf{n}_M \end{bmatrix} \tag{13.7}$$

whereby the partial vectors $\mathbf{z}_m$, $\mathbf{s}_m$ and $\mathbf{n}_m$ have the spatial dimension N. The index m represents the time $1, \ldots, M$. Thus the covariance matrix receives the block Toeplitz form:

$$
\mathbf{Q} = \begin{bmatrix}
\mathbf{Q}_0 & \mathbf{Q}_1 & \cdot & \cdot & \mathbf{Q}_{M-1} \\
\mathbf{Q}_1^* & \mathbf{Q}_0 & \mathbf{Q}_1 & \cdot & \cdot \\
\cdot & \mathbf{Q}_1^* & \cdot & \cdot & \cdot \\
\cdot & & \cdot & \cdot & \cdot \\
\mathbf{Q}_{M-1}^* & \cdot & \cdot & \cdot & \mathbf{Q}_0
\end{bmatrix}
\tag{13.8}
$$

The inverse matrix of $\mathbf{Q}$ is hermitian, but has no Toeplitz form. It has the structure:

$$
\mathbf{K} = \mathbf{Q}^{-1} = \begin{bmatrix}
\mathbf{K}_0 & \mathbf{K}_1 & \cdot & \cdot & \mathbf{K}_{M-1} \\
\mathbf{K}_1^* & \mathbf{K}_0 & \mathbf{K}_1 & \cdot & \cdot \\
\cdot & \mathbf{K}_1^* & \cdot & \cdot & \cdot \\
\cdot & & \cdot & \cdot & \cdot \\
\mathbf{K}_{M-1}^* & \cdot & \cdot & \cdot & \mathbf{K}_0
\end{bmatrix}
\tag{13.9}
$$

For large dimensions of the covariance matrix $\mathbf{Q}$ the inversion to $\mathbf{Q}^{-1}$ can lead to numerical problems. Moreover, this operation and also the necessary matrix multiplication $\mathbf{z}\mathbf{Q}^{-1}$ is computationally intensive. So we have a high processing load and cost even with a known $\mathbf{Q}^{-1}$.

The structure of the optimal processor is shown in Figure 13.5. Here a complex multiplication with a matrix of dimension $[NM, NM]$ is necessary.

13.4 Necessary degrees of freedom

For the development of suboptimal realisable concepts we have to determine the necessary spatial and temporal degrees of freedom. For this aim an eigenvalue analysis of the matrix $\mathbf{Q}$ is helpful and informative. The total number of the eigenvalues is NM. It has been shown [8] that, under the condition of Nyquist sampling in space and time, for the number L of dominant eigenvalues for the space-time covariance matrix of the clutter follows as a rule of thumb:

$$
L = M + N - 1
\tag{13.10}
$$

This number L gives the dimension of the orthogonal space or the number of eigenvectors occupied by the clutter signals. The complementary space with the dimension $NM - L$ is the orthogonal space in which the target signal may be detected. Signal and jammer space relations and projections into a corresponding space have been discussed already in chapter 10, section 10.2.3. If the number of dominant eigenvalues is smaller than the number of the available degrees of freedom an effective clutter suppression and signal detection can be expected.

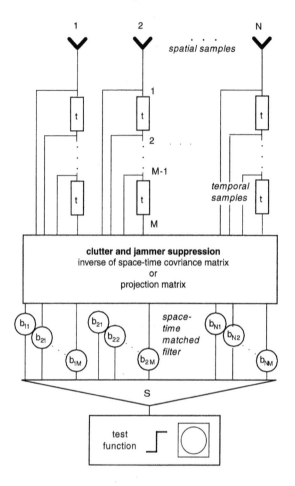

Figure 13.5   Block diagram for optimal STAP processor (courtesy R. Klemm) © IEE 1998 [8]

A clutter model was developed by taking into account the geometry, internal fluctuations, platform movement and inaccuracies. The matrix **Q** may be computed and the processing gain can be derived according to equation 13.6. An example of the distribution of the eigenvalues with $N = 24$ and $M = 3$ is shown in Figure 13.6. In practice the lower curve (x) for the case of a formed beam for transmission and reception applies, which suggests good clutter suppression since the eigenvalues are favourably concentrated thus leaving a sufficient dimensional signal space.

Additional spatial degrees of freedom are required if there are nearfield scatterers which introduce to the receiving array antenna clutter signals with a certain Doppler shift from directions other than that given by equation 13.4.

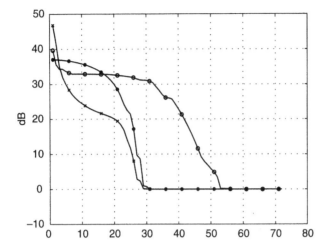

Figure 13.6 *Distribution of eigenvalues, N = 24, M = 3, omnidirectional sensors (circled line), directive sensors receiving (starred line), directive sensors receiving and transmitting (crossed line) (courtesy R. Klemm), © IEE 1998 [8]*

13.5 Suboptimal concept with FIR filter: reduction of time dimension

The complexity and expense of the processor must be reduced by developing a simplified and suboptimal processing structure. The optimum procedure will be a reference for comparison of the respective gain functions according to equation 13.6.

An initial option is suggested [8] as being replacing the suppression of the interference or clutter signals of **z** by a least-squares FIR filter (finite impulse response filter) instead of the matrix multiplication of equation 13.5. The received signal series would be convolved with this filter function as explained in chapter 2 section 2.4. This concept is especially useful if the signal sequence is temporally long (high value for M). This is the case especially for airborne radars in a mode with high PRF.

Now we want to derive an expression for such an FIR filter and we start from the basic equation 13.5:

$$y = \mathbf{z}^* \mathbf{Q}^{-1} \mathbf{s}$$

First, the signal vector **s** applies to a certain direction for beamforming and a certain Doppler frequency. It can be written with a matrix **B**, with dimension $[NM,M]$ for beamforming, and the Doppler vector **d**.

$\mathbf{s}_m$ and **b** are column vectors of the length N, column matrix **d** has elements representing the Doppler signals $d_m = \exp(j\psi m)$ for $m = 1, \ldots, M$. The phase difference $\psi = 2\pi f_d T$ (T = pulse period) corresponds to the Doppler frequency f_d

of the targets which shall be detected:

$$s = \begin{bmatrix} s_1 \\ s_2 \\ \vdots \\ s_M \end{bmatrix} = \begin{bmatrix} b & & & \\ & b & & \\ & & \ddots & \\ & & & b \end{bmatrix} \begin{bmatrix} d_1 \\ d_2 \\ \vdots \\ d_M \end{bmatrix} = Bd \tag{13.11}$$

In order to receive the FIR filter, one replaces matrix d by the column matrix e, also with length M:

$$e = \begin{bmatrix} 1 \\ 0 \\ \vdots \\ 0 \end{bmatrix}$$

This corresponds to a prediction error filter, estimating the actual signal from the previous ones and subtracting the estimate from the measured signal. This principle of estimation and subtraction was discussed in chapter 3, section 3.5.3. In this case we are looking for signals from the direction defined by beamforming vector b, combined with a two-dimensional suppression of clutter signals, but without the final Doppler filtering. Then the rule for filtering the signal z^* from clutter becomes:

$$z^* Q^{-1} Be = z^* \begin{bmatrix} K_0 b \\ K_1^* b \\ \vdots \\ K_{M-1}^* b \end{bmatrix} \tag{13.12}$$

Each partial matrix $K_m^* b$ becomes a column of the length N. The complete filter vector has the length NM. So we reduce the number of complex multiplications for the clutter suppression from $(NM)^2$ to NM. This filter has the characteristic of minimising the output clutter interference power for a steady data stream in time, for which the input interference process is characterised by Q. The estimation of the actual interference signals takes place by means of the past clutter interference signals taking into account their correlation in time and space. The estimated clutter interference signals are then subtracted from the actual measured signals. From this a temporal sequence of signals without clutter interference results, which is then processed by a Doppler filter bank.

In contrast, multiplication with the complete matrix Q^{-1} [NM, NM] according to equation 13.5 results in mutual interference cancellation for the whole $N \times M$ signal field or block, thus without the assumption of a temporal stationary data stream.

The above mentioned estimation of the actual clutter signal from the past or surrounding samples is possible if the sample rate fulfils the sampling theorem discussed in chapter 6 equation 6.21 (Nyquist condition) in space and time [7]. In the Doppler domain the pulse frequency PRF has to be higher than the maximum clutter Doppler

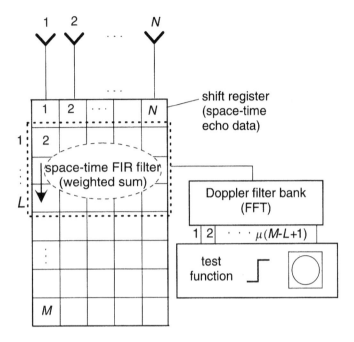

Figure 13.7   Block diagram FIR-STAP (courtesy R. Klemm) © IEE 1998 [8]

frequency resulting from the platform motion. In space the antenna elements must be on a grid for an unambiguous beam in space, that is an element spacing $d = \lambda/2$ (see chapter 4). This statement explains the basic function of STAP, which is equally valid for forward or sideways-looking arrays.

For a further simplification we may reduce the value M, the number of temporal samples for the FIR filter, to a lower value L. It can be selected independent of the pulse number M for the following Doppler filtering. From computations and simulations satisfactory results for L equal to only 2 or 3 have already resulted [8]. The coefficients of the FIR filter have to be computed and applied for each beam direction individually.

The configuration with the FIR filter then replaces matrix multiplication and needs only a following Doppler filter bank, realised by an FFT. This concept is shown in Figure 13.7.

13.6   Suboptimal concept with subarrays: reduction of spatial dimension

A further reduction of the complexity may be achieved by a spatial transformation [8] to reduce the number of spatial degrees of freedom or the number of antenna channels. This is especially of interest for arrays with larger element numbers. This

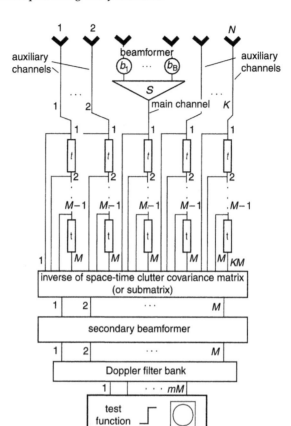

Figure 13.8   Symmetric auxiliary sensor processor (courtesy R. Klemm) © IEE
1998 [8]

transform is of the form:

$$
\mathbf{T} = \begin{bmatrix} \mathbf{T}_s & & & \\ & \mathbf{T}_s & 0 & \\ & 0 & \ddots & \\ & & & \mathbf{T}_s \end{bmatrix}
\tag{13.13}
$$

There are M submatrices $\mathbf{T}_s[N, K]$, therefore matrix $\mathbf{T}$ has the dimension $[NM, KM]$. Each has as many columns K as spatial channels are required after the transform and each column contains the weighting elements to form a subarray.

For linear arrays a symmetrical combination of the antenna channels according to Figure 13.8 has proven to be a very effective suboptimal processing configuration. The central antenna elements are summed by a beamformer with weights $b_i(i = 3, \ldots, N - 2)$, the outer antenna elements serve as auxiliary channels, two on each side. Thus the dimension of the covariance matrix is substantially reduced from

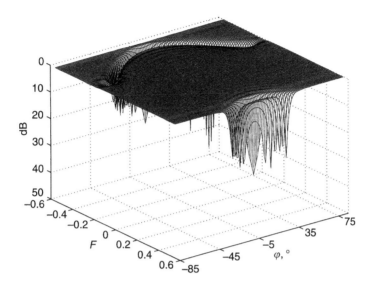

Figure 13.9   Filter response above sin φ, fd plane (forward looking) (courtesy R. Klemm) © IEE 1998 [8]

N to $K = 5$. In this case the transform submatrix has the dimension $[N, 5]$. For the matrix $\mathbf{T}_s$ we have:

$$\mathbf{T}_s = \begin{bmatrix} 1 & 0 & 0 & 0 & 0 \\ 0 & 1 & 0 & 0 & 0 \\ \cdot & \cdot & b_3 & \cdot & \cdot \\ \cdot & \cdot & \vdots & \cdot & \cdot \\ \cdot & \cdot & b_{N-2} & \cdot & \cdot \\ 0 & 0 & 0 & 1 & 0 \\ 0 & 0 & 0 & 0 & 1 \end{bmatrix} \tag{13.14}$$

This configuration for a reduction of the spatial channels may be combined with the FIR filter concept of the previous section. So the dimension of the filter for clutter suppression reduces to $K \times L = 5 \times 3$, which gives a good chance for realisation.

The effectiveness of this simplified concept may be compared with the optimum solution from the respective gain curves. Figure 13.9 shows the gain above the (φ, f_d) plane. In comparison with Figure 13.3 we recognise a narrow ditch above the clutter ellipse. All gain curves are normalised to $CNR \times NM$, with CNR for the clutter-to-noise ratio at the input.

As an example, in Figure 13.10 gain curves are shown as a cut for $\varphi = const$ through the gain function above the (φ, f_d) plane. Here, a gain comparison is made between the optimum processor, computed for $N = 24$ and $M = 24$, and the sub-array processor with five subarrays. There are only very small differences in the neighbourhood of the clutter notch.

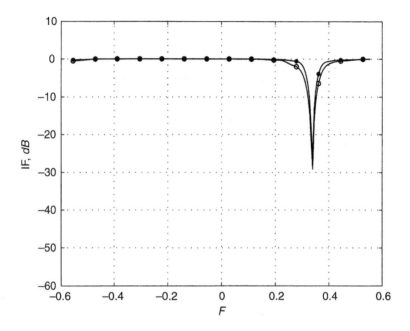

*Figure 13.10   Gain curve for auxiliary sensor concept (K = 5) (circled line), opti-
mum processor (starred line), (courtesy R. Klemm), © IEE 1998 [8]*

A future multifunction radar for airborne applications will have a planar array at
the aircraft nose looking forward. The array will probably be divided into subarrays to
achieve adaptive jammer suppression, as explained in chapter 10. All subarrays sum
up neighbouring receiving channels. The phase shifters in each channel are steered
for the desired look direction. Based on this subarray architecture STAP processing
is possible as illustrated in Figure 13.11. By combining some neighbouring subarrays
to super-subarrays, according to the example above, an adequate number of spatial
degrees of freedom may be provided. The gain has been computed [8] for different subarray architectures and subarray numbers. A typical example is shown in Figure 13.12:
a narrow notch in the gain function is achieved for a number of super-subarrays of
eight or 16. These have been formed by combination out of 32 subarrays. But for only
four super-subarrays the gain curve is insufficient, because there are by projection on
the horizontal plane only two degrees of freedom available for the azimuth direction.

In practice for beamforming a tapering function (Taylor, Bayliss function, see
chapter 4) is applied to achieve low sidelobes. For this case the gain function has
also been computed [7] and this is shown in Figure 13.13. A narrow notch is also
achieved here.

Conventional processing would first form a beam in the look direction, followed
by a Doppler filter bank. For a forward-looking array the gain function has been
computed [8] for comparison; this is also shown in Figure 13.13. We recognise a much
broader notch for the conventional beamformer and Doppler filter, so the detection

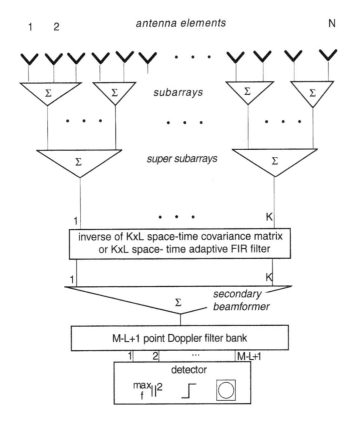

Figure 13.11   Subarray FIR filter concept with super subarrays (courtesy R. Klemm)
© IEE 1998 [8]

especially of slow targets with their Doppler spectrum near the clutter frequencies
would be severely degraded compared to STAP.

There have been published suggestions [9] for applying STAP processing to antennas with monopulse capability, which means using only the Σ and Δ beams. This
corresponds to only two spatial degrees of freedom. For forward-looking STAP the
number of degrees of freedom is too low to give acceptable results, as demonstrated
in Figure 13.12, where with four subarrays a broad notch appears. Only for the side-
looking case, section 13.8, can satisfying results be expected. But if degrading effects
from nearfield scatterers, channel mismatch or internal motion occur then the Σ and Δ
beam concept will not be a sufficient basis for making use of the advantages of STAP.

13.7   Adaptive processing

The matrix $\mathbf{Q}$ can be computed from knowledge of the antenna geometry and the flight
parameters (platform movement) for the respective range. Then efficient filtering is

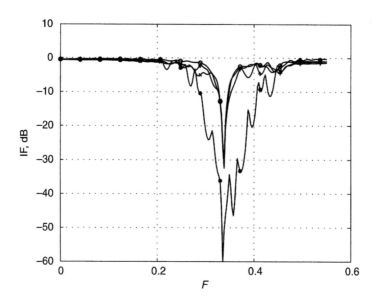

Figure 13.12   Planar forward-looking array, gain with subarrays, 32 subarrays (circled line), 16 subarrays (crossed line), 8 subarrays (plus line), 4 subarrays (starred line), (courtesy R. Klemm), © IEE 1998 [8]

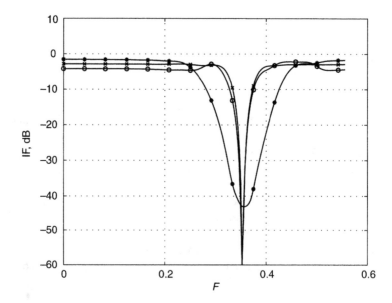

Figure 13.13   Linear array with tapering, optimum processing (crossed line), STAP FIR filter (circled line), beamformer (starred line), Doppler filter (plus line) (courtesy R. Klemm) © IEE 1998 [8]

achievable, although it should be aimed at an adaptive operation in which an estimate of $\mathbf{Q}$ is computed and renewed from measured data. The principles discussed in chapter 10 apply. Because the number of data snapshots or training data (secondary data) should be at least twice the dimension of the matrix $\mathbf{Q}$, it is favourable for the adaptation process to use only as few degrees of freedom as adequate. This reduction of matrix dimension is accomplished by the suggestion of suboptimal concepts discussed above, using the FIR filter and subarrays.

This statement is especially important in nonstationary clutter scenarios, because in this case the estimation of $\mathbf{Q}$ can be performed only in the direct neighbourhood of the cell under test and therefore with only few training data sets.

Apart from the suppression of ground and weather clutter the suppression of jammers can also be included. Additional spatial degrees of freedom have to be provided if common processing for clutter and jammer suppression is to be achieved. Another possibility is cascading jammer suppression and STAP. Jammer suppression is performed by spatial processing only, for example with 16 subarrays the jammer suppression is performed as suggested in chapter 10. Then in a second step five super-subarrays are formed providing the necessary spatial channels for STAP processing.

A special problem arises if jammers are illuminating clutter reflectors. This is called multipath jamming or hot clutter, and additional degrees of freedom are necessary [10,11]. From this adaptation auxiliary beams looking to the jammers are formed. These beams give an estimate for the jammer signal which has to be subtracted after adequate weighting for the direction with the illuminated clutter cell. This is automatically accomplished by estimating $\mathbf{Q}$ and applying $\mathbf{Q}^{-1}$. The weighting will also apply fast time components, that is for neighbouring range elements, to compensate for multipath delay effects.

Other degrading effects which cause a decorrelation and therefore widening of the filter ditch are:

- channel mismatch
- wide signal bandwidth
- range ambiguity
- internal clutter motion

Channel matching will be improved for future systems by IF sampling and digital lowpass filtering as described in chapter 6. Wide signal bandwidth will be handled by extending the filtering process in the fast time domain (with sampling according to bandwidth), or in other words, the range dimension. Broadband beamforming is discussed in chapter 5, section 2.

13.8 Sideways-looking radar

Reconnaissance systems with a linear antenna array along the flight vector, e.g. on a missile, a drone or an aircraft, were built or conceived and have been studied by several authors [8]. According to Figure 13.2 the velocity vector is now $\mathbf{v} = (0, v_y, 0)$.

The Doppler shift in direction r^0 is now:

$$f_d = \frac{2v_y}{c} \sqrt{1 - \left(\frac{h}{r}\right)^2} \sin\varphi \qquad (13.15)$$

Thus the clutter spectrum of ground clutter is concentrated above a rising straight line in the $(\sin\varphi, f_d)$ plane. This clutter power distribution is then still modulated with the transmit beam pattern and the receiving element pattern.

The associated gain for the suppression of this clutter distribution according to equation 13.6 shows a narrow linear ditch. Similar conclusions as for the forward-looking case are valid with respect to the degrees of freedom in space and time.

13.9   Conclusions

To detect slowly moving targets, which always have their spectrum near the clutter spectrum, the ditch of the processing gain function, necessary for clutter suppression, must be as narrow as possible, matched to the clutter spectral width. This is best accomplished by STAP processing with sufficient degrees of freedom. These spatial and temporal degrees of freedom can be provided for future airborne phased-array radars by an adequate number of subarrays or super-subarrays and an FIR filter in the time domain. The final Doppler filter bank then follows.

13.10   References

1 BRENNAN, L. E., MALLETT, J. D., and REED, I. S.: 'Adaptive arrays in airborne MTI', *IEEE Trans. Antennas Propag.*, 1976, **24** (5), pp. 607–615
2 KLEMM, R.: 'Adaptive clutter suppression for airborne phased array radar', *IEE Proc.*, Part F, Radar Signal Process, 1983, **130** (1), pp. 125–132
3 WARD, J. T.: 'Space-time adaptive processing for airborne radar'. Lincoln Laboratory technical report 1015, Lexington, MA, USA, 1994
4 RICHARDSON, P. G.: 'Analysis of the adaptive space time adaptive procressing technique', *IEE Proc. Radar Sonar Navig.*, 1994, **141** (4), pp. 187–195
5 RICHARDSON, P. G., and HAYWARD, S. D.: 'Adaptive space-time processing for forward looking radar'. Proceedings of IEEE international conference on *Radar*, Alexandria, VA, USA, 1995, pp. 629–634
6 KLEMM, R.: 'Adaptive airborne MTI: comparison of sideways and forward looking radar'. IEEE international conference on *Radar*, Alexandria, VA, USA, 1995, pp. 614–618
7 KLEMM, R.: 'Adaptive airborne MTI with tapered antenna arrays', *IEE Proc. Radar Sonar Navig.*, 1998, **145** (1), pp. 3–8
8 KLEMM, R.: 'Space-time adaptive processing – principles and applications' (IEE, London, UK, 1998)

9 WANG, H., ZHANG, Y., and ZHANG, Q.: 'An improved and affordable space-time adaptive processing approach', Proceedings of CIE international conference on *Radar*, Beijing, China, October 1996, pp. 72–77

10 FANTE, R., and TORRES, J.: 'Cancellation of diffuse jammer multipath by an airborne radar', *IEEE Trans. Aerosp. Electron. Syst.*, 1995, **31** (2), pp. 805–820

11 MARSHALL, D., and GAVEL, R.: 'Simultaneous mitigation of multipath jamming and ground clutter'. Proceedings of *Adaptive sensor array processing workshop*, MIT Lincoln Laboratory, 1996, vol.1, pp. 193–239.

Chapter 14

Synthetic aperture radar with active phased arrays

Synthetic aperture radar (SAR) with coherent focussing has already been proposed, in 1953 by C. W. Sherwin [1]. It allows imaging of the ground scene in all weather conditions with high resolution using a radar on board a flying platform or satellite. The first actual systems applied optical processing for image generation. Since then many developments have achieved finer resolution and faster processing of received SAR signals by digital processing.

New and extended capabilities can be achieved by applying an active phased-array radar for SAR operation, including array signal processing methods discussed in the preceding chapters. Opportunities are particularly good for the task of detecting, locating and imaging moving objects. The suppression of jamming may also be achieved.

14.1 Basic principle of SAR

As an introduction, the principle of conventional SAR with a single antenna and receiving channel will be explained.

For imaging of a ground scene a high resolution in range (in the direction of the line of sight) and cross range (orthogonal to the line of sight and parallel to the flight path), both independent of range, is necessary and achievable. Resolution values of the order of metres or even sometimes decimetres are required. Range resolution is achieved by applying an appropriate signal modulation to the transmitted pulse, as with other radar systems. This may be a short pulse with a length according to the required range resolution. More popular is linear frequency modulation (chirp) with a frequency shift determined by the required bandwidth reciprocal to the range resolution, as discussed in chapter 7. Then, by pulse compression, the required range resolution is achieved.

For the cross range or tangential dimension x an extremely narrow beam width Θ_s, produced by a synthetic array, is necessary. Because the cross-range resolution,

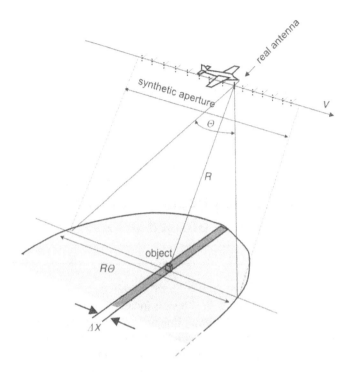

Figure 14.1   Principle and geometry of synthetic aperture radar

given by $\Delta x = R\Theta_s$, should be independent of range R, the beamwidth Θ_s should be proportional to $1/R$. This requirement can be fulfilled by a radar on board an airplane as shown in Figure 14.1. The antenna of this radar is pointing with its beam axis orthogonal to the flight direction to the ground scene and continuously illuminates with its footprint a strip parallel to the flight path. Pulses are transmitted at a certain pulse recurrence frequency, PRF, as is usual with other radars, and the echoes are received and stored. Now the trick is to process the series of stored echo signals as if they were all received at the same instant of time. Because these echoes are received during a certain flight path they correspond to virtual antenna elements at those positions, where the pulses have been transmitted and received, corresponding to the respective positions of the airplane on its flight path. These virtual antenna positions (dashed in Figure 14.1) thus form a synthetic aperture. The length of this linear aperture may be hundreds or even thousands of metres, dependent on the system parameters. Of course, such a long aperture could never be realised with a real aperture.

This synthetic aperture may be as long as required for cross-range resolution. The maximum length is given by the extension of the footprint produced by the real beam illuminating the ground. Single objects are illuminated during the passage of this footprint and their stored echoes may be used to form the synthetic array.

The beamwidth of the real antenna is constant and therefore the footprint extension becomes proportional to range R.

The antenna at the airplane, in Figure 14.1 indicated for simplicity as a dipole, is usually a reflector antenna with a certain dimension L in the flight direction, resulting in a real 3 dB beamwidth, Θ, of approximately:

$$\Theta = \frac{\lambda}{L} \tag{14.1}$$

The resolution Δx follows with the synthetic aperture length $L_S = R\Theta$ and taking into account the two-way transmit and receive beamforming:

$$\Delta x = R\frac{\lambda}{2R\Theta} = \frac{L}{2} \tag{14.2}$$

The cross-range resolution is therefore independent of range as required and is given by the length of the real aperture.

The time corresponding to the synthetic aperture, T_{sa}, depends on the airplane's velocity v:

$$T_{sa} = \frac{R\Theta}{v} = \frac{R\lambda}{2\,\Delta x v} \tag{14.3}$$

T_{sa} increases with range R and is reciprocal to Δx. The length of the synthetic aperture is:

$$L_{sa} = vT_{sa}$$

The signal processing for generating the image is performed in two steps. The received signals are sampled according to the bandwidth and stored in a two-dimensional memory as shown in Figure 14.2. The two dimensions are azimuth and range. In the time scale we may term them slow time for sampling with the PRF and fast time for sampling according to the bandwidth or range resolution, respectively. The echo samples following a transmit pulse are stored as columns in the memory as indicated in Figure 14.2. Each column represents one radar period.

Objects wandering through the illuminating beam change their line of sight distance: it decreases from a maximum value when entering the beam or footprint to a minimum value at the boresight position and then increases again until leaving the beam.

According to Figure 14.1 we observe the range development for a single object as indicated in Figure 14.3, showing the geometry in a plane determined by the line of sight to the object and the airplane's flight vector. The object has coordinates x, y in a coordinate system fixed to the flying platform. The boresight distance to an object is assumed to be R_0. It may be passed at time $t = 0$. On the right-hand side of the diagram ($x > 0$) we have $t < 0$. The direction to the object is given by the angle from boresight, $\varphi(t)$. The time dependent range, $R(t)$, is then given by:

$$R(t) = \sqrt{x(t)^2 + R_0^2} = \sqrt{(vt)^2 + R_0^2}$$

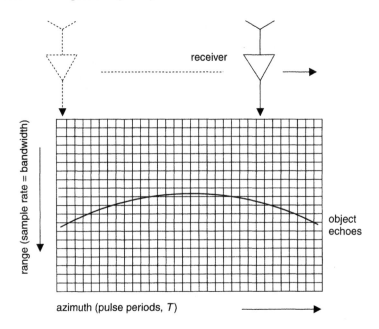

Figure 14.2   Received echoes in the memory field

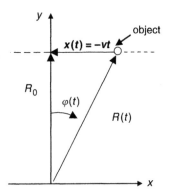

Figure 14.3   Geometry for SAR passing an object

This may be approximated using a Taylor series (for small x: $\sqrt{1+x} \approx 1 + x/2$):

$$R(t) = R_0 + \frac{1}{2}\frac{x(t)^2}{R_0} + \cdots \tag{14.4}$$

After the application of pulse compression the echoes of a single object are therefore on a parabolic curve, as shown in Figure 14.2. For the radial velocity follows:

$$\dot{R}(t) = -v\frac{x(t)}{R(t)} = -vu(t) \quad \text{with } u(t) = \sin\varphi(t) \tag{14.5}$$

The $-$ sign follows with Figure 14.3: for positive values of x the value of R is decreasing. Together with $w(t) = y/R(t) = R_0/R(t) = \cos \varphi(t)$ we have the directional unity vector $\mathbf{u}$:

$$\mathbf{u}(t) = \begin{bmatrix} u(t) \\ w(t) \end{bmatrix} = \begin{bmatrix} \sin \varphi \\ \cos \varphi \end{bmatrix} \tag{14.6}$$

The distance change causes an approximately linear decreasing Doppler frequency shift. The maximum value of the Doppler frequency occurs at entering the beam. At the 3 dB points of the real antenna beam the velocity component of scatterers at the ground relative to the flying platform is $\pm v \sin (\Theta/2)$. Taking into account equation 14.2 the Doppler frequency at these points is given by:

$$f_{D_3\,\text{dB}} = \pm \frac{2v}{\lambda} \sin \left(\frac{\Theta}{2} \right) \approx \pm \frac{2v}{\lambda} \frac{\Theta}{2} = \pm \frac{v}{2\Delta x} \tag{14.7}$$

To avoid azimuth ambiguities, the sampling frequency with respect to the azimuth (slow time) has to be greater than or equal to the Doppler bandwidth $B_D = v/\Delta x$. The maximum Doppler frequency entering over the real antenna sidelobes is determined by the flight velocity v:

$$\left| f_{D_{\text{max}}} \right| = \frac{2v}{\lambda}$$

For an object passing the boresight position at $t = 0$ we can write the expected signal function dependent on time with the two-way pattern of the real antenna, $D(u(t))$, and the signal amplitude, a, by:

$$s(t) = a \exp \left(-j \frac{4\pi}{\lambda} R(t) \right) D(u((t))) \tag{14.8}$$

We will use the abbreviation $\beta = 2\pi/\lambda$. With equation 14.4 and a_1 containing a fixed phase term at R_0 we get:

$$s(t) \approx a_1 \exp \left(-j\beta \frac{v^2}{R_0} t^2 \right) D(u(t)) \tag{14.9}$$

This is the equation of a linear frequency-modulated signal (azimuth chirp), because of the quadratic phase with time. For image generation by azimuth compression the received signal belonging to a range element has to be filtered or convolved with the complex conjugate function $s^*(t)$. All objects produce the same signal series with a shift in time corresponding to their respective azimuth positions and will be compressed to image points at their azimuth positions as desired. This is, in principle, a continuous process, new signal columns enter the memory (Figure 14.2) and the oldest are deleted. For each range element the signals are passing the filter function $s^*(t)$ for azimuth compression, generating the image of the observed strip. This filter operation is discussed in principle in chapter 2, section 2.4.

The convolution of the received signal with $s^*(t)$ may be achieved more economically by using a Fourier transform (FFT) as described in chapter 2, section 2.4, which results in a processing for blocks of received data.

Before the azimuth compression is performed the range walk according to the parabolic distance function has to be corrected. According to equation 14.4 and 14.5 the values $R(t)$ and $\dot{R}(t)$ are coupled by $x(t)$. In other words, to each Doppler frequency shift given by $\dot{R}(t)$ belongs a certain range displacement $R - R_0$. The Doppler shift is computed by an FFT of signal segments in the azimuth dimension. The corresponding range displacement is applied to these frequency parts of the signals for range walk correction. After back transformation into the time domain the range walk correction is achieved.

The width of the illuminated strip is determined by the real beamwidth in elevation of the SAR antenna aboard. The maximum range is additionally limited by the radar's transmit power and the other radar parameters according to the radar equation. Because the signal integration time is proportional to R, we have to note for the signal-to-noise-ratio: SNR $\sim 1/R^3$.

So far we have assumed an ideal flight in a straight line with known and constant velocity v above ground. In practice there are deviations from this ideal flight path by manoeuvres and wind influences and turbulences. The focusing of the synthetic aperture for each range element requires phase accuracies corresponding to only fractions of the wavelength. Therefore additional focusing procedures are necessary to achieve high-resolution imaging. The actual flight path may be determined by differential GPS system and/or a precise inertial navigation system. Autofocus techniques may be applied additionally [2].

The relations discussed so far shall now be illustrated by a parameter example for a typical modern SAR system. We assume as requirements:

$$\Delta x = 0.5\,\mathrm{m}, \qquad \lambda = 0.03\,\mathrm{m}, \qquad R = 50\,\mathrm{km}, \qquad v = 200\,\mathrm{m/s}$$

Then follows with equation 14.3 for the aperture time:

$$T_{sa} = 7.5\,\mathrm{s}$$

with equation 14.4 for the range walk:

$$\Delta R = R(T_{sa}/2) - R_0 = 5.625\,\mathrm{m}$$

with equation 14.7 for the Doppler shift of fixed objects at the 3 dB beam points:

$$f_{D3\,\mathrm{dB}} = \pm 200\,\mathrm{Hz}$$

The range walk amounts therefore to about 11 range cells and has to be corrected. The pulse period for an unambiguous range is 330 μs. This represents an azimuth sampling frequency of 3 kHz, which is well above the 400 Hz as the Doppler bandwidth. The number of signals in the azimuth dimension within the 3 dB width of the real beam is $N = 22\,522$. The length of the synthetic aperture for maximum range vT_{sa} is 1500 m.

14.2 Problems of moving-target detection and location

Fixed objects wandering through the real beam cause a signal with a linear decreasing Doppler shift (down chirp) according to equation 14.9. The zero crossing point determines, after azimuth compression, the azimuth position of the image point. This is shown in Figure 14.4. A moving approaching object with a velocity vector orthogonal to our flight path will show an additional Doppler shift. This is indicated in Figure 14.4 by the dashed fat line. This signal is similar, to a large extent, to the echo signal from a fixed object which is shifted in time or azimuth. Therefore, by azimuth compression the moving object will be displayed with a displaced azimuth position. Pictures with trains or cars located by this displacement beside railways or streets have been generated. The reason for this effect is the lack of an independent direction estimation. For the so far discussed SAR with one antenna and receiving channel the azimuth position is estimated only from the course of the Doppler shift with time and there is, as discussed, a coupling of Doppler shift and azimuth position. For moving targets with velocity components also parallel to the flight vector there will be a change of the resulting Doppler slope and there is no effective azimuth compression because of the mismatch between the received signal and the reference $s^*(t)$.

If only moving targets are to be detected then the frequency band which is occupied by the random ensemble of fixed objects or clutter, indicated by the hatched strip, must be suppressed by signal processing. The width of this clutter band is determined by the antenna pattern of the real antenna, with its main lobe and also the sidelobes. A clutter suppression filter would have to suppress at least the frequency band of the main lobe clutter. This filter would then pass only the signal of a moving target with a velocity high enough so that its main part of the Doppler spectrum is above or below the clutter band.

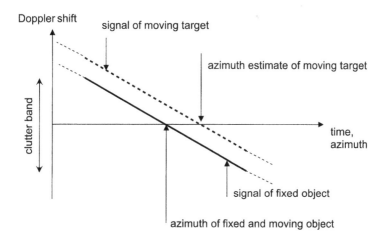

Figure 14.4   Doppler shift above azimuth for fixed and moving target

The detection of slowly moving targets in particular, e.g. vehicles on the ground, is impossible with such a one-channel SAR. The signal spectrum would be almost completely suppressed by the clutter filter. If there are moving targets with a high cross section then these could pass the detection threshold but could not be distinguished from fixed targets. Moreover the image would be blurred because the focusing is not matched properly. A correct azimuth location would also be impossible.

14.3   Clutter suppression with multichannel array radar

With an active receiving array oriented along the flight path the direction of arrival of the received wavefronts may be estimated with great accuracy, as described in chapter 11. An array divided into four subarrays, as shown in Figure 14.5, proved suitable. The phase shifters within the transmit/receive modules for each antenna element are all steered for the desired beam direction. After summing the signals from a subarray the receiving channels, which consist of downconverters and bandpass amplifiers up to analogue-to-digital conversion, follow. Processing for moving-target detection will be discussed in the following.

According to equation 14.5 the Doppler frequency shift of fixed objects is dependent on the direction u:

$$f_D(u) = \frac{2\dot{R}}{\lambda} = \frac{2v}{\lambda} u \qquad (14.10)$$

In each individual direction u only the Doppler frequency of clutter echoes has to be suppressed and not the whole clutter band. Using this approach an essential improvement for the detection of moving targets can be expected. This corresponds to the methods and results of space-time processing in chapter 13.

As demonstrated by the parameter example above, the number of echo signals, M, is very large. A transformation of the echo signals, divided into suitable data

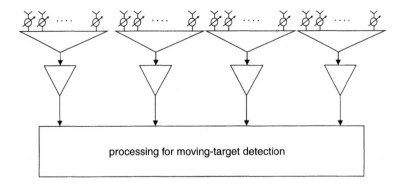

Figure 14.5   Array with subarrays for moving-target detection

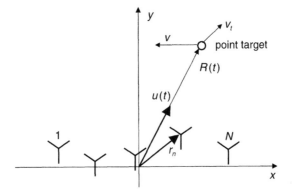

Figure 14.6   Geometry of receiving array for SAR

blocks, into the frequency domain by an FFT is therefore suitable in this case for further processing because the spectral lines become mutually independent.

Following the developments of Ender in Reference 3, first the received signal vector for a point target at the output of N antenna elements is given. The geometry is sketched in Figure 14.6.

The scalar product $\mathbf{ur}_n$ of the antenna element (or subaperture phase centre) coordinates $\mathbf{r}_n = (r_{n_x}, r_{n_y})$ with the unit vector $\mathbf{u}$ (the length of this vector is normalised to one) in the direction of arrival determines the relative phases at the antenna elements. The N antenna elements have the two-way directional pattern $D_n(\mathbf{u})$. The signal vector at the antenna elements (antenna vector) for a normalised signal or wavefront from direction $\mathbf{u}$ is then given by:

$$\mathbf{a}(\mathbf{u}) = (D_n(\mathbf{u}) \exp{(j\beta \mathbf{ur}_n)})_{n=1}^{N} \qquad (14.11)$$

The signal produced by a fixed point target at the ground according to equation 14.9 is then:

$$\mathbf{s}(t) = s_0 \exp{\left(-j\beta \frac{v^2}{R_0} t^2\right)} \mathbf{a}(\mathbf{u}(t)) \qquad (14.12)$$

We have to emphasise that $\mathbf{s}(t)$ is now an N-dimensional vector. For a moving target crossing the y axis at time $t = 0$ and with velocity vector $v_t = (v_x, v_y)$, that is with velocity components v_x (along or parallel to the flight vector) and v_y (cross to the flight vector), the signal vector $\mathbf{s}_t$ becomes:

$$\mathbf{s}_t(t) = s_{t_0} \exp{\left(-j\beta\left(2v_y t + \frac{(v - v_x)^2}{R_0} t^2\right)\right)} \mathbf{a}(\mathbf{u}(t)) \qquad (14.13)$$

The vector (v_x, v_y) adds to the velocity vector $\mathbf{v}$ of the radar platform.

Within each range element within the corresponding strip a continuous reflectivity on the ground can be described by a function $p(t)$. This would be measured by an infinitesimally narrow fixed boresight beam onboard our radar platform and thus scanning the ground by flying along the strip.

Therefore the received clutter process, $\mathbf{w}(t)$, results from a convolution of $\mathbf{s}(t)$ with $p(t)$ and the added receiver noise vector $\mathbf{n}$:

$$\mathbf{w}(t) = \mathbf{s}(t)^* p(t) + \mathbf{n}(t) \tag{14.14}$$

The clutter process $p(t)$ may be described by its autocorrelation function. This may be transformed with the Wiener Khintchine relation (chapter 2, equation 2.35) into the frequency domain resulting in the spectral power density $P(\omega)$. This is a scalar function. The Fourier transforms of $\mathbf{s}(t)$, $\mathbf{w}(t)$ and $\mathbf{n}(t)$ are $\mathbf{S}(\omega)$, $\mathbf{W}(\omega)$ and $\mathbf{N}(\omega)$, respectively, which are all column vectors with N components. From equation 14.14 follows for the cross-spectral matrix describing the N-dimensional received clutter process in the frequency domain $\mathbf{W}(\omega)$:

$$\mathbf{C}_w(\omega) = \mathbf{S}(\omega) P(\omega) \mathbf{S}^*(\omega) + \mathbf{C}_N(\omega) \tag{14.15}$$

The matrix $\mathbf{C}_w(\omega)$ describes the passing clutter scene within the illuminated strip at the output of the array's receiving channels for one range element.

All time functions are sampled with the radar's pulse period T. If M samples in the time domain are applied we get also M spectral lines with an unambiguous interval $\Delta\omega = 2\pi/T$ and a spectral resolution of $2\pi/MT$ and frequencies $2\pi m/MT$ for $m = 1, \ldots, M$. We would have to compute M matrices $\mathbf{C}_w(\omega_m)$ with dimension $[N, N]$.

A signal from a moving target is already given by equation 14.13. A parameter set ϑ shall describe the target's motion:

$$\mathbf{s}(t, \vartheta) = b \, \exp\left(-j\beta R(t, \vartheta)\right) \mathbf{a}(\mathbf{u}(t, \vartheta)) \tag{14.16}$$

The received or measured signal vector (N components), corresponding to one range element, from clutter, noise and possibly from moving targets, is denoted $\mathbf{z}(t)$ and its Fourier transform by $\mathbf{Z}(\omega)$. Supposing the components of $\mathbf{Z}$ are jointly Gaussian distributed because of the receiver noise, then the test variable T according to the likelihood ratio test for the detection of a target with the parameter set ϑ (chapter 3 section 3.3, equation 3.50) is given by:

$$T(\mathbf{z}_1, \ldots, \mathbf{z}_M) = \left| \sum_{m=1}^{M} \mathbf{S}(\omega_m, \vartheta)^* \mathbf{C}_w^{-1}(\omega_m) \mathbf{Z}(\omega_m) \right| \tag{14.17}$$

For $T > \eta$ a moving target is accepted, with η as the decision threshold. The test contributions from all m frequencies are summed up with this equation, because a moving target shows a Doppler frequency variation during illumination by the real aperture. The magnitude is taken because of the unknown phase of the received signal.

Clutter suppression is accomplished in equation 14.17 by the operation $\mathbf{C}_w^{-1}(\omega_m)\mathbf{Z}(\omega_m)$. This is a well-known expression, discussed in chapter 3 and already applied for clutter or other interference suppression. The result from $\mathbf{C}_w^{-1}\mathbf{Z}$ is integrated over all Doppler frequencies of interest as is usual with matched filtering.

For white noise and large M the frequency components are mutually independent for different receiving channels or frequencies and we may assume for the cross-spectral matrix for noise $\mathbf{C}_N(\omega) = \sigma^2\mathbf{I}$. By application of the matrix inversion formula equation 2.17 we get from equation 14.15, keeping in mind the dependence of $\mathbf{C}_w$ on ω:

$$\mathbf{C}_w^{-1} = \mathbf{C}_N^{-1} - \mathbf{C}_N^{-1}\mathbf{S}(\mathbf{S}^*\mathbf{C}_N^{-1}\mathbf{S} + P^{-1})^{-1}\mathbf{S}^*\mathbf{C}_N^{-1}$$

$$\mathbf{C}_w^{-1} = \frac{1}{\sigma^2}\mathbf{I} - \frac{1}{\sigma^2}\mathbf{I}\mathbf{S}\left(\frac{P\mathbf{S}^*\mathbf{S} + \sigma^2}{P\sigma^2}\right)^{-1}\mathbf{S}^*\frac{1}{\sigma^2}\mathbf{I}$$

$$= \frac{1}{\sigma^2}\left(\mathbf{I} - \frac{\mathbf{S}\mathbf{S}^*}{\sigma^2/P + \mathbf{S}^*\mathbf{S}}\right)$$

For a high clutter level compared to noise the ratio σ^2/P tends to zero and $\mathbf{C}_w^{-1}$ simplifies finally to:

$$\mathbf{C}_w^{-1} = \frac{1}{\sigma^2}\left(\mathbf{I} - \frac{\mathbf{S}\mathbf{S}^*}{\mathbf{S}^*\mathbf{S}}\right) \qquad (14.18)$$

This is a projection matrix in a space orthogonal to the clutter space defined by column vector $\mathbf{S}$. The relationship between the inverse of an interference covariance and a projection matrix has already been discussed in chapter 10 in connection with jammer suppression.

For explanation: if $\mathbf{Z}(\omega)$ contains a clutter component $\mathbf{S}$ we get:

$$\mathbf{C}_w^{-1}\mathbf{S} = \frac{1}{\sigma^2}\left(\mathbf{I} - \frac{\mathbf{S}\mathbf{S}^*}{\mathbf{S}^*\mathbf{S}}\right)\mathbf{S} = \frac{1}{\sigma^2}\left(\mathbf{S} - \frac{\mathbf{S}(\mathbf{S}^*\mathbf{S})}{\mathbf{S}^*\mathbf{S}}\right) = \frac{1}{\sigma^2}\left(\mathbf{S} - \mathbf{S}\right) = 0$$

For each Doppler frequency ω a narrow notch in the array pattern is produced by this projection for the specific direction of the respective clutter vector $\mathbf{S}(\omega)$.

The signal-to-clutter-plus-noise ratio (SNCR) is generally given according to chapter 3 equation 3.85 by:

$$\mathrm{SNCR} = \mathbf{S}^*\mathbf{C}^{-1}\mathbf{S}$$

and in this case for a signal with parameter set ϑ by:

$$\mathrm{SNCR}_{filter}(\vartheta) = \sum_{m=1}^{M}\mathbf{S}^*(\omega_m, \vartheta)\,\mathbf{C}_w^{-1}(\omega_m)\,\mathbf{S}(\omega_m, \vartheta) \qquad (14.19)$$

The signal vector $\mathbf{S}(\omega_m, \vartheta)$, perhaps from a moving target, may be proportional to any antenna vector or direction of arrival (DOA) vector $\mathbf{a}(\mathbf{u})$. Only those signals at frequency ω_m with an $\mathbf{a}(\mathbf{u})$ corresponding to the clutter frequency are suppressed by

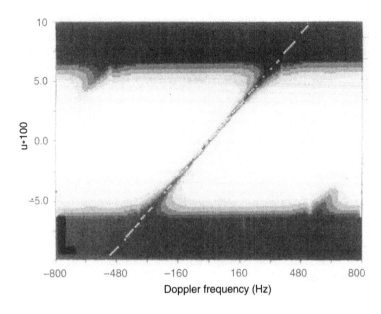

Figure 14.7   Space-frequency filter characteristic for moving target detection, white 0 dB, each grey level −5 dB (courtesy J. Ender)

$C_w^{-1}(\omega_m)$, all others are passed and may contribute to a detection. The resulting filter function may be expressed by:

$$f(u, \omega) = \mathbf{a}^*(\mathbf{u})\,\mathbf{C}_w^{-1}(\omega)\,\mathbf{a}(\mathbf{u}) \qquad (14.20)$$

This two-dimensional filter function is shown as an example in Figure 14.7. The contour plot shows a deep ditch for the clutter signals. During the scanning and sampling process by the real radar beam a target would deliver contributions for the summation according to equation 14.17, dependent on its path through this (u, ω) plane. Fixed targets would wander along the ditch and be suppressed. This corresponds to the space-time filter discussed in chapter 13.

The matrix $\mathbf{C}_w$ may be computed with equations 14.12 and 14.15 assuming a model for $P(\omega)$ and a known platform velocity above ground. If there are unknown influences to the platform motion, the matrix $\mathbf{C}_w$ has to be estimated from measurements of clutter signals:

$$\hat{\mathbf{C}}_w(\omega_m) = \frac{1}{K}\sum_{k=1}^{K}\mathbf{Z}_k(\omega_m)\,\mathbf{Z}_k^*(\omega_m) \quad \text{for } m = 1, \dots, M \qquad (14.21)$$

This is the sample matrix inversion (SMI) method discussed in chapter 10. The samples may be taken from a set of K range elements. To improve the numerical stability a diagonally-loaded matrix may be applied as discussed in chapter 10 section 10.4.1.

The $\mathbf{Z}_k(\omega_m)$ are computed from data segments long enough, for example $M = 256$ or 512, to achieve mutually uncorrelated values $\mathbf{Z}_k(\omega_m)$ and $\mathbf{Z}_k(\omega_n)$ for $m \neq n$.

The matrix $\hat{\mathbf{C}}_w(\omega_m)$, with dimension $[N, N]$, may be represented by its N eigenvalues $\lambda_n(\omega_m)$ in the diagonal matrix $\Lambda(\omega_m)$ and the eigenvector matrix $\mathbf{X}(\omega_m)$ with column vectors $\mathbf{x}_n(\omega_m)$, as discussed in chapter 10, section 10.2.3, equation 10.13:

$$\hat{\mathbf{C}}_w(\omega_m) = \mathbf{X}(\omega_m)\,\Lambda(\omega_m)\,\mathbf{X}(\omega_m)^* \tag{14.22}$$

The dominating eigenvalue λ_1 corresponds to the ground clutter. This is demonstrated in Figure 14.8 for the $N = 4$ channels of the experimental AER system, described in section 14.8. These four channels correspond to the antenna subarrays as shown in Figure 14.5; the courses of the four eigenvalues with the Doppler frequency are shown. The ratio of λ_1 to λ_2 indicates the achievable clutter suppression. The eigenvector $\mathbf{x}_1(\omega_m)$ connected to the dominating eigenvalue $\lambda_1(\omega_m)$ corresponds to the clutter model signal of equation 14.12. The important difference is: the eigenvector is evaluated from real clutter measurements and therefore includes all deviations of the antenna and receiving system from the model, such as antenna element coupling, inaccuracies and mutual mismatch of receiving channels and analogue-to-digital conversion.

In Figure 14.9 the relative phases are shown within the unambiguous interval $-\pi, \ldots, \pi$ of the four components of eigenvector $\mathbf{x}_1$. There are small deviations from the ideal linear phase function assumed in the model of equation 14.12. These relative phases corresponding to the respective directions are not influenced by the

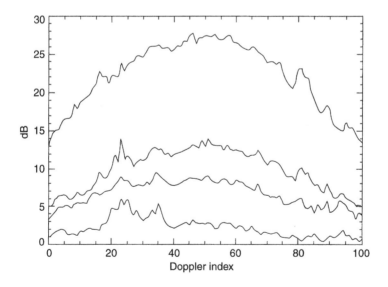

Figure 14.8   Eigenvalues dependent on Doppler frequency for $N = 4$ (courtesy J. Ender)

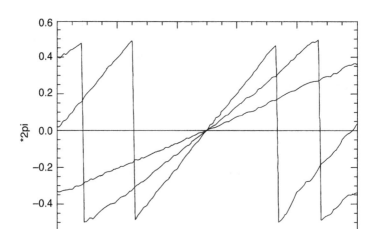

Figure 14.9   Relative phases of the elements of eigenvector x_1 (courtesy J. Ender)

Doppler frequency shift itself, because this shift is small compared to the carrier frequency, in this case 10 GHz.

Because the clutter Doppler frequency ω_k is coupled to a corresponding direction u_k according to equation 14.10, we can use the set of $x_1(\omega_k)$ as estimates of the calibrated direction of arrival (DOA) vectors for the set of directions, corresponding to the resolution given by the spectral resolution with M signal samples, with a_k as a scaling factor:

$$\hat{a}(u_k) = a_k \mathbf{x}_1(\omega_k) \quad k = 1, \ldots, M \tag{14.23}$$

This set of DOA vectors is also called the array manifold, because it describes the behaviour of the array for all directions of interest.

Because the spectral power density of the clutter area at the edges and in the sidelobe region may be smaller than that from a moving target, it is possible that by the adaptation process the target may be cancelled. This effect may be avoided by the application of coherent subspace transformation [20]. It is assumed that by a transform matrix $\mathbf{T}$ to get $\mathbf{Y} = \mathbf{TZ}$ it is possible to obtain for all frequencies the same covariance matrix and therefore the same clutter subspace. Now data at all frequencies contribute to the estimation of the covariance matrix and the target influence is minimised. This transformation may be realised for example by the matrix:

$$\mathbf{T}(f) = diag(a_n(\mathbf{u}(f_0))/a_n(\mathbf{u}(f)))_{n=1}^{N}$$

All data vectors $\mathbf{Z}$ are transformed to a centre frequency f_0, using the array manifold $\mathbf{a}(\mathbf{u})$ from the above measurement and the actual $\mathbf{u}(f)$ from the flight parameters.

14.4 Target location

A moving target may be detected during the illumination by the real beam. It then travels through the (u,ω) plane as indicated in Figure 14.7. If there are several detections they may be combined by a kind of tracking process to one target report.

We may assume moving-target detection within at least one Doppler cell at frequency ω. The measured signal vector $\mathbf{Z}(\omega)$ which resulted in the detection shall now also be used for the direction estimation. As discussed in chapter 11 section 11.4.1, to determine $\hat{u}$ we have to maximise, using equation 11.21, the expression:

$$\frac{|\mathbf{a}^*(u)\mathbf{C}_w^{-1}(\omega)\mathbf{Z}(\omega)|^2}{\mathbf{a}^*(u)\mathbf{C}_w^{-1}(\omega)\mathbf{a}(u)} = \text{maximum for } u = \hat{u} \qquad (14.24)$$

$\mathbf{C}_w^{-1}$ suppresses the ground clutter and $\mathbf{a}(u)$ is given by equation 14.11 and acts as the beam-steering vector. It has to be scanned computationally through all directions u to find the maximum. An equivalent and more efficient procedure is applying a monopulse procedure with a difference beam according to equation 11.15. This procedure assumes a well-calibrated receiving array with $\mathbf{a}(u)$ describing properly the real beam-steering vectors (array manifold). An elegant method developed by Ender [3] is using the array manifold given in the preceding section by the set $\mathbf{x}_1(\omega_k)$ from equations 14.21 and 14.22. These vectors may be used instead of the vectors $\mathbf{a}(u)$.

An alternative processing [3] results from superresolution methods: according to equation 12.22 in chapter 12 for a measured signal vector $\mathbf{Z}$ at a certain frequency the estimator S is maximised by variation or selection of $\omega_k (k = 1, \ldots, M)$ and the corresponding $\mathbf{x}_1(\omega_k)$:

$$S(\omega_k) = \frac{1}{\mathbf{Z}^*(\mathbf{I} - \mathbf{x}_1(\omega_k)\mathbf{x}_1^*(\omega_k))\mathbf{Z}} \qquad (14.25)$$

The expression $(\mathbf{I} - \mathbf{x}_1(\omega_k)\mathbf{x}_1^*(\omega_k))$ is the projection into the space orthogonal to the clutter space at frequency ω_k. If the denominator tends to zero the best matching between $\mathbf{Z}$ and $\mathbf{x}_1(\omega_k)$ is found. The direction $\hat{u}$ coupled to ω_k according to equation 14.23 is determined. An example of the course of S as a function of the Doppler frequency is given in Figure 14.10.

14.5 SAR/MTI processing

In Figure 14.11 we find summarised the processing scheme for imaging of ground-scene and moving-target detection [8,9]. The incoming continuous data stream is compressed in range with the range reference function. Using the azimuth FFT the data are transformed into the Doppler frequency domain; range walk correction is performed at this stage. The centroid is determined for the transformed clutter data. By the airplane's drift movement and by the antenna steering direction this centroid frequency may be shifted away from Doppler frequency zero. The width of the clutter band is determined from the airplane's velocity and the antenna pattern.

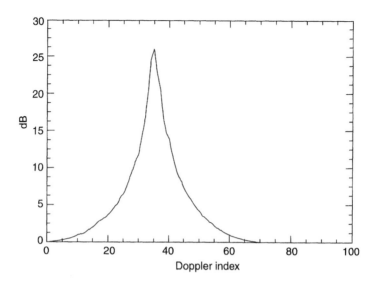

Figure 14.10   Response of estimator S for high-resolution direction estimation (courtesy J. Ender)

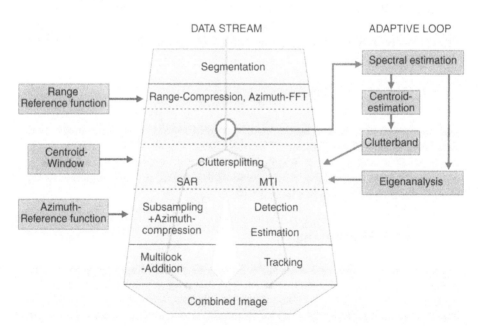

Figure 14.11   SAR/MTI processing scheme (courtesy J. Ender)

*Figure 14.12   Moving targets above a ground scene, detected and located with AER
system (courtesy J. Ender) © IEE 1999 [12]*

On the left-hand branch of the block diagram the fixed ground scene echoes are
preserved by a filter passing the clutter band. They lead by azimuth focusing to the
image of the fixed ground scene. Within the MTI branch the spectral matrix C_w is
estimated by using all available range samples. The clutter subspace is computed
from the eigenvector x_1 corresponding to the largest eigenvalue λ_1. In Figure 14.8
the course of λ_1 over ω illustrates the clutter band. If λ_1 shows peaks outside the
clutterband these are caused by moving targets with high speed. If $z^*(\omega)x_1(\omega)$ for
frequencies ω outside the clutter band exceeds a detection threshold then moving
targets are accepted and reported. The subspace spanned by x_2 corresponding to the
second largest eigenvalue λ_2 is orthogonal to the clutter subspace and contains the
slow targets within the clutter band. After detection the positions of the moving targets
are estimated. All target reports are combined and displayed optionally above the
ground scene. An example for the measured results achieved with the AER system
is given in Figure 14.12: cars on a motorway junction are detected and correctly
positioned.

14.6   Jammer suppression

A noise jammer would mask the ground image corresponding to the additional noise
introduced into the receiving channel, especially if the jamming signals are received

Figure 14.13   SAR image with jamming and with jammer attenuation (courtesy J. Ender)

via the real main beam. A strong jammer would also be effective in the sidelobe region and image masking would be suffered within the disturbed azimuth section.

An SAR system with an active receiving array and several subarrays offers the potential for suppressing the jamming signal, using the spatial degrees of freedom, according to the discussion in chapter 10.

Early experiments with a jammer have been performed with the AER system and described by Ender [10]. In Figure 14.13 a jammed image is shown in the upper part and in the lower part the same scene with jammer attenuation. The jammer-to-noise ratio in the received signals was about 30 dB in the centre of the main beam.

The special problem with SAR signals is to preserve the image quality despite the jammer suppression. Because the time corresponding to the signal series used to generate an image is quite long, i.e. several seconds, all fluctuations of the receiving gain with respect to echoes from a certain object would deteriorate the image. In other words, the point-spreading function, determining the image of a point target, would be broadened and gain increased sidelobes.

In principle, the methods for spatial suppression of jammers described in chapter 10 could be applied. The data vector $\mathbf{z}(t)$ $[N,1]$ has to be multiplied with a filter matrix $\mathbf{F}$ $[N, N]$ before image generation by beamforming with vector $\mathbf{a}(u)$, given by equation 14.11, and convolution with the azimuth chirp $s^*(t)$, given by equation 14.9.

According to equation 10.6 the product $\mathbf{w}^* = \mathbf{a}^*\mathbf{F}$ may be computed first to obtain a beamforming vector $\mathbf{w}^*$ with spatial notches in the jammer directions.

The filter matrix $\mathbf{F}$ should be determined for an adaptive operation from measured jammer signals, since the directions of jammers change with time, at least by the movement of our radar platform. Jammers may be at different ranges resulting in an individual rate of change for the directions. The jamming signals are described by the covariance matrix $\mathbf{Q}$ at the output of the N receiving channels. According to equation 10.5 a direct solution for the filter matrix $\mathbf{F}$ is given by $\mathbf{F} = \mathbf{Q}^{-1}$. Because $\mathbf{Q}$ is unknown we can apply only an estimated sample covariance matrix $\hat{\mathbf{Q}}$. This estimated $\hat{\mathbf{Q}}$ should be computed from as many range elements as possible outside ground clutter, e.g. at high ranges and containing predominantly jamming signals. As discussed in section 10.4.1 the statistical fluctuations of the beamforming weight vector $\mathbf{w}$ depend on the number of samples S for the estimation of $\hat{\mathbf{Q}}$. To counter these fluctuation effects the LSMI method (loaded sample matrix inversion) may be applied given by equation 10.18. Another approach [10] tries to find a compromise for a minimum quadratic difference between the image by vectors $\mathbf{w}$ and $\mathbf{a}$ on the one hand and the residual jammer power on the other hand. The result of this approach is close to the LSMI method.

A further approach is to apply a projection matrix derived from the jammer subspace according to equation 10.16. This has been used to evaluate Figure 14.13 by offline processing of the measured data. As a result the width of the jammed strip is reduced considerably.

Ender also suggests [10] a reconstruction of the image-generating signal by an approach using the conditional expectation. The measured signal is first, for the purpose of jammer suppression, transformed into the subspace orthogonal to the jammer space. By this transformation a certain deformation may occur. This deformation is corrected afterwards, leading to a point-spreading function with low sidelobes comparable to that without jammer cancellation. This method has been tested up to now with simulations.

14.7 Object height by interferometry

The image of the ground scene discussed so far is two dimensional. We may get an impression of the height of single elevated objects such as buildings, towers or trees by the individual radar shadow. A radar height measurement is possible by an additional antenna at a suitable vertical distance to the main receiving antenna. Both antennas deliver complex-valued images of the scenes S_1 and S_2. Now the product $S_{if} = S_1 S_2^*$ is formed. The magnitude of S_{if} gives again the two-dimensional intensity image of the ground scene. But the phase argument Φ of S_{if} is dependent on the range differences and these depend on the target's height h. This evaluation of the differential phase is generally named interferometry [13–19]; the geometry is shown in Figure 14.14. The phase difference Φ is given by:

$$\Phi = \frac{2\pi}{\lambda}(r_1 - r_2) = \frac{2\pi}{\lambda}\left[\left(r_2^2 + B^2 + 2r_2 B \cos \Theta\right)^{1/2} - r_2\right] \qquad (14.26)$$

Known values are H, B, r_2, and Φ mod 2π.

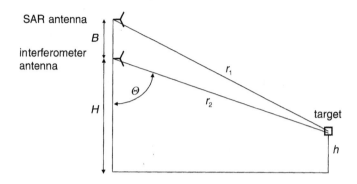

Figure 14.14   Geometry for interferometric SAR

From equation 14.26 we get:

$$r_1 = r_2 + \frac{\Phi\lambda}{2\pi}$$

and:

$$r_1^2 = r_2^2 + B^2 + 2r_2 B \cos\Theta$$

$$\Theta = \cos^{-1}\left(\frac{r_1^2 - r_2^2 - B^2}{2r_2 B}\right)$$

This leads finally to the relation for the target height :

$$h = H - r_2 \cos\Theta \tag{14.27}$$

For the whole image the height h is computed to obtain a three-dimensional target scene. An example achieved with the AER system is shown in Figure 14.15 with a perspective presentation of a volcanic mountain. By interferometric SAR a digital elevation map is generated with the actual height values for trees and buildings.

The baselength would be selected about $B = 1$ m. Then for a range of $r_1 = 10$ km a height accuracy of about 1 m is achievable [18]. The problem with two antennas is an ambiguity of the phase by multiples of 2π. This may be resolved to a certain extent by a procedure called phase unwrapping [19]. One method proposes linking the phases of neighbouring pixels (resolution cells) along a suitable chosen integration path. The assumption is used that these phases cannot differ by more than π because the steps in height are generally limited.

The height estimation accuracy is inversely proportional to the baselength B. But a long baseline increases the number of phase ambiguities or the fringe frequency. A solution is the addition of a third antenna. Now three different baselength values are available. A small baseline resolves the ambiguity and the long baseline improves

Figure 14.15 Interferometric 3D image of a volcanic mountain scene, measured with AER system of FGAN (courtesy J. Ender, L. Rößing)

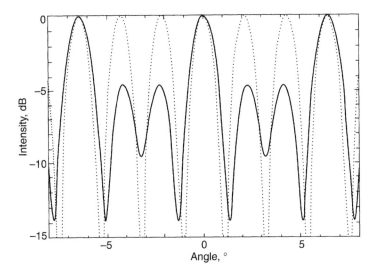

Figure 14.16 Array pattern for two antenna elements (dotted line) and three antenna elements (solid line) with baseline ratio 1 : 2 : 3 (courtesy L. Rößing, J. Ender)

the resolution. A choice has to be made for the positions of the antennas. For minimal sidelobes of this three-element array the baseline ratio $1 : 2 : 3$ has been suggested [11]. In Figure 14.16 the resultant pattern is compared with the pattern for the two antennas of the usual interferometer.

The three antennas forming a vertical array may also be used directly to estimate the elevation angle of each individual pixel. For pixel v with complex reflectivity c_v, receiver noise vector $\mathbf{n}_v$ and unit vector $\mathbf{u}_v$ to the pixel v we get for the three-element vector $\mathbf{z}_v$ [11]:

$$\mathbf{z}_v = c_v \cdot \mathbf{a}_v(\mathbf{u}_v) + \mathbf{n}_v \tag{14.28}$$

The unit vector $\mathbf{u}_v$ now describes the target elevation; $\mathbf{a}(\mathbf{u}_v)$ is the direction-of-arrival vector for the antenna positions $\mathbf{r}_i$ and factors k_i for the two-way characteristics of the ith transmit-receive path:

$$\mathbf{a}(\mathbf{u}_v) = \begin{bmatrix} k_1(\mathbf{u}_v) \exp\left(j\beta(\mathbf{u}_v \cdot \mathbf{r}_1)\right) \\ k_2(\mathbf{u}_v) \exp\left(j\beta(\mathbf{u}_v \cdot \mathbf{r}_2)\right) \\ k_3(\mathbf{u}_v) \exp\left(j\beta(\mathbf{u}_v \cdot \mathbf{r}_3)\right) \end{bmatrix} \tag{14.29}$$

The factors k_i of vector $\mathbf{k}$ must be determined by some kind of calibration. The same method as for azimuth estimation described in section 14.4 may be applied. From the estimated covariance matrix the eigenvector, according to the maximum or principal eigenvalue λ_1, which predominates alone above noise, gives the vector $\mathbf{k}$ or the array manifold. With this knowledge we may again apply for the direction estimation the monopulse procedure as discussed in sections 11.4.1 and 14.4 with equation 14.24. The estimated $\Delta\hat{u}$ from the flat earth can be used to compute by a geometric operation the height of each pixel [11].

The covariance matrix may be estimated for an individual pixel. If there are two scatterers at different heights we will get two eigenvalues λ_1, λ_2, both distinctly above the noise level. With both corresponding eigenvectors $\mathbf{x}_1, \mathbf{x}_2$ the MUSIC method for superresolution, discussed in chapter 12 section 12.4, may be applied to estimate the height of both scatterers. For example, the echoes from the road on a bridge and from the surface of the valley below can be separated.

14.8   Experimental system AER and results

An experimental phased-array SAR system called AER (experimental airborne radar) [7] has been under development at FGAN since 1989. The aim has been to verify by experiments the possibilities offered by an active receiving array, discussed in the preceding sections.

Additionally to the signal-processing possibilities, a limited scan capability of $\pm 8°$ from boresight should make it possible to operate in spotlight mode. With this mode the beam is steered in the opposite direction to the flight movement to illuminate a target area longer than is possible with a fixed beam. According to equation 14.3 an improved cross-range resolution Δx then results from the increased illumination or aperture time T_{sa} and therefore a correspondingly increased synthetic aperture length. Also, using azimuth steering, a squinted forward-looking mode is possible.

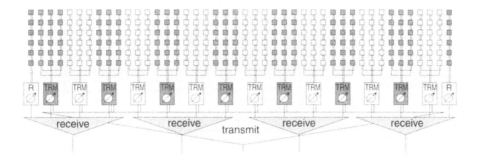

Figure 14.17   Structure of array antenna for AER with subarrays (courtesy J. Ender)

The system operates at X band (10 GHz). The rectangular antenna array with its structure is shown in Figure 14.17. As antenna elements rectangular patches have been chosen, and the array has 264 antenna elements which are all fed alternatively at two points for selection of polarisation (vertical or horizontal). Six elements are combined by a column feed network for forming the elevation pattern with a beam width of 16°. There are 14 subarrays, each with three columns which have been combined in the horizontal direction for cost saving. This limits the azimuth scan capability but corresponds to the requirement of scanning only ±8°. Each of these subarrays is combined with a transmit/receive module, although the edge columns have only a receive module. For the purpose of MTI processing these triple subarrays are combined on the next level to four super-subarrays with corresponding receiving channels. By using a switched combiner/splitter network a four-channel polarimetric mode is possible. With a second array receiving antenna an interferometric mode is achieved and applied to measurements.

The system's maximum range requirement is 20 km and this was found to be sufficient for the experimental verification and demonstration of the proposed procedures. The transmit power is generated by 14 transmit/receive modules. The total peak power from all modules together is 80 W. The transmit/receive module has a low-noise receiving amplifier, separate four-bit phase shifters for transmit and receive, a polarisation switch, a limiter and an interface for the digital control signals. The transmit power is applied with an amplitude taper to reduce the sidelobes against backfolding of the clutter spectrum. The transmit polarisation can be changed from pulse to pulse. A second and third receiving antenna have been built for interferometry. The antenna can be steered mechanically in elevation by a navigational computer to adjust the antenna to the desired strip on the ground.

The resolution is chosen 1 m × 1 m, therefore the signal bandwidth is 150 MHz. The transmit signal is modulated with linear or nonlinear chirp. The waveform is generated by a direct digital synthesiser. The I and Q components of the received signals from the four subarray channels are each analogue-to-digital converted with eight bits at a rate of 160 MHz. The data are first stored in a 1 Gbyte RAM memory and then spooled to hard disks.

*Figure 14.18   Antenna array for AER and complete radar subsystem (courtesy
J. Ender)*

The system was first tested by experiments with a van on motorways on high
bridges and looking into valleys. An example of the resulting images is shown in
Figure 14.20.

For flight tests the array antenna is mounted along the flight direction looking
sideways out of the side door of a C 160 Transall aircraft as the flying platform.
Several successful measurement campaigns have been performed afterwards with
this aircraft as carrier [9–11]. Some examples of results have already been given here
in Figures 14.12, 14.13 and 14.15.

14.9   Summary and motivation for SAR with active phased-array antennas

The potential of an active phased-array radar system for SAR application has been
discussed in the preceding sections. One interesting possibility is the detection and
location of moving targets, even if they are slowly moving. The lower limit for
detection is a radial velocity of only about 1 m/s, that is the velocity of pedestrians.
Subsequent accurate location by a monopulse procedure is possible.

A spotlight mode, steering the antenna arbitrarily to a scene of interest to achieve
a long synthetic aperture, results in improved cross-range resolution. An autofocus
procedure by correlation analysis is also practicable [2].

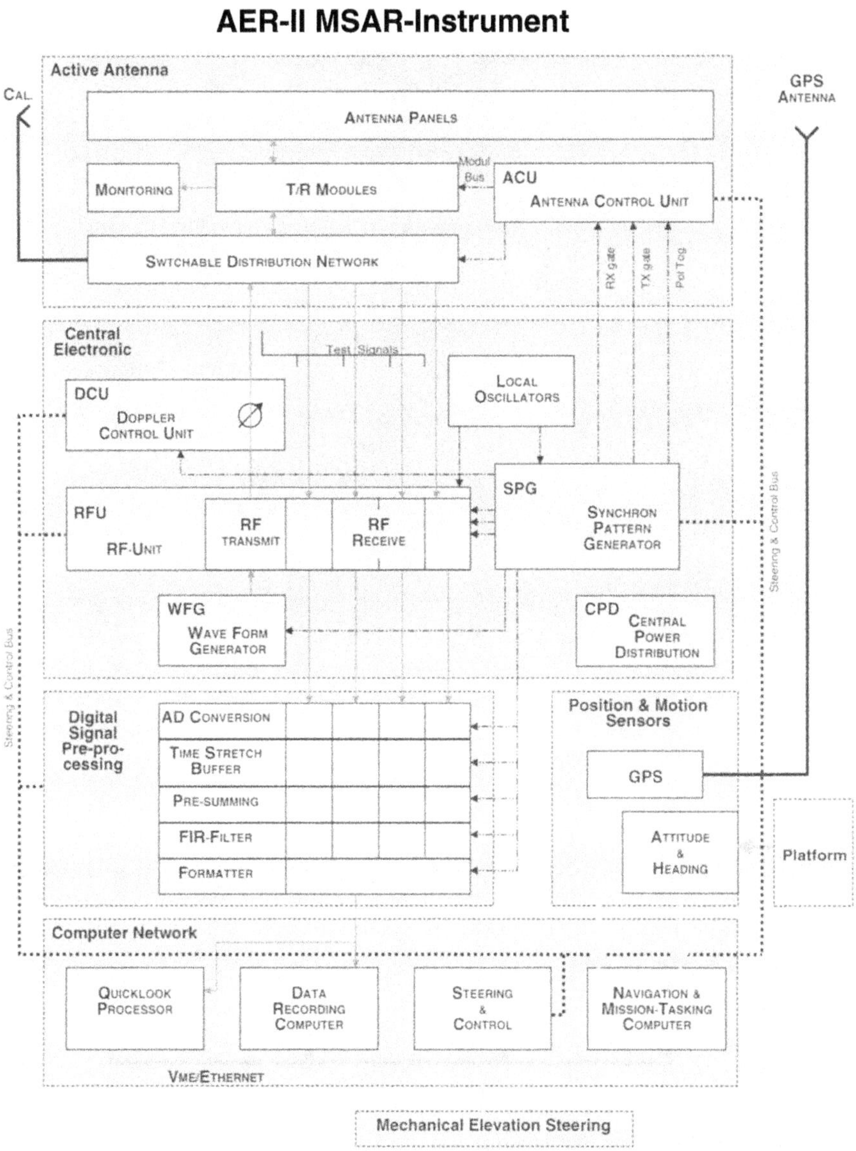

AER-II MSAR-Instrument

Figure 14.19   Block diagram of AER-II MSAR instrument (courtesy J. Ender)

Suppression of jammer signals may be achieved by the methods of adaptive beamforming, thus reducing the extent of the jammed area or radial strip.

Using additional antennas (one or better two with suitable vertical positions) the height of each pixel can be determined by interferometric processing to achieve a three-dimensional image of the ground scene.

Figure 14.20   SAR image of Mosel valley, measured with AER from a van on a motorway bridge (courtesy J. Ender)

All these methods have been developed in theory and validated by experiments with the AER system. Apart from these methods resulting from the application of array signal-processing methods, there are also some additional operational advantages with active phased arrays for SAR:

- a high mean time between failure (MTBF) by the application of solid-state transmit amplifiers
- high redundancy by multiple transmit/receiving modules and therefore a distributed transmit power generation, resulting in a graceful degradation in case of failure of individual modules
- avoiding distribution losses between a central transmitter and the array antenna elements

Beam agility offers:

- selection of specific targets and scenes for imaging
- compensation of platform movements (roll, pitch and yaw)
- improved resolution with spotlight mode
- elevation scan to image several strips at the same time in a multiplex operation
- multiplex combination of SAR and scan-MTI operation

Problems in realising SAR with an active phased-array antenna are to be expected in the following technical areas: producing the necessary transmit/receive modules with reasonable cost, calibration and mutual matching of the channels, achieving a high bandwidth for fine range resolution, realising the processing algorithms for real-time operation. All these problems will be solved in the future and the advantages of a multichannel SAR with an active phased array will be used for improved exploration and surveillance for civil and military tasks.

14.10   References

1 SHERWIN, C. W., RUINA, P., and RAWCLIFFE, R. D.: 'Some early developments in synthetic aperture radar systems', *IRE Trans. on Military Electronics*, April 1962, **6**, pp. 111–115

2 ENDER, J.: 'Analyse der Selbstfokussierung eines Radars mit synthetischer Apertur unter Verwendung einer Gruppenantenne'. Dissertation, Ruhr-Universität Bochum, 1991

3 ENDER, J.: 'Signal processing for multi channel SAR applied to the experimental SAR system AER'. Proceedings of International conference on *Radar*, Paris, France, 1994, pp. 220–225

4 ENDER, J.: 'Azimutpositionierung bewegter Ziele mit Mehrkanal-SAR'. Proceedings of URSI-conference, Klein-Heubach, Germany, 1994, S. 715–724

5 ENDER, J.: 'Detection and estimation of moving target signals by multi-channel SAR', *AEÜ*, 1996, **50**, (2), pp. 150–156

6 ENDER, J.: 'Detection and estimation of moving target signals by multi-channel SAR'. Conference proceedings *EUSAR'96*, Königswinter, Germany, 1996, pp. 411–417

7 ENDER, J.: 'The airborne experimental multi-channel SAR-system AER II'. Conference proceedings *EUSAR'96*, Königswinter, Germany, 1996, pp. 49–52

8 ENDER, J., and RÖßING, L.: 'Mehrkanal-SAR-Verfahren: Mit dem flugzeuggetragenen System "AER II" erzielte experimentelle Ergebnisse'. Proceedings 9, *Radar* symposium der DGON, Stuttgart, Germany, 1997, pp. 309–320

9 ENDER, J.: 'Experimental results achieved with the airborne multi-channel SAR system AER II'. Conference proceedings *EUSAR'98*, Friedrichshafen, Germany, 1998, pp. 315–318

10 ENDER, J.: 'Anti-jamming adaptive filtering for SAR imaging'. Proceedings international *Radar* symposium der DGON, München, Germany, 1998, pp. 1403–1413

11 RÖßING, L., and ENDER, J.: 'Advanced SAR interferometry techniques with AER II'. International *Radar* symposium der DGON, München, Germany, 1998, pp. 1261–1269

12 ENDER, J.: 'Space-time processing for multi-channel synthetic aperture radar', Electron. Commun. Eng. J. 1999, **11** (1), pp. 29–38

13 GOLDSTEIN, R. M., ZEBKER, H. A., and WERNER, C. L.: 'Satellite radar interferometry: two-dimensional phase unwrapping', *Radio Sci.*, 1988, **23** (4), pp. 713–720

14 RODRIGUEZ, E., and MARTIN, J. M.: 'Theory and design of interferometric SARs', *IEE Proc. F, Radar Signal Process.*, 1992, **139** (2), pp. 147–159

15 MASSONET, D., ROSSI, M., and CARMONA, C., *et al.* 'The displacement field of the Landers earthquake mapped by radar interferometry', *Nature*, July 1993, **364**, pp. 138–142

16 MASSONET, D., BRIOLE, P., and ARNAUD, A.: 'Deflation of Mount Etna monitored by spaceborne radar interferometry', *Nature*, June 1995, **375**, pp. 567–570

17 CURRIE, A., BAKER, C. J., BULLOCL, R., EDWARDS, R., and GRIFFITHS, H. D.: 'High resolution 3-D SAR imaging'. Proceedings of *Radar 95* conference, Washington, USA, May 1995, pp. 468–472

18 GRIFFITH, H. D.: 'Interferometric synthetic aperture radar', *Electron. Communi. Eng. J*, 1995, pp. 247–256

19 GRIFFITH, H. D., and WILKINSON, A. J.: 'Improvements in phase unwrapping algorithms for interferometric SAR'. Proceedings of international conference on *Radar*, Paris, France, 1994, pp. 342–349

20 ENDER, J.: 'Subspace tansformation techniques applied to multi-channel SAR/MTI'. IGARSS conference, Hamburg, Germany, 1999, Session BB2

Chapter 15

Inverse synthetic aperture radar (ISAR)

15.1 Introduction: basic principles

In chapter 14 we have seen high-resolution images of the ground scene produced by a radar flying on an elevated platform and thereby forming a long synthetic aperture. This long aperture resulted in a very fine azimuth or cross-range resolution. For ground-based radar systems it is often also highly desirable to get images of detected targets, for example of the observed aircraft. This could be a valuable contribution to a classification of flying targets.

Early ideas on this topic have emerged in the open literature since 1969: Brown and Fredricks [1] gave a discussion on range Doppler imaging and Kock [2] suggested an improved gain for a ground radar by a synthetic aperture beam for detecting and perhaps resolving aircraft. The author [3] presented in 1973 radar experiments with aircraft targets which demonstrated the feasibility of target imaging in one dimension (cross range). Then followed several publications discussing the method more comprehensively and presenting further experimental results, even for two-dimensional images [4–7].

The generation of a radar target image may be confined to targets which are already detected, acquisitioned and tracked by our multifunction radar. Then it is possible to transmit to a selected flying target a suitable pulse series, using beam agility and the tracking capability of a phased-array radar. The basic idea is explained in Figure 15.1.

The radar transmits a regular pulse series to the target and echoes are received and written in a memory. Now we reverse the relative movement in our mind: we assume the target fixed in space and the radar moving in the reverse direction. With the stored echoes we can form by processing a synthetic aperture specialised for the selected target. The flight path of the radar, produced by the target's movement, forms the synthetic aperture or array. By forming a cluster of very narrow beams with this synthetic array the cross-section elements on the target will be resolved correspondingly. Because the target is within the nearfield of the synthetic array we have additionally to apply a focusing operation determined by the target's range and its velocity. The name for this method is inverse synthetic aperture radar or ISAR. The high resolution

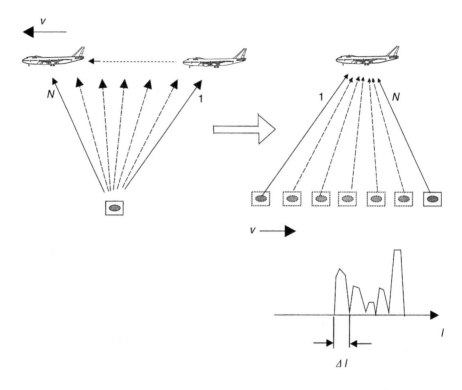

Figure 15.1   Basic principle for inverse synthetic aperture (ISAR)

given by this method is in that dimension, which is orthogonal to the line of sight or cross range. By this method we get a one-dimensional cross-range image with a certain resolution Δl, as indicated in Figure 15.1. The signal bandwidth of radar systems, e.g. of the order of 1 MHz, determines the range resolution, e.g. 150 m, which means that an aircraft target is completely covered by only one range cell.

Another equivalent explanation of the ISAR method results from the target motion: the target echoes will have a mean Doppler frequency given by the radial velocity of the aircraft. By movement straight ahead the target shows additionally a certain rotation with respect to the radar site, with the rotation axis orthogonal to the line of sight. This causes all cross-section elements to rotate with respect to the target centre and show a different velocity in the direction to our radar, which is proportional to the distance from the rotation axis, and a corresponding differential Doppler frequency. By a corresponding focusing or matched filtering with a high Doppler resolution cross-range resolution is achieved. Only if the target flies exactly on a circle around the radar is there no change of aspect angle and no Doppler shift at all, and ISAR could not be applied.

To achieve a two-dimensional target image we also need a corresponding range resolution by applying a high signal bandwidth. We will first discuss cross-range resolution by ISAR alone.

15.2 Synthetic aperture and beamforming by target motion

Now we will first analyse the geometric relation to recognise some basic conditions for generating such a radar cross-range image. The flight path of an aircraft may be nearly tangential to our radar site. According to Figures 15.1 and 15.2, the synthetic array is formed with antenna points $-N, \ldots, N$ and distance $d = vT$ between these points, with $v = $ velocity of the aircraft and $T = $ radar pulse period. To the array point k corresponds the received signal z_k. The range of the target from the array centre $k = 0$ may be R. To form a beam in direction Θ the phase along the path r has to be compensated for transmit and receive. For the element at position x_k we get:

$$r^2(x) = R^2 + x^2 - 2Rx \sin \Theta$$

$$r(x) = R\left(1 + \frac{x^2}{R^2} - 2\frac{x}{R} \sin \Theta\right)^{1/2} \tag{15.1}$$

A sum beam is formed in direction Θ by processing the received signals by the equation:

$$S(\Theta) = \sum_{k=-N}^{N} z_k \exp\left[-\frac{4\pi j}{\lambda}(R - r(x_k, \Theta))\right] \tag{15.2}$$

For $x \ll R$ we may approximate equation 15.1, using $\sqrt{1 + x} \approx 1 + x/2$:

$$r \approx R + \frac{x^2}{2R} - x \sin \Theta \quad \text{or} \quad R - r \approx -\frac{x^2}{2R} + x \sin \Theta \tag{15.3}$$

$$S(\Theta) \approx S_1(\Theta) = \sum_{k=-N}^{N} z_k \exp\left[\frac{4\pi j}{\lambda}\frac{x_k^2}{2R}\right] \exp\left[-\frac{4\pi j}{\lambda} x_k \sin \Theta\right] \tag{15.4}$$

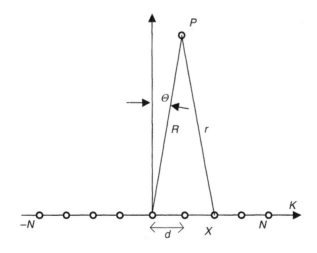

Figure 15.2   Geometry for synthetic array for nearly tangential flight path

The first factor represents the nearfield phase correction with respect to the target position at point P. This focusing operation is fortunately independent of Θ. The second factor represents the Fourier transform for forming the sum beam out of the signals z_k after the focusing correction. The angular resolution by the beam S_1 can be derived from equation 15.4. The beam pattern is, according to equation 4.11, for a steering direction Θ_0 and with $u = \sin \Theta$ given by:

$$S_1(u/u_0) = \sum_{k=-N}^{N} \exp\left[\frac{4\pi j}{\lambda}(u - u_0)\right]$$

$$= (2N + 1)\frac{\sin\left[(2N + 1)2\pi d/\lambda(u - u_0)\right]}{(2N + 1)\sin\left[2\pi d/\lambda(u - u_0)\right]} \tag{15.5}$$

The first zero appears at:

$$u_1 - u_0 = \frac{\lambda}{2d(2N + 1)}$$

As in section 4.2.1 we get for the half-power beamwidth Δu by setting $u - u_0 = \Delta u/2$:

$$(2N + 1)\frac{2\pi d}{\lambda}\frac{\Delta u}{2} = 1.39$$

$$\Delta u = \frac{1.39}{\pi}\frac{\lambda}{(2N + 1)d} = 0.442\frac{\lambda}{L} \tag{15.6}$$

with L for the length of the synthetic aperture. In comparison to equation 4.17 we have for the beamwidth now only half of the value, because the phase increments of the transmit and receive path are both effective. The cross-range resolution along an axis orthogonal to the line of sight (LOS) and in the plane given by this LOS and the flight vector, that is usually the horizontal plane, is then given by:

$$\Delta l = R\Delta\Theta \approx R\Delta u = 0.442\frac{R\lambda}{(2N + 1)vT} \tag{15.7}$$

15.3   Target cross-range image

The Fourier transform of the focused signals according to equation 15.4 gives the target signal contributions within a set of $2N + 1$ beams. The cross range scale Δl is given by equation 15.7. In other words, the target is divided into a series of cross-section elements of length Δl, representing the image. The scale is given by the target's range and velocity, known from the tracking procedure.

We estimate the angular sector Φ within which we observe the aircraft:

$$\Phi = 2\tan^{-1}\left(\frac{(2N + 1)vT}{2R}\right) \approx \frac{(2N + 1)vT}{R} \tag{15.8}$$

Then equation 15.7 may be expressed by:

$$\Delta l \approx 0.442 \frac{\lambda}{\Phi} \tag{15.9}$$

For a lateral resolution of, for example, $\Delta l = 10\lambda$ we need an observation sector of only $\Phi = 0.0442$ or $2.53°$. For an X band ground radar this would mean a resolution of 0.3 m. On the other hand Φ is the amount of the aircraft rotation with respect to the line of sight necessary to allow this reasonable cross range imaging. If we observe a target on a curved flight path, the scale Δl may be derived from Φ, known from the tracking procedure.

As for SAR images, we may have also for ISAR target images a resolution independent of range.

We may get a further feeling for the conditions for ISAR imaging by assuming realistic parameter values:

$$v = 250 \,\mathrm{m/s}$$
$$R = 50 \,\mathrm{km}$$
$$\lambda = 0.1 \,\mathrm{m}$$

To achieve for example a resolution $\Delta l = 1$ m we need an observation angular sector as above of $\Phi = 0.0442$ and the aperture length follows with equation 15.8 to:

$$L = R\Phi = (2N + 1)vT$$
$$= 5 \cdot 10^4 \cdot 0.0442$$
$$= 2210 \,\mathrm{m}$$

The total observation time $L/v = 2210/250 = 8.84$ s has to be covered by $2N + 1$ pulses at a period T.

Next we have to choose N or T: by the Fourier transform a set of $2N+1$ beams are given with an angular separation according to equation 15.5 given by $\lambda/(2d(2N + 1))$. The periodic cross-range repetition interval CRI or unambiguous interval is therefore given by:

$$\mathrm{CRI} = \frac{\lambda R}{2d} = \frac{\lambda R}{2vT} \tag{15.10}$$

This gives a maximum value for T, because the CRI must be larger than the possible target dimension. It depends on R. For our parameter examples and a CRI > 1000 m follows:

$$T \leq \frac{\lambda R}{2v\mathrm{CRI}} = \frac{0.1 \cdot 5 \cdot 10^4}{250 \cdot 1000} = 20 \,\mathrm{ms}$$

The minimum number of pulses then follows from the total observation time given above to:

$$2N + 1 = 8.84/0.02 = 442$$

We recognise that for achieving a reasonable cross-range image the angular sector is relatively narrow, the aperture length or the flight path is several kilometres long, the measurement time is several seconds and the pulse period may be relatively long compared to usual radar operation.

With this parameterised feeling we recognise further that focusing with the first factor in equation 15.4 is absolutely necessary because the maximum phase correction $\Delta\varphi$ for $k = -N, +N$, is for our parameter example given by:

$$\Delta\varphi = \frac{4\pi x_N^2}{2\lambda R} = 244 \cdot 2\pi$$

15.4   Focusing

15.4.1   Range focusing and range walk

We have assumed so far with respect to focusing and beamforming that all reflecting points of a target are distributed on a line orthogonal to the LOS. This is of course not the case for airplanes. The question arises whether the range focusing factor in equation 15.4, optimised for range R to the target centre, is valid for all contributing reflection points of the target. By assuming a range difference ΔR from R a certain defocusing may be caused. This follows from the expression for the focused sum beam and echoes with range difference ΔR. We assume for simplicity $\Theta = 0$ for the target direction:

$$S(0, \Delta R) = \sum_{k=-N}^{N} \exp\left[-\frac{2\pi j}{\lambda}\frac{x_k^2}{R + \Delta R}\right]\exp\left[\frac{2\pi j}{\lambda}\frac{x_k^2}{R}\right]$$

$$= \sum_{k=-N}^{N} \exp\left[\frac{2\pi j}{\lambda}x_k^2\frac{-R + R + \Delta R}{(R + \Delta R)R}\right]$$

$$\approx \sum_{k=-N}^{N} \exp\left[\frac{2\pi j}{\lambda}x_k^2\frac{\Delta R}{R^2}\right]$$

For the above chosen parameters the function $S(0, \Delta R)/S(0, 0)$ is presented in Figure 15.3. The values x_k are given according to our parameters by:

$$x_k = vTk = dk \quad \text{with } d = 5\,\text{m}$$

The decay down to the half power, that is $S(0, \Delta R)/S(0, 0) = 0.707$, is seen for $\Delta R > 80\,\text{m}$. This is well above the expected target dimension of about 30–50 m and we may use therefore the same focusing for the whole target.

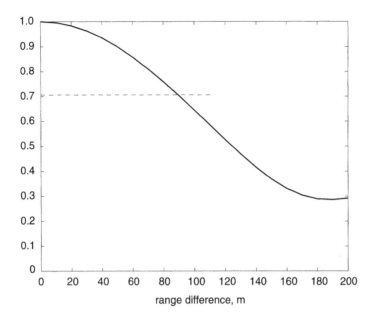

Figure 15.3 *Focusing sensitivity dependent on range difference* ΔR *given by the normalised sum beam, 442 pulses, $d = 5\,m$, $R = 50\,km$*

The next question to discuss is a possible range walk during the relatively long observation time. Up to now we have assumed a bandwidth usual for a multifunction radar of the order of 1 MHz; then the range resolution is 150 m. For a tangential flying target the range increase at the edges of the synthetic aperture, that is for $k = -N, +N$, is given for our assumed parameters by:

$$\delta R = \left(50000^2 + 1105^2\right)^{1/2} - 50\,000 = 12\,\text{m}$$

So δR is small compared to the range cell extent and all echoes of the target will remain, in this case, within one range cell. The target's range cell has been selected by the target-detection and tracking process.

The situation is more complicated if the target is not flying boresight to our radar. Then there is a certain angle Θ under which the synthetic array is observing the target. All effects discussed for scanning with linear arrays as presented in chapter 4 apply. A beam broadening by the factor $1/\cos \Theta$ reduces the resolution and has to be compensated for by a longer observation time or path. The increasing or decreasing of range will cause a change of the target's range cell or bin. Then for each observation point the range cell with maximum target amplitude may be selected.

The remaining problem is to know the flight path and target movement with sufficient accuracy. With SAR systems one could use the inertial navigation aids on the flying platform for focusing the image. With ISAR this possibility is generally not available and we have to estimate the flight parameters with our own radar measurements.

15.4.2   Focusing with straight-line assumption

A first possibility is to assume, after detection and tracking, an exact straight flight path. Then R is measured by the radar and v is roughly known by the tracking process. A fine estimate v is derived by a variation to maximise the synthetic sum beam outputs. This assumption may be justified by the inertia of the flight path with respect to an acceleration within the observation interval of some seconds. There will be enough opportunities for a selected target to repeat the procedure and select the best image or to form an average from the series of images. With this method initial experiments have been performed [3], an example is given in Figure 15.4. The resolved echo amplitude distribution representing the cross-section distribution above the cross-range dimension is shown. The estimated velocity is $v = 745$ km/h, the range is $R = 77$ km. The measured aircraft, a tangential flying Boeing 737, with its length of 30 m, fits the main evaluated cross-section elements from the measurement, as indicated by the aircraft sketch.

Another approach for focusing is to assume a polynomial of a suitable degree for the overall phase history of the target. By confining ourselves to flight paths which are approximately straight the phase polynomial is used only up to quadratic terms [6].

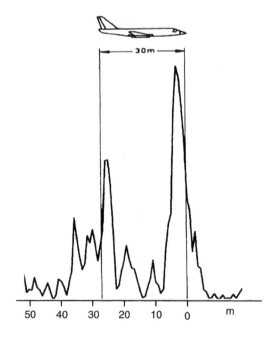

Figure 15.4   Cross-range resolution of tangential flying Boeing 737, horizontal axis in metres

In the next section we discuss possibilities for focusing with any kind of phase history.

15.4.3 Focusing for arbitrary flight paths

15.4.3.1 Iteration with Fourier transform and Kalman filter

We have discussed the angle of necessary rotation to form the synthetic aperture as being only about 2° to 3°. Therefore targets flying along a turn may produce this rotation in a shorter time in favour of less random effects like wind shear. Ender [7] has developed a focusing method for ISAR data of airplanes by combining position estimates and spectral information from the echoes in a Kalman filter. By applying an interpolation the fine range history is estimated and used for a first motion compensation. The variation of the centre frequencies of the spectra of data sections indicates the remaining motion compensation error. These are fed back to the tracking filter for improved iteration, and the next iterations therefore start with improved motion compensation. The expected spectral resolution will be higher and therefore longer data segments may be evaluated. The iteration is stopped if the Fourier resolution becomes smaller than the remaining fast or uncorrelated motion compensation error. For the experiments performed with the ELRA system (chapter 17) the final resolution was achieved after two to four iterations. The linear scale for the image has to be computed from the flight-path parameters corresponding to equation 15.9.

According to experiments with the ELRA system difficulties only arose with this focusing method in the transition phase from a straight to a curved flight.

With Figures 15.5 and 15.6 we see examples from these experiments with curved flight paths. The cross-range resolution is generally 1 m. The sketch of the aircraft indicates the position of the measured target (Boeing 737). Figure 15.5 gives an example for a boresight image and Figure 15.6 for a lateral image of the front view. Figure 15.7 gives a view from behind. In this case four jet engines are the predominating reflectors (Boeing 747).

Figure 15.5   Cross-range resolution of Boeing 737 from boresight, measured with ELRA system (courtesy J. Ender)

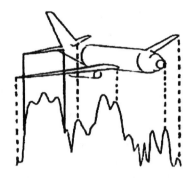

Figure 15.6   Cross-range resolution of Boeing 737, aside from the front view, measured with ELRA system (courtesy J. Ender)

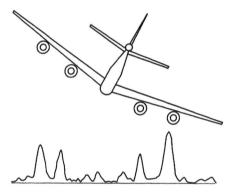

Figure 15.7   Cross-range resolution of Boeing 747 from behind, measured with ELRA system (courtesy J. Ender)

15.4.3.2   Doppler shift of a target and its scatterers

We will now discuss, using simple geometry, the course of the Doppler frequency of an observed target. This situation has already been presented in chapter 8 in section 8.6.5 with Figure 8.18. Now we are interested in the width of the instantaneous Doppler spectrum of a complex target, modelled with two scattering centres at the edges.

The target centre starts with its movement at $t = 0$ at $x = x_0$ as the distance from the boresight point. The target with length l may have two scatterers at the end points $\pm l/2$ away from the centre, indicated in the Figure 15.8 by the solid squares. We compute the Doppler frequency as a function of time for the target centre and the two scatterers by observing the respective ranges $r(t)$, $r_1(t)$, $r_2(t)$:

$$r(t) = R\left[1 + \left(\frac{x_0 - vt}{R}\right)^2\right]^{1/2} \qquad r_1(t) = R\left[1 + \left(\frac{x_0 - l/2 - vt}{R}\right)^2\right]^{1/2}$$

$$r_2(t) = R\left[1 + \left(\frac{x_0 + l/2 - vt}{R}\right)^2\right]^{1/2} \tag{15.11}$$

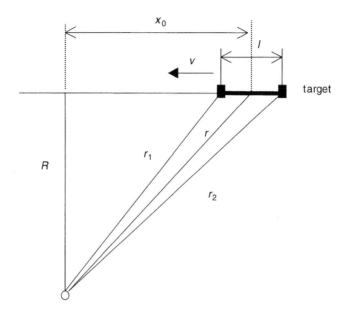

Figure 15.8 Target with two scatterers on tangential flight path

The Doppler frequencies are given by:

$$f_d(t) = -\frac{2}{\lambda}\frac{dr(t)}{dt}$$

For the central point we get:

$$f_d(t) = -\frac{2}{\lambda}\frac{R}{2[1 + (x_0 - vt/R)^2]^{1/2}}\frac{2(x_0 - vt)}{R}\left(\frac{-v}{R}\right) \approx \frac{2}{\lambda R}(x_0 v - v^2 t)$$

(15.12)

The approximation is valid only for $x_0 \ll R$. This corresponds to the central section of Figure 8.18, where the frequency change is nearly linear.

The observation time t goes from $t = 0$ to $t = 2x_0/v$, forming a synthetic aperture of length $2x_0$. With our selected parameters the total Doppler range is with equation 15.12:

$$\Delta f_d = \frac{4x_0 v}{\lambda R} = \frac{4 \cdot 1105 \cdot 250}{0.1 \cdot 50000} = 221\,\text{Hz}$$

The Doppler frequency of the target centre will decrease linearly with time from 110.5 to $-110.5\,\text{Hz}$. From this knowledge follows a pulse recurrence frequency PRF of at least 220 Hz or a pulse period of about 5 ms.

For a flight path section with a certain angle to the line of sight LOS the velocity vector v may be decomposed into a tangential and a radial vector v_t and v_r, respectively. The Doppler frequency for the tangential velocity vector has just been

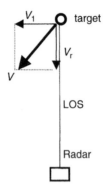

Figure 15.9   Flight path velocity components

discussed. The radial velocity vector causes a corresponding constant Doppler shift of the linear frequency function. The PRF has to be selected correspondingly higher.

The other two Doppler frequencies for the two scatterers are given by:

$$f_{d_1}(t) = \frac{2}{\lambda R}((x_0 - l/2)v - v^2 t)$$

$$f_{d_2}(t) = \frac{2}{\lambda R}((x_0 + l/2)v - v^2 t)$$

(15.13)

The scatterers at the target edges, as indicated in Figure 15.8, produce Doppler frequencies according to equation 15.13 with a difference δf_d above or below the Doppler frequency from the target centre. With a total target length of $l = 15\,\text{m}$ we get:

$$\delta f_d = \frac{lv}{\lambda R} = \frac{15 \cdot 250}{0.1 \cdot 50000} = 0.75\,\text{Hz}$$

Therefore, the instantaneous bandwidth is only 1.5 Hz and very small compared to the total Doppler frequency variation of 221 Hz.

If there is a certain target motion divergent from the idealised straight path without acceleration then the instantaneous bandwidth still remains at about 1.5 Hz, but the function of the mean instantaneous frequency $f_{d_m}(t)$ with time will be somehow nonlinear. If we wanted to know this function $f_{d_m}(t)$ the motion-compensation phase could be computed by integration with time.

15.4.3.3   Focusing with Wigner-Ville distribution

For focusing we have to estimate the function $f_{d_m}(t)$. This is a problem of analysing the time-frequency distribution of the received signal.

One possible method would be the short-time Fourier transform which is applied to a series of time-windowed sections of the signal. The discrete time variable is taken as the series of time instants which corresponds to the centre of the windows.

A narrow window then will give a good time resolution but a poor frequency resolution or *vice versa*.

The Wigner-Ville distribution (WVD) [8] has been suggested more recently for this focusing task [9]. It results in a relatively high resolution in time and frequency compared to the short-time Fourier transform. On the other hand, the WVD has some other problems because it is a nonlinear transformation (it is called bilinear). Cross products appear when several signals and noise are present. The possible distortion depends on the relative power of signal and noise. In the ISAR application we can assume that the target is already tracked and the signal-to-noise ratio is sufficiently large and well above noise. Because we expect a relatively narrow spectrum around the varying centre frequency, as discussed above, the WVD is a very suitable analysis tool. The WVD of a complex analytical signal $z(t)$ is defined by:

$$W(t, f) = \int_{-\infty}^{+\infty} z\left(t + \frac{\tau}{2}\right) z^*\left(t - \frac{\tau}{2}\right) \exp\left(-j2\pi f\tau\right) d\tau \tag{15.14}$$

Because we have available only a finite data series within a time interval Δ we can form only a windowed version of equation 15.14 or a so-called pseudo WVD. A weighting function $g(\tau)$ of length Δ, which is zero outside our time window, may also be introduced. So we get:

$$W_w(t, f) = \int_{-\Delta/2}^{+\Delta/2} g(\tau) z\left(t + \frac{\tau}{2}\right) z^*\left(t - \frac{\tau}{2}\right) \exp\left(-j2\pi f\tau\right) d\tau \tag{15.15}$$

We introduce a small transform: $\tilde{\tau} = \tau/2$. With this follows:

$$W_w(t, f) = 2\int_{-\Delta/4}^{+\Delta/4} g(\tilde{\tau}) z(t + \tilde{\tau}) z^*(t - \tilde{\tau}) \exp\left(-j2\pi f 2\tilde{\tau}\right) d\tilde{\tau} \tag{15.16}$$

Because we have to process sampled data equation 15.16 becomes the discrete Wigner-Ville distribution (DWVD):

$$W(n, k) = 2\sum_{m=-N/2}^{N/2} z(n + m) z^*(n - m) \exp\left(-j2\pi m 2k/(N + 1)\right) \tag{15.17}$$

Index n corresponds to time nT_s, the data assumed to be sampled at a period T_s or sampling frequency $f_s = 1/T_s$ according to the signal bandwidth. Index m corresponds to the time shift $\tau = 2mT_s$ within the time window of length $N + 1$ (N even number) and index k corresponds to frequency $2k/T(N + 1)$. The unambiguous frequency representation is, because of the factor 2 before k in equation 15.17, between 0 and $f_s/2$.

The WVD has several very useful properties [8] in relation to an ISAR focusing application:

- $W(t, f)$ is real and can represent the variation in energy
- the first moment with respect to f yields the instantaneous frequency $f_{d_m}(t)$:

$$f_{d_m}(t) = \frac{\int_{f=0}^{\infty} f W(f, t) \, df}{\int_{f=0}^{\infty} W(f, t) \, df} \tag{15.18}$$

- the spectral extent of the WVD gives an indication of the signal bandwidth
- the WVD is time and frequency invariant; shifts in time or frequency are followed by the WVD

For illustration we give some examples of the WVD for a sampled signal series, computed with a MATLAB program using equation 15.17. The total number of samples, taken at a sampling frequency f_s, is chosen as $n_1 = 450$. The window slides along this series with a width of $N + 1 = 41$. In Figure 15.10 we first have a signal with constant Doppler frequency $f_d = 0.25 \, f_s$.

Figure 15.11 shows the WVD of a linear ascending frequency (LFM, chirp) up to half the sampling frequency f_s. In Figure 15.11 we use a contour plot as an alternative presentation. In Figure 15.12 the frequency is oscillating according to an assumed

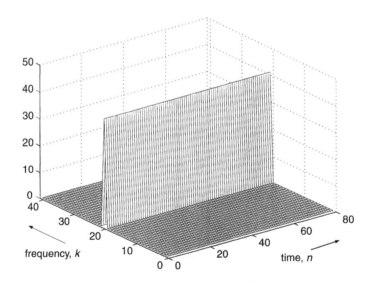

Figure 15.10   WVD of constant Doppler signal

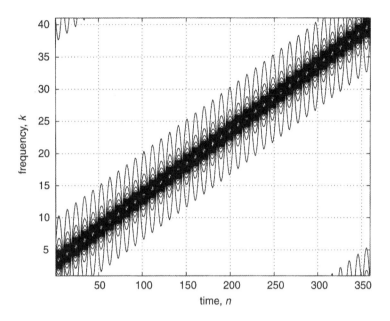

Figure 15.11   WVD of linear frequency modulation

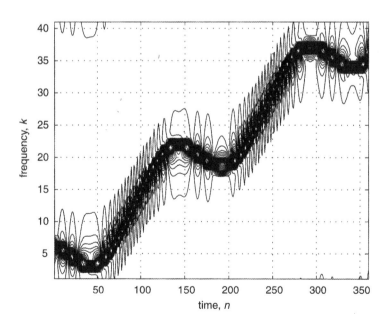

Figure 15.12   WVD for LFM and additional phase oscillation

signal given by:

$$z = \exp\left[j\pi \cdot 0.5 \cdot f_m \cdot n^2 + j \cdot 1.5 \cdot 2\pi \sin\left(2\pi n/150\right)\right] \qquad (15.19)$$

The WVD follows with its maximum all these frequency variations, as we require.

After determination of the frequency function according to equation 15.18 the focusing phase is derived by integration:

$$\varphi_c(t_k) = 2\pi \int_{\tau=0}^{t_k} f_{d_m}(\tau)\, d\tau \qquad (15.20)$$

This focusing phase is used for forming the set of narrow beams for image generation according to equation 15.4 and using for the scale equation 15.7:

$$S_1(\Theta) = \sum_{k=-N}^{N} z_k \exp\left(-j\varphi_{c_k}\right) \exp\left(-\frac{4\pi j}{\lambda} x_k \sin\Theta\right) \qquad (15.21)$$

For extended targets a multidimensional WVD has been studied [12] and applied to experimental data to use also the directional information available from an active receiving array.

15.5   High range resolution

We have seen a possible cross-range resolution of the order of 1 m. To get a two-dimensional image it would be desirable to match the range and cross-range resolution. This means, as discussed in chapter 7, a required signal bandwidth of 150 MHz. This would be the frequency shift of a frequency-modulated pulse (linear or nonlinear), or we could apply a phase-coded pulse with a subpulse length of only 6.6 ns. This high bandwidth is generally not available with multifunction radar systems and not required for the other functions such as target search and tracking.

An elegant solution is a synthetic high bandwidth with stepped frequency pulses [5,10].

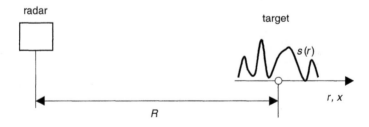

Figure 15.13   Target to be resolved in range

Instead of a single pulse a series of $2N + 1$ pulses is transmitted. The frequency for the kth transmitted pulse is chosen by:

$$f_k = f_0 + k\Delta f \quad k = -N, \ldots, 0, \ldots, N \tag{15.22}$$

The total used frequency band is $B = 2N\Delta f$ around f_0. The pulse length τ is chosen to cover the whole target extent in range, that is $\tau \geq 2\Delta r/c$, with Δr the target range extent and c velocity of light. The target centre may be at range R, the scattering elements of the target may have a distance r relative to this centre. The scattering amplitude is described by a function $s(r)$.

The received echoes are mixed down with the same frequency as transmitted and we get the complex amplitudes at frequency $k(-N, N)$:

$$S_k = S_{I_k} + j S_{Q_k}$$
$$= A \int dr\, s(r) \exp\left[-j\frac{4\pi}{c}(f_0 + k\,\Delta f)(R + r)\right] \tag{15.23}$$

given by integration along the target extent.

The reconstruction of $s(r)$ is possible by a Fourier transform. The result is denoted by $\hat{s}(x)$, with x also the range from the target centre:

$$\hat{s}(x) = \sum_{k=-N}^{N} \exp\left[j\frac{4\pi}{c}k\Delta f x\right] S_k \tag{15.24}$$

We will now discuss the properties of this reconstruction, with respect to resolution and the unambiguous range window.

We apply equation 15.23 in equation 15.24:

$$\hat{s}(x) = \sum_{k=-N}^{N} \exp\left[j\frac{4\pi}{c}k\Delta f x\right] \int dr\, s(r) \exp\left[-j\frac{4\pi}{c}(f_0 + k\Delta f)(R + r)\right] \tag{15.25}$$

The target range R results only in a phase factor. For simplicity we assume $R = 0$. We rewrite equation 15.25:

$$\hat{s}(x) = \int dr\, s(r) \exp\left[-j\frac{4\pi}{c}f_0 r\right] \sum_{k=-N}^{N} \exp\left[j\frac{4\pi}{c}k\Delta f(x - r)\right] \tag{15.26}$$

The sum expression is again using the formula given by equation 4.5 in chapter 4:

$$\hat{s}(x) = \int dr\, s(r) \exp\left[-j\frac{4\pi}{c}f_0 r\right] \frac{\sin\left[(2N + 1)2\pi\,\Delta f(x - r)/c\right]}{\sin\left[2\pi\,\Delta f(x - r)/c\right]}$$
$$\hat{s}(x) = (2N + 1) \int dr\, s(r) \exp\left[-j\frac{4\pi}{c}f_0 r\right] \frac{\sin\left[(2N + 1)2\pi\,\Delta f(x - r)/c\right]}{(2N + 1)\sin\left[2\pi\,\Delta f(x - r)/c\right]} \tag{15.27}$$

This equation performs a convolution of $s(r)$ and the point-spreading function given by the last term, which is approximately the function $\sin(\xi)/\xi$ with $\xi = (2N + 1)2\pi \Delta f (x - r)/c$. The second term results in an additional phase rotation which we may neglect, because we are interested only in the amplitude information.

Therefore, each target point, given by a delta impulse $\delta(r)$, has as its image a $\sin(\xi)/\xi$ function. The 3 dB width of this image, δr, is given by:

$$2\pi(2N + 1) \Delta f \, \delta r/2c = 1.39$$

or:

$$\delta r = 0.442 \frac{c}{(2N + 1) \Delta f} \approx \frac{c}{2B} \qquad (15.28)$$

The unambiguous range window follows also from equation 15.26: the exp sum expression is the same if for every k holds:

$$\exp\left[j\frac{4\pi}{c}k \Delta f (r + \Delta r)\right] = \exp\left[j\frac{4\pi}{c}k\Delta f(r)\right]$$

resulting in:

$$\frac{4\pi}{c}\Delta f \Delta r = 2\pi$$

or:

$$\Delta r = \frac{c}{2\Delta f} \qquad (15.29)$$

With equations 15.28 and 15.29 we can choose the radar parameters for the high-resolution mode.

For example we choose:

$$\delta r = 1\,\text{m}$$

$$\Delta r = 100\,\text{m}$$

Then follows:

$$B = 150\,\text{MHz}$$

$$\Delta f = 1.5\,\text{MHz}$$

This means that we have to transmit for each range profile a burst of $2N + 1 = 101$ pulses. Each burst corresponds to one pulse for cross-range focusing, discussed in section 15.3. The pulse length matched to the range window Δr may be chosen $\tau = 1\,\mu\text{s}$. Because we know the target range R from the tracking procedure we can apply an ambiguous pulse recurrence frequency (PRF) to achieve a short frame time for generating the complete range profile. The time schedule is sketched in Figure 15.14.

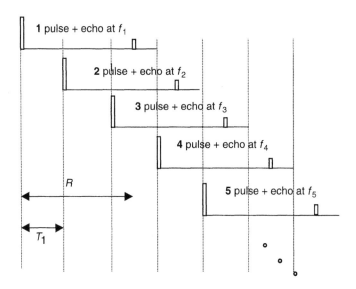

Figure 15.14   Time schedule for high-range resolution by synthetic high bandwidth

The solid-state transmit amplifiers of a multifunction array radar allow an arbitrary waveform, described by pulse length and pulse period. The target is illuminated for all pulses k with increasing frequencies f_k with subinterval length T_1, as indicated in the figure. The *LO* frequency has to be increased, as mentioned above, accordingly during reception of the echo within interval k.

We may choose for our example $T_1 = 5\,\mu s$ and the total frame time for one range profile would be $T = 0.505$ ms. The range variation within this short interval would be for a radial velocity of 300 m/s only 0.15 m, small compared to the assumed high-resolution range bin of 1 m.

After forming the high-resolution range profiles a range alignment procedure generally has to follow. For this alignment a crosscorrelation between neighbouring profiles may be used to find the range shift for correction [10]. Then the cross-range resolution is evaluated for each range element according to sections 15.3 and 15.4. The result then is the desired two-dimensional image.

An example of a two-dimensional image, evaluated under favourable conditions, is given in Figure 15.15 [13]. It shows an aircraft target of opportunity (Learjet), the darkness representing the resolved target reflectivity. The measurement has been performed with a high-performance tracking radar at K_u band. The target was at a range of about 190 km, flying on a smooth right turn.

Instead of the discussed series of pulses with a stepped frequency according to equation 15.22, a series of pulses representing segments of an LFM pulse could be transmitted [14]. These partial chirps make use of the available signal bandwidth. The advantage would be resolution enhancement with only a few pulses. For targets with higher relative velocities motion compensation is necessary.

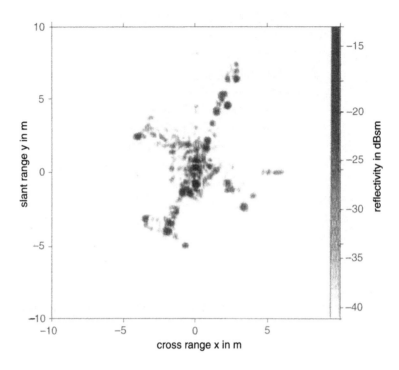

*Figure 15.15   Two-dimensional radar image of aircraft, K_u band, resolution 0.25 m
for both dimensions (courtesy K. Magura)*

15.6   Alternative image planes

As discussed above the image plane contains the LOS (range resolution) and is orthogonal to the rotation axis of the target with respect to our radar (cross-range resolution).
This has been produced so far by a more or less straight horizontal flight path resulting in a vertical rotation axis. The image plane is therefore horizontal and we get the
target image as seen from above the target.

In some cases the target produces rotations around another axis. In the case of a
ship target the pitch, roll and yaw may be evaluated for respective images [5]. The
rule given above applies in all cases. Examples for these cases are:

- if a ship is approaching our radar the pitch allows an image in the length/height
  dimension (range/cross range), as seen from broadside
- if a ship is approaching our radar the yaw allows an image in the length/breadth
  dimension (range/cross range), as seen from above
- if a ship is passing broadside to our radar the roll allows an image in the
  breadth/height (range/cross range) dimension, as seen in a front view

Generally, the movements will be superposed resulting in a rotation axis varying
with time. The interpretation of the corresponding image series is an interesting and
demanding task of its own.

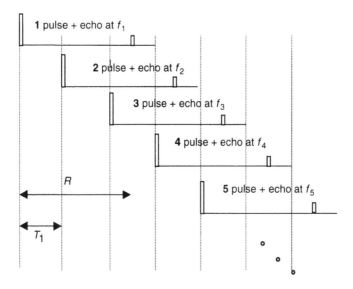

Figure 15.14 Time schedule for high-range resolution by synthetic high bandwidth

The solid-state transmit amplifiers of a multifunction array radar allow an arbitrary waveform, described by pulse length and pulse period. The target is illuminated for all pulses k with increasing frequencies f_k with subinterval length T_1, as indicated in the figure. The *LO* frequency has to be increased, as mentioned above, accordingly during reception of the echo within interval k.

We may choose for our example $T_1 = 5\,\mu s$ and the total frame time for one range profile would be $T = 0.505\,ms$. The range variation within this short interval would be for a radial velocity of 300 m/s only 0.15 m, small compared to the assumed high-resolution range bin of 1 m.

After forming the high-resolution range profiles a range alignment procedure generally has to follow. For this alignment a crosscorrelation between neighbouring profiles may be used to find the range shift for correction [10]. Then the cross-range resolution is evaluated for each range element according to sections 15.3 and 15.4. The result then is the desired two-dimensional image.

An example of a two-dimensional image, evaluated under favourable conditions, is given in Figure 15.15 [13]. It shows an aircraft target of opportunity (Learjet), the darkness representing the resolved target reflectivity. The measurement has been performed with a high-performance tracking radar at K_u band. The target was at a range of about 190 km, flying on a smooth right turn.

Instead of the discussed series of pulses with a stepped frequency according to equation 15.22, a series of pulses representing segments of an LFM pulse could be transmitted [14]. These partial chirps make use of the available signal bandwidth. The advantage would be resolution enhancement with only a few pulses. For targets with higher relative velocities motion compensation is necessary.

9 BARBAROSSA, S., and FARINA, A.: 'Detection and imaging of moving objects with synthetic aperture radar, part 2: joint time-frequency analysis by Wigner-Ville distribution', *IEE Proc. F, Radar Signal Process.* 1992, **139** (1), pp. 89–97

10 WEHNER, D. R.: 'High resolution radar' (Artech House, Inc., Norwood, MA, USA, 1995, 2nd edn)

11 RUTTENBERG, K., and CHANZIT, L.: 'High resolution by means of pulse to pulse frequency shifting'. Proceedings of IEEE *EASCON*, 1968, pp. 47–51

12 RIECK, W.: 'Time-frequency distribution of multi-channel SAR-data for auto-focussing of moving targets'. Proceedings of IEE international Conference on *Radar*, Edinburgh, UK, 1997, pp. 224–228

13 MAGURA, K., and RUSTEMEIER, M.: 'Nichtkooperative Zielidentifizierung von Flugobjekten durch Radarabbildung', *Frequenz*, 1996, **50** (7–8), pp. 147–156

14 BERENS, P.: 'SAR with ultra-high range resolution using synthetic bandwidth'. Proceedings of *IGARSS* 99, 1999, Session CC05

Chapter 16
Target classification

Classification and possibly identification of observed radar targets is a natural requirement for radar systems applied for air or maritime traffic control and defence. In all cases without target information, for example provided by secondary surveillance radar (SSR) or with identification of friend or foe (IFF), the radar echo signal has to be evaluated. For air surveillance it is of great importance to distinguish as fast as possible between targets of interest, such as airplanes, and false targets like birds or flocks of birds. By this distinction the formation and processing of false tracks, thereby wasting computer and radar time and power is avoided.

Possibilities and some results for radar imaging of targets have been discussed in chapters 14 and 15. Target imaging is, of course, the most advantageous method for classification. In cases where this is not possible, other methods for classification may be applied.

Attempts have also been made to use the echo series of a target for target classification. The echo series of airplanes is influenced by typical internal motion, vibration or rotation. Vibration and rotation effects cause a signal fluctuation and modulation in contrast to an ideal point target. In particular, the rotation of propeller and jet engine cause modulation of the echo signal. These fluctuation and spectral effects can be used if the radar waveform is chosen suitably to perform adequate sampling. With a phased-array multifunction radar we can use the freedom of waveform selection and dwell time to evaluate these effects adequately.

In an era of more advanced radar techniques, incoherent methods have initially been tried for target classification, for example by observing the propeller modulation [1]. Here, coherent techniques will be applied, evaluating the produced Doppler spectrum.

16.1 Fluctuation effects

Extensive target measurements have been performed by FGAN-FFM with several coherent radar systems and a digital recording system [2,3]. From the results some proposals for target classification were derived [4].

We suggested the separate evaluation of amplitude and phase fluctuation. As fluctuation estimators g were applied the ratios of standard deviation to mean of amplitude and phase, respectively.

The echo series with N signals out of one target dwell is given by:

$$z(n) = a_n \exp\left(j\Phi_n\right) \tag{16.1}$$

This signal was extracted out of the recorded data window, which was given by a range gate and an angular sector where the rotating radar antenna sampled the selected target with a series of radar periods. The echo series for evaluation of a target was selected from the respective range bin.

For the amplitude we get for mean and standard deviation, corresponding to equations 3.31 and 3.33:

$$\bar{a} = \frac{1}{N}\sum_{1}^{N} a_n$$

$$\sigma_a^2 = \left[\frac{1}{N-1}\left(\sum_{n=1}^{N} a_i^2 - N\bar{a}^2\right)\right]$$

and for the fluctuation estimator:

$$g_a = \sigma_a/\bar{a} \tag{16.2}$$

For determining the phase-fluctuation estimator the phase difference of consecutive signals was first computed:

$$z(n) = a_n \exp\left(j\Delta\Phi_n\right) \tag{16.3}$$

All targets generally show a slowly varying Doppler frequency shift f_d. The phase difference $\Delta\Phi$ is determined by f_d and the radar pulse period T: $\Delta\Phi = f_d T$. If this phase difference is nearly constant then there is little fluctuation.

To avoid the problem of the 2π step we applied averaging of the components of the phasor $p = \exp\left(j\Delta\Phi_i\right) = x_i + jy_i$. The variance of the phase difference is expressed by:

$$\sigma_p^2 = \frac{1}{N-1}\left[\sum_{n=1}^{N}(x_i^2 + y_i^2) - \frac{1}{N}\left(\sum_{n=1}^{N} x_i\right)^2 - \frac{1}{N}\left(\sum_{n=1}^{N} y_i\right)^2\right]$$

σ_p is related to the mean of p:

$$\bar{p} = \left(\left(\frac{1}{N}\sum_{n=1}^{N} x_i\right)^2 + \left(\frac{1}{N}\sum_{n=1}^{N} y_i\right)^2\right)^{1/2}$$

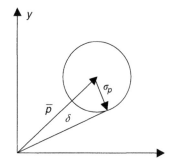

Figure 16.1   Phase fluctuation angle

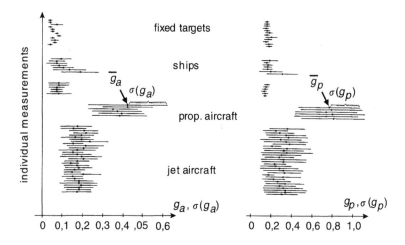

Figure 16.2   Amplitude and phase fluctuation estimators g_a and g_p with their mean and standard deviation for different types of target, each bar represents a measurement series © IEE 1977 [4]

As illustrated in Figure 16.1, σ_p indicates the radius of a fluctuation circle around the mean $\bar{p}$. The phase fluctuation estimator:

$$g_p = \frac{\sigma_p}{\bar{p}} \tag{16.4}$$

can be explained as $\sin \delta \approx \delta$, expressing the estimator g_p as an angle.

Both estimators have been applied to various measured target data [4]. The results are summarised in Figure 16.2. On the left-hand side we see the amplitude fluctuation estimator g_a and on the right-hand side the phase fluctuation estimator g_p, both with their standard deviation within the measurement series, for various targets. Each horizontal bar corresponds to a series of scans of the observed target, and a decision between target types may be made by a threshold comparison. A distinction between

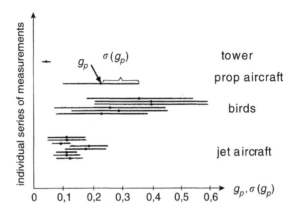

Figure 16.3   Phase fluctuation estimator g_p with mean and standard deviation for different types of target, each bar represents a measurement series © IEE 1977 [4]

propeller and jet airplanes is possible with a probability of 0.94 for correctly deciding for a jet airplane and 0.1 for incorrectly rejecting it.

Birds often show irregular movements and wing beats. Their high corresponding phase fluctuation may be used for recognition. Measurement results are given in Figure 16.3 for comparison with aircraft echo fluctuation.

16.2   Doppler spectrum evaluation

A part of the radar pulse reflected from an inbound flying aircraft target is scattered back from the blades of the jet engines. The exploitation of jet-engine modulation (JEM) seems to be an effective contribution to target classification. This phenomenon is observable only within an angular sector of about 60° from the nose-on direction. The rotating fans create with their blades a periodic modulation of the reflected echoes. These modulation effects can be seen in the spectrum of the received signal series by certain frequency lines. Spectral analysis procedures are therefore the main tools for using JEM [5].

The received echo signal may have amplitude and phase modulation. These modulations will have a fundamental period given by the rotation period, additionally, there may be modulations from the rotating fans creating harmonics depending on the number of blades.

We will discuss a simple model to illustrate the generated spectrum and the evaluation methods.

The received signal is given with the carrier frequency f_0 and amplitude modulation $a(t)$ and phase modulation $\Phi(t)$ by:

$$r(t) = a(t) \exp(j\Phi(t)) \exp(j2\pi f_0 t) = z(t) \exp(j2\pi f_0 t) \qquad (16.5)$$

After mixing the signal down to baseband we have to consider only the complex modulation $z(t)$.

We assume for convenience that we have compensated for the Doppler shift resulting from the radial velocity of the aircraft. The elemental scatterers will generate a modulation in phase and amplitude with a frequency given by the rotation rate and the number of blades. By superposition of all these scattering contributions their frequency content will be unchanged.

The scattering elements may change their range in a sinusoidal manner. The model signal is composed of amplitude and phase modulation. We assume two stages of the turbine with k_1 and k_2 fan blades, respectively. The rotation of the turbine is at the fundamental frequency f_T and the blades produce the fundamental frequencies $f_T k_1$ and $f_T k_2$:

$$
\begin{aligned}
z_m(t) = a_0 &+ a_1 \exp\left(j 2\pi f_T k_1 t\right) + a_2 \exp\left(j 2\pi f_T k_2 t\right) \\
&+ b_1 \exp\left(j 2\pi c_1 \sin\left(2\pi f_T k_1 t\right)\right) + b_2 \exp\left(j 2\pi c_2 \sin\left(2\pi f_T k_2 t\right)\right) \\
&+ b_3 \exp\left(j 2\pi c_1 \sin\left(2\pi f_T k_1 t\right) + j 2\pi c_2 \sin\left(2\pi f_T k_2 t\right)\right)
\end{aligned}
\tag{16.6}
$$

The motivation for this set-up is the following:

- a_0 is a constant term
- a_1 is the amplitude at frequency $f_T k_1$
- a_2 is the amplitude at frequency $f_T k_2$
- b_1 is the amplitude of a term with sinusoidal phase modulation at frequency $f_T k_1$, the amplitude of this phase modulation is c_1; c_1 depends on the range variation with respect to the carrier wave length, higher frequencies result in a higher value for c_1; the corresponding function is given for k_2
- the third row has the sum of both phases

We assume for the two-stage jet engine a first stage having a 21-blade fan and a second stage having a 33-blade fan. The fundamental rotation frequency f_T is assumed to be 100 Hz [5].

With a MATLAB program the spectrum of z_m is computed for a variety of parameter assumptions. The number of time samples is selected $N = 1024$ during one period $T = 1/f_T$. By the FFT the spectrum $F(f)$ is computed up to frequency $N f_T$, that is for our parameters from 0 in steps of 100 Hz up to 102.4 kHz.

We first assume only amplitude modulation according to the first row of equation 16.6. The result is given in Figure 16.4. There are only the two corresponding spectral lines at 2.1 and 3.3 kHz. Next we assume for Figure 16.5 only phase modulation at frequency k_1. The factor c_1 is chosen to be equal to 3, which means that the phase varies across the phase interval $3 \times 2\pi$. Now we have harmonics at multiples of frequency k_1. One observes an increasing number of harmonics with an increasing phase interval. We observe the same for frequency k_2 in Figure 16.6 with a factor $c_2 = 2$. Then we add the sinusoidal phases for both frequencies according to the third row of equation 16.6. Now, as shown in Figure 16.7, many spectral lines

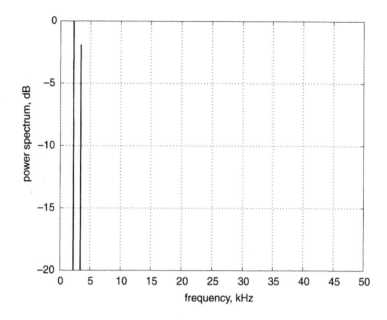

Figure 16.4   Model spectrum of amplitude modulation for two fans

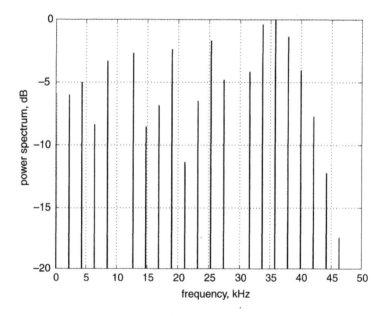

Figure 16.5   Model spectrum by phase modulation at frequence k_1

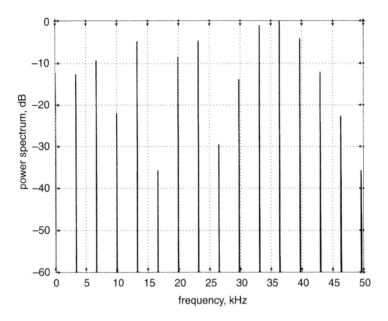

Figure 16.6   Model spectrum by phase modulation at frequency k_2

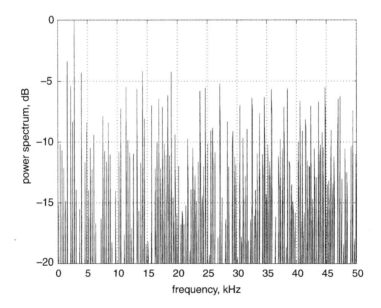

Figure 16.7   Summation of phase modulations k_1, k_2

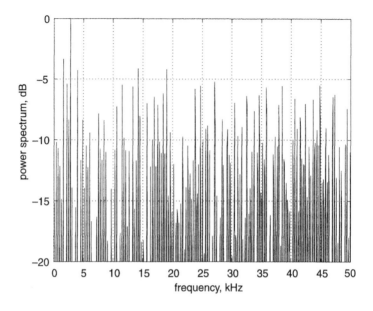

Figure 16.8   Combination of all three types of modulation

appear and the spectrum becomes more noisy. Finally, all model signals are summed up leading to the spectrum shown in Figure 16.8. Each spectrum is normalised to its respective maximum value. The harmonic lines from both frequencies k_1 and k_2 are now superposed together with the noisy spectrum from the added phase modulation signal. This spectrum is processed with the so-called cepstrum, which is defined for the spectrum $F(f)$ and the inverse Fourier transform as:

$$C(\tau) = FFT^{-1}(\log|F(f)|) \tag{16.7}$$

In Figure 16.9 we see the cepstrum according to equation 16.7. It shows a clear peak at a time $\tau = 333$ of 1000 units (10 μs). This corresponds to the one third rotation symmetry because the common divisor of 21 and 33 is just 3.

In Figures 16.10–16.13 finally two experimental measured JEM spectra and their corresponding cepstra are shown. In principle, these look similar to our model spectrum. The generation of all the spectral lines has heuristically been explained by the combination of spectral lines according to the fan-blade numbers and their ratio. So we can expect an individual spectrum according to the respective engine type. For a real measured spectrum the inverse problem has to be solved to find the spectral harmonics fitting together. For efficient classification of aircraft with jet engines the set of measured and recorded spectral patterns for the different aircraft types must be available as a reference basis. For obvious reasons these have not been published.

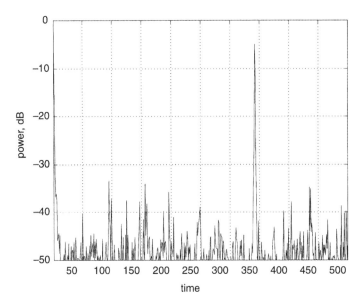

Figure 16.9   Cepstrum for the spectrum in Figure 16.8

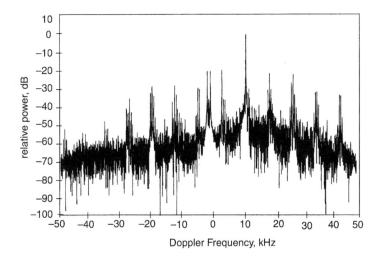

Figure 16.10   Example 1, measured JEM spectrum (courtesy H. Schneider FGAN-FHR)

The precondition for the evaluation of the JEM spectrum is a sufficiently high pulse recurrence frequency for recognising the higher-order spectral lines up to about 20–50 kHz. With a solid-state multifunction radar with its waveform agility this should be possible to establish. The observation time has to be at least some rotation periods of the engine, which means some 10 ms.

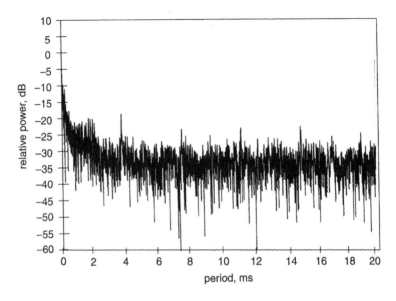

Figure 16.11   Example 1, cepstrum to the spectrum of Figure 16.10 (courtesy H. Schneider, FGAN-FHR)

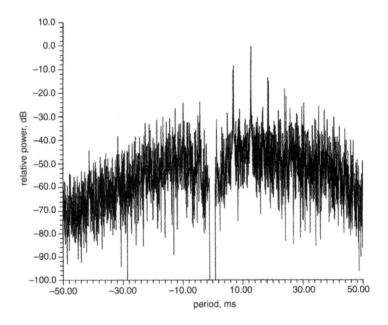

Figure 16.12   Example 2, measured JEM power spectrum (courtesy H. Schneider FGAN-FHR)

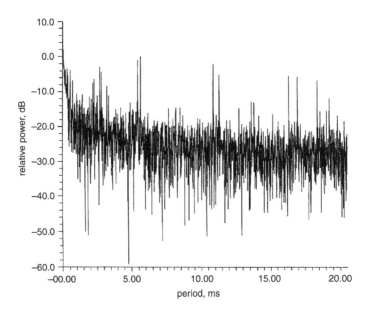

Figure 16.13   Example 2, cepstrum to the spectrum of Figure 16.10 (courtesy H. Schneider, FGAN-FHR)

16.3   References

1 KERR, D. E., and GOLDSTEIN, H.: 'Radar targets and echoes', in 'Propagation of short radio waves, vol. 13' (MIT Radiation Laboratory Series, McGraw-Hill Book Company, New York–Toronto–London, 1951) chapter 6, pp. 539–543

2 WIRTH, W. D.: 'Clutter- und Signalspektren aus Radicord-Aufnahmen und die Bewegtzielentdeckung durch Impulsradar', *Nach. Tech. Z.*, 1968, **12**, pp. 759–765 (in German)

3 V. SCHLACHTA, K.: 'Characteristics of echoes from aircraft and clutter derived from digital recorded data'. AGARD *Radar* conference proceedings, Istanbul, Turkey, 1970, pp. 5.1–20

4 V. SCHLACHTA, K.: 'A contribution to target classification'. Proceedings of IEE international conference on *radar*, London, UK, October 1977, pp. 135–1439

5 BELL, M. R., and GRUBBS, R. A.: 'JEM modeling and measurement for radar target identification', *IEEE Trans. Aerosp. Electron. Syst.*, 1993, **29** (1), pp. 73–87

Chapter 17

Experimental phased-array system ELRA

Research work in the area of phased-array radar has been performed at the research institute FGAN-FFM (now FGAN-FHR) since 1970. The phased-array system ELRA (electronic steerable radar) serves as an experimental basis. The first successful radar operation with the tracking of flying targets dates back to the year 1975. Since then many modifications and further developments corresponding to advances in theoretical knowledge and in technology have been installed in the areas of hardware and software. Most of the new concepts described in this book have already been implemented in the ELRA system or are under preparation for implementation. The main effort has been concentrated in the areas of signal and data processing and system control.

The development of this experimental system has proved to be important for the verification of concepts and theoretical proposals and the recognition of new problems and questions. It was also, last but not least, necessary and useful for the demonstration of the system to interested visitors and personalities who are responsible for the determination of targets for the radar research work in Germany. The demonstrations were especially important for the motivation and encouragement of industrial developments in the field of phased-array radar.

The modular system concept offers the possibility of exchanging and modernising individual parts of hardware and software, which has been done several times since the beginning of the project. Only the two antenna cabins on the roof of the research institute of FGAN (formerly FFM and now FHR), shown in Figure 17.1, are used all the time. The system development was performed in steps, by increasing the number of antenna modules and increasing the number and efficiency of the radar functions and of the available waveforms. In this sense there will be no final complete system during the next few years, as long as there are new concepts to be implemented, verified and demonstrated.

The final aim is an adaptive radar system which is self adapting against varying disturbance signals such as clutter and jamming from multiple sources, selecting appropriate waveforms for the different tasks and distributing the available radar energy in an economic way to achieve secure detection and tracking of all targets of

Figure 17.1   ELRA antenna cabins for receive and transmit at FGAN

interest. Target recognition procedures will be applied according to the latest results and technical possibilities in this field.

In the following the basic concepts and system components will be described as an illustration for the variety of techniques applied.

17.1   System overview

The block diagram of the ELRA system in Figure 17.2 gives an overview of the main system components. At the beginning of the project we decided to use separate active array antennas for transmitting and receiving. Each antenna element has its own transmit or receive channel.

The following aspects led to the decision for separate arrays:

- independent distribution of antenna elements for receiving and transmitting on the aperture planes
- independent sidelobes for receiving and transmitting, which multiply for the two-way radar function and thus avoid single high effective sidelobes
- independent technologies for transmit and receive channels
- reduction of coupling problems between transmitting and receiving systems
- avoidance of the cost for transmit/receive switches
- cost minimisation by independently selecting the number of modules to achieve a certain radar range: because the transmit channels are more expensive compared to the receivers a higher number of receiver channels is advantageous; the product of element numbers determines the gain product and the range according to the radar equation discussed in chapter 4 section 4.5.2

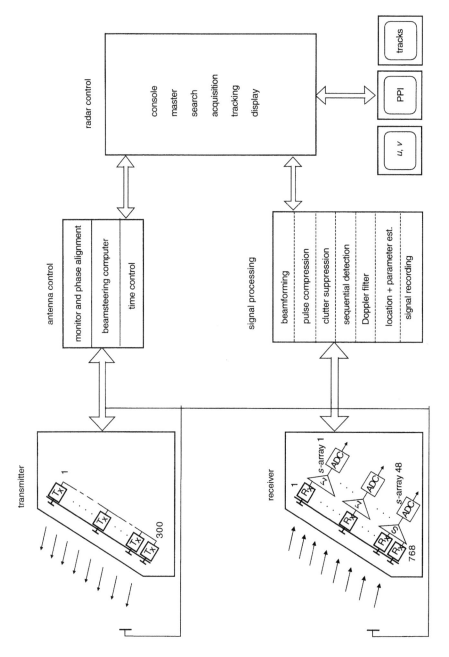

Figure 17.2   Block diagram for experimental phased-array system ELRA

The following advantages resulted from the choice of active arrays:

- application of solid-state devices for the transmitting channels for distributed generation of the total transmit power, as discussed already in chapter 4 section 4.5
- avoiding high-voltage devices
- possibility for a stepwise extension of the number of transmit and receive modules
- high operational safety and only graceful degradation in the case of antenna module failures, without an additional stand-by transmitter
- high flexibility and rich possibilities for array processing and beamforming on the receiving side

The antennas are connected to the control subsystems for timing, phase shifter control and monitoring. The whole system is of course coherent; the RF reference signal and all clock pulses are derived from a stable crystal-controlled oscillator. On the receiving side the signals are digitised after forming subarrays. The digitised signals are applied to the signal-processing block, comprising beamforming for the sum and two difference beams (azimuth and elevation), followed by pulse compression, clutter suppression, sequential detection, Doppler filtering and target location. Part of the final signal processing and especially the radar control are performed by a net of microcomputers (PCs).

17.2   Antenna parameter selection

The parameters of the ELRA system have been chosen according to the tasks of an air surveillance radar. The radar range should provide enough targets of opportunity from the usual air traffic, of the order of 20–50 airplanes, for the different experiments. Therefore the maximum range was chosen as 200 km. To achieve low propagation attenuation the carrier frequency band was selected within the S band (2–4 GHz). Finally, the frequency, allocated by the responsible authority, is 2.72 GHz or a wavelength of $\lambda = 11$ cm. The approved bandwidth is 1 MHz, therefore experiments with frequency agility were prohibited. The achievable range resolution is, with this bandwidth, 150 m. The 3 dB width of the pencil beam was selected at about $2°$ as a typical and common value for air surveillance. The planar antennas may be scanned $120°$ in azimuth. By tilting the antenna planes back by $30°$ the elevation can be scanned from 0 to $90°$.

To save on cost by installing only a moderate number of transmit and receive modules a randomly-thinned antenna element distribution was chosen. The required pencil beams are produced by planar circular arrays. Low near-in sidelobes are achieved by a thinning taper according to the Taylor weighting.

The transmit array has the following main parameters:

- planar circular aperture with a diameter of 3.1 m (this corresponds to 28 λ)
- 300 antenna elements with a thinned density-tapered distribution on a regular $\lambda/2$ grid; this corresponds to a filling factor of only 12 per cent in comparison with the fully filled array; with this element number a mean sidelobe level of -25 dB is achievable

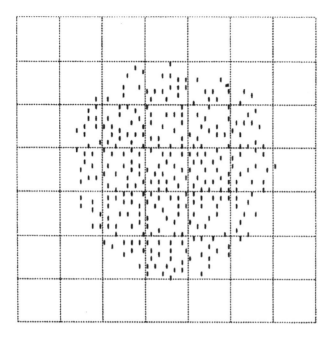

Figure 17.3   Distribution of antenna elements for ELRA transmit array

- the 300 transmit power amplifiers each deliver 10 W peak or 3 W mean power
- the final amplifiers are driven by an active distribution network with preamplifiers
- the pulse length is variable between 1 µs minimum and 640 µs maximum
- the pulse repetition frequency is variable between 250 Hz and 10 kHz
- the phase shifters before the final amplifiers have a resolution of three bit; this is sufficient for the achievable sidelobe level with this randomly-thinned array
- the antenna elements are etched vertical microstrip dipoles resulting in vertical polarisation

The distribution of the antenna elements for transmitting is shown in Figure 17.3. The main parameters of the receive array are the following:

- planar circular array with a diameter of 4.1 m (corresponding to 37 λ)
- 768 antenna elements with a randomly-thinned density-tapered distribution on a regular λ/16 grid; the element density also follows a Taylor function; the filling factor in relation to a regular spaced λ/2 array is 18 per cent
- the 3 dB beamwidth is about 2°
- the mean sidelobe level is −29 dB
- 768 active receive modules are connected to the antenna elements
- antenna elements are vertical dipoles for vertical polarisation, realised as etched microstrip devices; 200 antenna elements are crossed dipoles for measurements with vertical and horizontal polarisation
- the receiving channels are grouped into 48 subarrays according to Figure 17.4

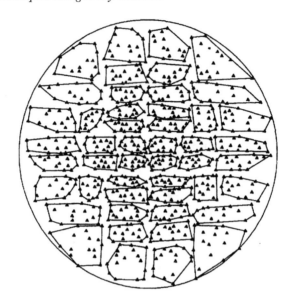

*Figure 17.4   Antenna element distribution of ELRA receive array, lines indicate
subarrays, two horizontal rows of subarrays contain crossed dipoles*

The main antenna and radar parameters together with a maximum pulse length 640 µs
and 16 integrated pulses result in a range of 200 km for targets with a radar cross
section of $\sigma = 1\,\text{m}^2$.

17.3   Antenna elements and modules

As antenna elements printed dipoles, etched using a microstrip technique on RT
duroid boards, are used. For transmitting the usual $\lambda/2$ dipoles with their arms
parallel to the ground plane are applied. They are directly connected to the transistor power amplifiers to avoid any losses from connecting cables. For the receiving
antenna, with a little increased element density, V-shaped dipoles are used to reduce
mutual coupling effects between neighbouring dipoles. Both antenna elements are
shown in Figure 17.5. All dipoles are mounted on the planar antenna ground planes
(aluminium plates) through vertical slots. The polarisation is therefore vertical.
192 receiving antenna elements, out of the total of 768, positioned in the centre two
horizontal rows of the subarrays, are crossed vertical and horizontal dipoles. Their
outputs can be selected by a fast pin-diode switch. This antenna element is shown
in Figure 17.6. With these dipoles a fan beam with horizontal polarisation may be
generated for special experiments.

The final transmit amplifiers are based on a bipolar microwave transistor and
a microstrip device, shown in Figure 17.5. Cooling by the aperture plane and
the surrounding air has proved to be sufficient. The maximum pulse length is

Figure 17.5   *Antenna modules for transmitting and receiving: dipoles with transmit power amplifier and low-noise receiver amplifier*

Figure 17.6   *Crossed dipoles for receiving of vertical and horizontal polarised signals, mounted on the antenna plane*

640 µs at an output power of 10 W, a duty cycle of 0.32 and an efficiency of 0.3. The amplifier is designed for class C operation, thus minimising amplitude variations caused by a varying input power by the attenuation variation of the steered phase shifters. The bandwidth of this amplifier is 100 MHz, which is

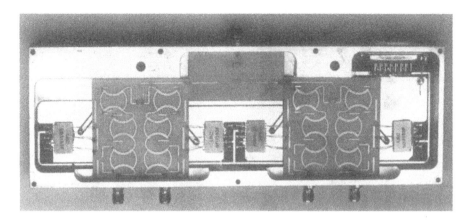

Figure 17.7   Box with four transmit phase shifters: three bits, pin-diode switches

more than sufficient for the fixed allowed frequency and the signal bandwidth of 1 MHz.

The phase of each transmit channel is steered by a pin-diode phase shifter prior to the final amplifier. The phase is quantised with three bits, that means 45° steps for the steered phase. This is adequate for the sidelobe level, achievable with the randomly-thinned array. The beam direction then can be steered with 0.1° accuracy. The phase steering is accomplished with microstrip branchline couplers and reflecting lines with their length adjusted for phase shifts of 45°, 90° and 180°. The transmission loss of the phase shifter is about 6 dB. The phase error is less than 10°. Four phase shifters are combined in one box, as shown in Figure 17.7.

The phase modulation of the transmit pulses for later pulse compression of the received signal is performed by one central phase shifter for the common transmit radio frequency (RF) reference signal. It is realised in the same manner as above but with six-bit resolution and is adjusted for a phase error of less than 2°.

The 768 receiving channels are equipped with a low-noise amplifier followed by a mixer with image frequency suppression, converting the signal down to the intermediate frequency (IF) of 30 MHz, followed by an IF preamplifier. The bandpass filter within the following IF amplifier is made using two different surface acoustic wave filters (SAW), one of which is matched to the subpulse length of 2 μs and the other to 10 μs [14]. This has proved to be a versatile solution for achieving an almost equal frequency filter characteristic for all receiving channels. The IF gain can be reduced for the short-range application by 20 dB to avoid signal limiting. The following mixer converts the IF signal down to baseband with orthogonal components I and Q.

Phase steering at the receiving side is achieved in conjunction with the baseband mixer. The reference signal is steered in its phase by a delay line with eight taps and a multiplex switch, steered by the three-bit phase-steering command. This economic and versatile solution avoids any amplitude variation by phase switching of the received signals.

17.4   Antenna control and monitoring

The main part of the antenna control is a special arithmetic unit for the serial computation of all steering commands for the phase shifters within the antenna modules. The main features of this unit are as follows:

- 0.2 µs computation time for each phase-steering command
- transmit and receive antenna may be controlled independently
- antenna focusing, selectable for near or farfield
- additional phase to broaden or shift the beam

The computed phase-steering commands and their resulting phase values have to compensate for the phase differences between the antenna channels caused by different cable length and inertial or transition phases of the amplifiers and filters. The main task, besides this phase correction, is to achieve a planar phase front for the desired beam direction, thus allowing the array to be focused into the farfield. For special experiments or measurements a special quadratic phase front is formed, focusing the antenna beam to an auxiliary antenna in the nearfield in front of the transmit or receive array antennas, respectively.

These auxiliary antennas are at a fixed and known position. They are used for automatic testing and monitoring of the antennas [8,13].

On the transmitter side a small part of the transmitted pulse is automatically coupled to the test antenna, so that the complete pulse, superposed from all channels, can be monitored. A suitable switching procedure for the phase shifters for the individual channel phases allows the extraction of the basic inertial channel phases and monitoring of the phase-shifter states.

For the N antenna elements with index i $(i = 1, \ldots, N)$ the positions on the antenna plane are given by their coordinates (x_i, y_i). For a desired beam direction (u, v) the steering phase is computed by:

$$\Phi_i = u x_i + v y_i + \Phi_{ic} + \Phi_{if} + \Phi_{ib} \qquad (17.1)$$

The phase correction values Φ_{ic} are used to focus the beam to the nearfield auxiliary antenna. The phase values Φ_{if} are applied to transform the focusing from the nearfield to the farfield. Finally, the phase values Φ_{ib} are applied optionally to achieve a certain beam forming.

By steering the phase of a selected transmit channel i to the states $0°$ and $180°$ the test signal at the auxiliary antenna results in two different signal vectors as shown in Figure 17.8. The phase of the difference vector gives the channel phase $-\Phi_{ic}$. All values Φ_{ic} are measured serially and represent the relative channel phases. Phase changes in the antenna channels caused by temperature effects, change of connecting cable length, amplifiers or the phase drift of other components are compensated for automatically by this method. During the measurement of one selected channel all the other channels have a fixed phase and contribute a superposed constant vector, cancelled by forming the difference vector.

Additionally, all the other phase states are selected for the channel i. In case of ELRA with three-bit phase resolution these are eight states at $45°$ steps. The results

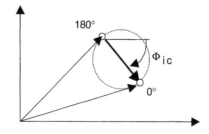

Figure 17.8   Measurement of channel phase

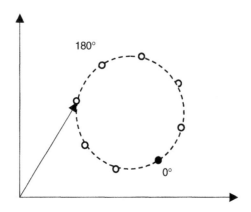

Figure 17.9   Phase states of 3-bit phase shifter

should appear approximately on a circle as sketched in Figure 17.9. Otherwise a failure of the phase shifter may be recognised.

On the receiver side a corresponding auxiliary antenna transmits a test pulse to all antenna elements. By switching the phase shifters and evaluating the sum beam output the basic inertial channel phases are derived and the phase shifters are checked accordingly. In Figure 17.10 some failure cases of the receiving channel are shown, which can be detected automatically by a corresponding diagnostic procedure. The channel amplitude is also monitored by this sequential phase steering. The diameter of the circle, estimated by a least-square fit to the measured points, gives the estimate of the relative channel gain. In Figure 17.11 the transmit antenna elements are plotted with vertical bars and the faulty elements are indicated by stars. This diagram is available online on a monitor.

The set of phase correction values is measured and stored for the selectable bandwidth (0.1 and 1 MHz) and the two receiver gain settings.

The measurement channel with the auxiliary antenna must have a sufficient dynamic range to cope with this superposed signal from all the other resting channels. It is realised with a ten-bit analogue-to-digital converter. To reduce the dynamic

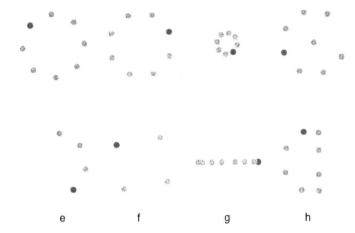

e   f   g   h

Figure 17.10   *Fault diagnoses from phase switching (courtesy W. Sander)*
   a correct element, dark point for zero phase
   b correct element, dark point for another zero phase
   c amplitude or gain too low
   d one phase state faulty
   e faulty phase shifter, 180° bit missing
   f faulty phase shifter, 45° bit missing
   g faulty receive channel, Q component missing
   h faulty receive channel, limiting of I component

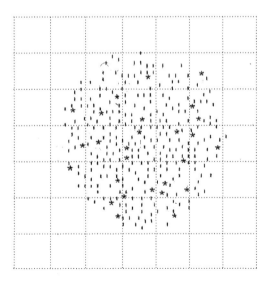

Figure 17.11   *Transmit antenna with faulty channels, marked by* ∗
   (courtesy W. Sander)

range a multiplexer selects the output of that subarray with the measured channel. To increase the signal-to-noise ratio further an integration of 64 succeeding measured signals is applied. For the transmit antenna the dynamic range is reduced by switching off a part of the antenna modules.

These measurement results are used for automatic failure analysis to facilitate the maintenance of the antenna system. Up to a tolerable number of failures there is no replacement necessary.

The nearfield-farfield transformation phase Φ_{if} may be computed in principle from the known position of the auxiliary antenna and its distance to the antenna elements. A more efficient focusing was achieved by a phase measurement performed once between all antenna elements and the fixed auxiliary antenna. The same dipole has been placed for this measurement into all element positions [8].

By scanning the antenna beam, focused on the nearfield, across the auxiliary antenna a cut through the antenna beam scan pattern is measured. This gives a versatile indication of the main beam width and the sidelobe level. Measurements with isolated targets in the farfield showed a good agreement with these nearfield measurements with respect to beamwidth and sidelobe level.

In Figure 17.12 we see the diagram of a single dipole, the scan pattern of the receiving antenna without focusing with the correction phases Φ_{ic} and with focusing.

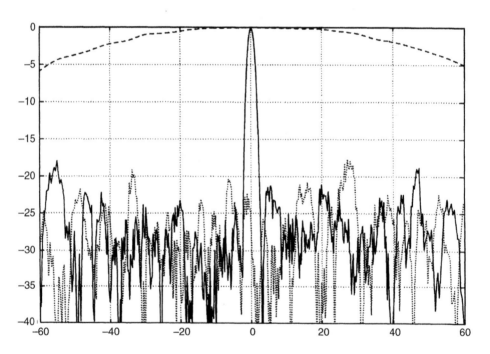

Figure 17.12   *Receiving antenna scan pattern (gain in dB above scan angle in °),*
measured with auxillary antenna (solid line), scan pattern with-
out focusing (line dotted), single element pattern (dashed line)
(courtesy W. Sander)

The ratio of the main beam peak-to-sidelobe level corresponds to the expected value according to the rule of thumb $1/N$, given in chapter 4 by equation 4.30.

This approach to antenna monitoring has the following advantages:

- by radiation coupling with the auxiliary antenna a coupling network for feeding a reference signal to all antenna channels is saved
- by phase-shifter steering of individual antenna channels selection switches are saved
- the monitoring also comprises the antenna elements
- coupling effects are included in the measurements; remark: the auxiliary antennas are in the nearfield of the whole array but in the farfield of single antenna elements and the subarrays

By the additional optional phase Φ_{ib} the beam may be formed by phase-only weighting. In particular a broadening of the transmit beam by a quadratic phase across the array is possible, as shown in Figure 17.13. The broadened transmit beam may be combined with a cluster of narrow receive beams stacked in elevation, according to chapter 5 section 5.3.6, for the purpose of accelerating the search function at shorter ranges. Measured beam patterns are shown in Figure 17.14. The reduced transmit gain, caused by this defocusing, is acceptable for this application. The angular resolution is maintained by the narrow receive beams.

Apart from the antenna monitoring, a sequence and timing control produces:

- transmit pulses with selectable phase modulation, length, period and number
- range gates steerable in width and position for the receiving signal
- clock and synchronising pulses for signal processing
- switching the processing path for the received signals

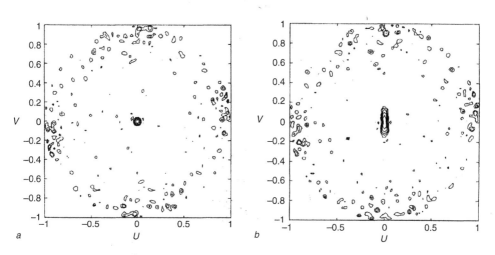

Figure 17.13 *Contour plots of the transmit beam above u,v plane (courtesy W. Sander)*

    a standard beam,

    b beam broadened in elevation by phase weighting

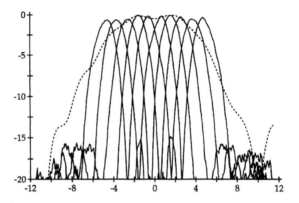

*Figure 17.14   Beam pattern in dB above angle in degrees, broadened transmit beam
(elevation) and cluster of narrow receive beams (courtesy W. Sander)*

17.5   Radar functions and waveforms of the ELRA system

The chosen parameters for the array antennas have been presented in section 17.2.
The waveforms have to be chosen suitably to achieve the desired target detection
performance. We required a detection range for targets of opportunity of about 200 km.
The well-known radar equation is used to select the remaining parameters.

For the signal-to-noise power ratio SNR we have with:

τ = pulse length	640 µs (selected)
P_T = total transmit power	300×10 W
G_T = transmit gain	$300\,\pi$
G_R = receiving gain	$768\,\pi$
λ = wavelength	0.11 m
σ = radar cross section	$2\,\text{m}^2$ (small aircraft)
R = range	to be determined
F = noise factor	2 (3 dB)
L = losses	4 (6 dB)

$$\text{SNR} = \frac{\tau P_T G_T G_R \lambda^2 \sigma}{(4\pi)^3 R^4 k T_0 F L} \tag{17.2}$$

We assume an integration of 16 received pulses, therefore an SNR of 1 or 0 dB
is sufficient for the single pulse. We solve equation 17.2 for R using the selected
parameters:

$$R^4 = \frac{640 \cdot 10^{-6} \cdot 10 \cdot 300 \cdot 300 \cdot \pi \cdot 768 \cdot \pi \cdot 0.11^2 \cdot 2}{1 \cdot (4\pi)^3 \cdot 4 \cdot 10^{-21} \cdot 2 \cdot 4} = 16.64 \cdot 10^{20}$$

or:

$$R = 202 \, \text{km}$$

With these parameters the ELRA system is, with respect to its range, comparable to a medium-range air surveillance system (ASR). The experimental results will therefore also be relevant for operational systems.

The new ability of a phased-array radar is its multifunction capability. The system has been developed to demonstrate these possibilities. The following functions are implemented:

(i) Search mode for target detection. Search volumes with different size and shape, and with individual frame rates, pulse period, pulse length, range resolution, detection method for:

- long range with low elevation
- medium range
- horizon
- short range

(ii) Acquisition for detected targets. Confirmation of target detection, parameter estimation and precise location in range r and direction (u,v) with monopulse for:

- long range
- medium range
- short range

(iii) Tracking of acquired targets (up to 40 targets):

- track initialisation
- individual tracking parameters for each target (tracking rate and waveform)
- α–β filter

(iv) Antenna monitoring and alignment.
(v) Test operations:

- signal recording for statistical analysis and target recognition
- antenna pattern measurements
- adaptive beam pattern
- generation of clutter maps

(vi) Display of functions and results:

- beam direction in (u,v)
- search, acquisition and tracking plots
- targets under tracking
- table of parameters for tracked targets on special request: position, velocity vector, track duration, estimated detection probability

- online statistics with estimated and updated duty factor for individual radar functions
- raw data oscilloscope

17.5.1   Available codes and transmit pulses

To perform the different radar tasks a set of coded pulses for transmitting has been implemented. The principles of pulse coding and compression have been discussed in chapter 7. The subpulse length can be chosen $\tau = 1$, 2 or 10 μs, dependent on the required range resolution, and the idea was to choose an adequate resolution for each individual radar function. For example, for long-range search with sequential detection a moderate resolution is appropriate, as discussed in chapter 9 section 9.1.5, and therefore $\tau = 10$ μs is applied. For a short-range search in clutter or for target location in the acquisition or tracking mode a higher range resolution is required and $\tau = 1$ or 2 μs is applied.

The total pulse length, determined by the code length, may be matched to the expected target cross section and range. Table 17.1 gives an overview of the variety of implemented codes. Subpulses of $\tau = 10$ μs are only applied for search.

The pulse length together with the selected number of pulses applied for a radar task allows energy management. For target acquisition and tracking the energy is, for

Table 17.1   *Implemented codes in ELRA Compression by:*
MF = matched filter
MFW = matched filter with Hamming weighting
LSMMF = least-mean-square mismatched filter

Modulation code	Code length	Subpulse μs	Pulse length μs	Compression by
—	1	10	10	MF
P3	16	10	160	MF, MFW
P3	32	10	320	MF, MFW
P3	64	10	640	MF, MFW
NLFM	64	10	640	MF
—	1	2	2	MF
Frank	16	1	16	MF
Frank	64	1	64	MF
P3	16	2	32	MF, MFW, LSMMF
P3	32	2	64	MF, MFW, LSMMF
P3	64	2	128	MF, MFW, LSMMF
P3	128	2	256	MF, MFW
P3	256	2	512	MF, MFW
NLFM	64	2	128	MF
NLFM	128	2	256	MF
NLFM	256	2	512	MF

example, increased by about 5 dB relative to the search energy in the same range to stabilise the target tracking. This is possible because it has to be applied to only a few beam positions; this means that all targets, when once detected, are then tracked with a high tenacity. If a track is lost because of target fading effects, a local search task will find that target again and tracking will continue. This ability is very important for reliable observation of air traffic, which is not achievable with the track-while-scan procedures of conventional mechanical rotating radars.

17.5.2 Time budget for a parameter example

As an illustrative example we give here a selection of waveforms for multifunction operation together with a suitable and sensible time budget. Of course, this selection may be changed by the operator or within an adaptive operation.

17.5.2.1 Search function

The search function has to cover the volume specified by the range $R = 200$ km and height $h = 20$ km and an azimuth sector of $120°$, typical for planar phased arrays. This is sketched as an elevation cut in Figure 17.15. The search function is divided into four subtasks.

For a *long-range search* up to $R = 200$ km with $\tau = 640$ μs sequential detection is applied. The reduction in mean test length compared to conventional tests is about 5 dB, as discussed in chapter 9. For the parameters discussed above we need, instead of the pulse number $N = 16$, only a mean test length $n_n = 5$ for a $P_d = 0.5$. For S search periods results a cumulative detection probability:

$$P_C = 1 - (1 - P_d)^S \tag{17.3}$$

For $S = 5$ results $P_C = 0.968$. The blind range is given by:

$$R_b = \frac{c\tau}{2} \tag{17.4}$$

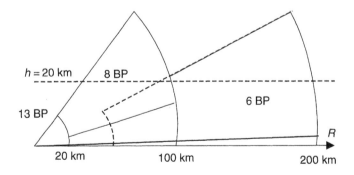

Figure 17.15 *Elevation coverage for multifunction search, the number of beam positions (BP) in elevation for the respective range is indicated*

and in this case $R_b = 96$ km. According to Figure 17.15 we need six beam positions (BP) for elevation coverage (beam width 2° to cover about 12°). In azimuth 60 BPs are necessary. With pulse period $T = 2$ ms we get a search time:

$$T_{LRS} = 6 \cdot 60 \cdot 5 \cdot 0.002 = 3.65 \text{ s}$$

The duty cycle is $\eta = 0.3$. For repetition of the long-range search we select a repetition period of $t_{LRS} = 15$ s. A radial flying target with velocity $v = 250$ m/s would approach 18.7 km within five search periods, which is less then ten per cent of the maximum range.

The *medium-range search* has to fill up the blind range of the long-range search. This is only necessary for areas where new targets may appear. All targets coming inbound from the maximum range will already have been detected and tracked. We have to search only for targets at low and high altitudes (h < 600 m or h > 20 km) or entering the observation volume from the sides. The necessary energy compared to a long-range search may be reduced by 2^4, according to the radar equation, and this is achieved by choosing $\tau = 128$ µs and a mean pulse number $n_n = 2$. The detection performance will then be the same as for 200 km. The new blind range is now 20 km. With $T = 0.8$ ms and eight BPs in elevation we get a search time of:

$$T_{MRS} = 8 \cdot 60 \cdot 2 \cdot 0.0008 = 0.7685 \text{ s}$$

The duty cycle is only $\eta \approx 0.16$. For the repetition period of this medium-range search we select $t_{MRS} = 7.5$ s.

Low altitudes are covered by the *coherent search* with a Doppler filter bank for clutter suppression. Therefore we select a higher fixed pulse number $N = 8$ and $\tau = 40$ µs. These values are adequate for a range of 100 km. With $T = 0.8$ ms this search requires a time of:

$$T_{CS} = 1 \cdot 60 \cdot 8 \cdot 0.0008 = 0.38 \text{ s}$$

The repetition period is chosen $t_{CS} = 5$ s. The duty factor is only $\eta \approx 0.05$.

Finally, the blind range of the medium-range search is covered by a *short-range search*. A pulse with $\tau = 2$ µs and $N = 3$ (MTI operation) is suggested. With $T = 0.2$ ms and $13 \cdot 60 = 780$ BPs the search is completed within:

$$T_{SRS} = 780 \cdot 3 \cdot 0.0002 = 0.468 \text{ s}$$

The repetition period is selected $t_{SRS} = 5$ s.

A relative search load with a basic search period of 15 s can now be derived:

long range:	$1 \cdot 3.65$ s	$= 3.65$ s
medium range:	$2 \cdot 0.768$ s	$= 1.536$ s
coherent search:	$3 \cdot 0.38$ s	$= 1.1404$ s
short range:	$3 \cdot 0.468$ s	$= 1.404$ s
	sum	$= 7.68$ s

So we end up with 51 per cent search load. The mean duty cycle is $\bar{\eta} = 0.19$, so we don't make use of the available mean power with $\eta = 0.3$. For different search tasks we have selected different dwell times and repetition periods in each beam position, matched to the individual functions. Here we recognise a fundamental advantage of beam agility.

17.5.2.2 Target location and tracking

Target location is applied after each search result for target acquisition and validation and afterwards for computer-controlled target tracking. The energy for each location task is increased by about 5 dB compared to the corresponding search task. Because the target range is now roughly known from the search results, there is no blind-range problem and the pulse width may be chosen up to a duty cycle of $\eta = 0.3$ at maximum range. For coherent Doppler processing for clutter suppression and Doppler frequency estimation we have to apply in all cases an adequate pulse series. The Doppler information gives the target radial velocity, which is used in the tracking procedure.

We select for target location at:

long range (150–200 km): $\tau = 640\,\mu s,$ $T = 2\,ms,$ $N = 12$
medium range (75–150 km): $\tau = 512\,\mu s,$ $T = 1.5\,ms,$ $N = 8$
short range (<75 km): $\tau = 16\,\mu s,$ $T = 0.6\,ms,$ $N = 16$

The resulting dwell times for these location tasks are 0.024, 0.012, 0.0096 s, respectively. For tracking ten targets within each area (range interval) at a medium period of 1 s we would need a time segment of 0.456 s within each s. That means a tracking load of 45.6 per cent. This would be compatible with our assumed search load. If more targets are to be tracked, the tracking rate could be reduced for targets flying straight ahead or the rate for search operations would be reduced by a priority control.

17.6 Digital signal processing

The task of signal processing is the detection, location and parameter estimation of targets within the specified observation space. The received echo signals are super-imposed by noise and external disturbance signals. After detection an estimation of target parameters (range, azimuth, elevation, Doppler frequency and amplitude), as accurate as possible follows. An overview of signal processing is given by the block diagram in Figure 17.16. Real-time processing with high data rates is achieved using special hardware, developed with standard digital logic devices, and signal processors (multiplier-accumulator devices MAC, FFT processors). Control and further offline processing is performed by programmable signal processors, which are programmed in assembler language.

Then microcomputers or PCs, programmed in a high-level language like C++, follow.

Signal processing starts with forming the receive beams. This is performed in two steps, according to the partial beamforming concept discussed in chapter 5 and

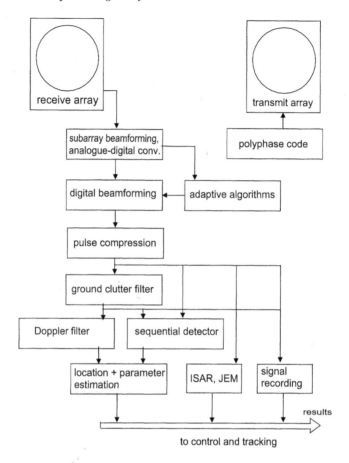

Figure 17.16   Block diagram for ELRA signal processing

sketched in Figure 5.2. First subarrays are generated by analogue summation of the baseband signals of the respective antenna channels. Allocation of the antenna channels to the 48 subarrays, each with 16 channels, is shown in Figure 17.4.

The output I and Q signals of the subarrays are digitised. Final beamforming is carried out by digital weighting and summing of the subarray outputs to generate the sum beam and two difference beams (for azimuth and elevation). If all weight values are equal to one we get the usual sum beam. By applying different signs to the right and left-hand halves of the array we get the azimuth difference beam. With different signs for the lower and upper half we get the elevation difference beam. This set of three standard beams is applied for usual radar operation and is realised by a fixed wired network of adders.

Arbitrary subarray weights may be used additionally for beamforming experiments. 128 different sets of weighting coefficients are precomputed and stored. They offer a high flexibility for beamforming experiments, especially for jammer suppression and superresolution.

If there is a disturbance by jammers the outputs of the subarrays may be used to compute adaptive weights for beamforming with adaptive jammer suppression, according to the principles and algorithms described in chapter 10 sections 10.3.2 and 10.4. Already performed offline adaptation experiments are described in chapter 10, section 10.6.

For each of the three beam outputs the received polyphase coded pulses are compressed. A digital correlator is applied for a code length of up to 256. The applied codes are described in chapter 7 and in section 17.5.

The next step is the suppression of fixed clutter by a recursive digital filter, discussed in chapter 8 section 8.1 with equation 8.5. Six such filters are necessary for the orthogonal signal components of the three beams. Up to 1024 range cells can be processed. For each range cell one memory element is provided for the estimated signal mean, which is recursively updated.

After the fixed clutter filter the Doppler processor follows. Its tasks are coherent signal integration, Doppler estimation and a final suppression of clutter. This processor is made with FFT elements, some special hardware and a PC. The filtering is performed according to the description in chapter 8 section 8.2. Two FFT elements work alternately overlapping in time. The capacity limit is given, with N_t = number of transmit pulses or radar periods and N_r = number of range cells, by the product $N_s * N_r \leq 16\,384$ for the processing of signals for one beam. For example, with 512 range cells up to 32 radar periods may be applied for the coherent search. The received signals according to a range cell may be weighted for sidelobe reduction by a window function, e.g. with a Hamming or Hanning function. For pulse numbers below an adequate binary number 2^n the missing signals are filled up with zeros (zero padding). By forming an average from the output powers of all filters the residual noise, clutter or jammer signal is estimated. The averaging process may be extended into the range dimension, resulting in a two-dimensional (Doppler and range) window for estimating the residual noise. This estimate then determines the detection threshold for achieving a nearly constant false-alarm rate (CFAR).

The suppression of weather and other Doppler-shifted clutter may be performed adaptively at the output of the filter bank, as described in chapter 8 section 8.3.

For the search function at long range outside of clutter areas a noncoherent sequential detector is applied as described in chapter 9 section 9.1.4. The test is performed for all range cells of the respective beam position. If there is the decision 'target present', the stored test function values λ_{in} are used to determine the range of the targets (up to 5). Against a varying noise level noise normalisation is applied to the received signal at the input of the sequential detector. A noise estimate is derived for the respective beam position in a time interval before transmitting. By dividing the signal by this estimate the noise level is kept nearly constant. The design signal of the sequential test is divided by the same estimate to maintain the target-detection probability at the required level. The mean test length will then increase accordingly, resulting in automatic energy management. The sequential detection is realised with only four digital signal processors (DSP) TMS320-C20. To avoid too long a test length in a beam position a truncation test length N_T can be selected.

For target location the sampled signals out of the sum beam are used for range estimation. Because the sampling raster is asynchronous to the received pulses the range position of the maximum of a target pulse is determined by interpolation. The direction estimates (azimuth, elevation) are determined by the monopulse procedure, discussed in chapter 11 section 11.1 with equation 11.15. If the angular differences to the beam direction are too large the target report is ignored to reduce the probability of a multiple acquisition of the same target out of neighbouring beam positions. The real part of the monopulse quotient is ideally zero for a single target. Deviations from this can be caused by multiple targets, discussed in chapter 11 section 11.6, or by a misalignment of the antenna. To avoid influences from clutter echoes the target location is performed after the filter against fixed clutter or the Doppler processor. The target location function is realised by a DSP coupled to a PC for control.

The ELRA system may be used for real-time recording of signals, especially from selected clutter areas and targets. For clutter signal recording a buffer memory (1 Mbyte) is available; then the data are transferred to mass storage. The evaluation results are used for the generation of a clutter map or for the evaluation of clutter spectra and probability-density distribution functions.

Target signals from selected aircraft may be recorded at a high data rate during continuous tracking. Then the evaluation of target spectra, phase fluctuation and ISAR images, as described in chapter 15, section 15.4.4 is possible.

17.7 Array signal processing

The outputs of the 48 subarrays are available for array-processing experiments. These data are transferred with an IEC bus to a microcomputer for offline evaluation. First experiments for adaptive beamforming, described in chapter 10, and superresolution have been performed.

The dynamic range of the subarray channels and their outputs is now extended from eight to 14 bit resolution after the analogue-to-digital conversion. Online adaptive beamforming is under development with the application of modern digital signal processors.

17.8 Control system, operation and display of functions and results

The first implementation of radar control by a single process computer was abandoned some years ago in favour of an expandable network of microcomputers (PCs). This concept was capable of fulfilling the requirements for real-time experimental multifunction operation of the ELRA system [13].

The microcomputers are based on Intel processors connected by an ethernet bus with a special protocol. The packet length may be chosen between eight and 63 bytes for short packets (commands) or up to 32 kbytes for long packets (data). The transfer rates are about 1280 packets/s or 14 packets/s, respectively. The maximum transfer rate results in 450 kbytes/s. Another bus is used to synchronise time-critical processes with a transfer rate of up to 8 Mbytes/s. The masterless computer network may be further extended; it now comprises eight PCs.

The complete radar function control has been divided into small tasks. Communication between tasks is performed only by messages and mailboxes. Functionally cooperating tasks are combined to task groups which can be executed independently from their network location. The distribution of task groups within the network is not adaptive but must be configured. Global data are stored in a data pool with access by the mailboxes. The complete software system can be tested on a single computer. For each subsystem the same software structure has been implemented. The operating system, up to now MS-DOS, has been supplemented by a resident multitasking system IPMX (interprocessor message exchange system) with integrated network drivers especially developed for the requirements of the ELRA system. Its main advantage is a short task switching time at minimum system overhead. The operating system is cooperative, which means that a new task with a higher priority will not interrupt a running task but will wait for the next possible interrupt. The radar control software RACOS (90 k instructions) comprises for each single computer at least a subsystem control task, keyboard/remote manager and eventually a data pool manager. Further task groups depend on the type of subsystem. The operating system will now be changed to LINUX.

The main applied programming language is Turbo Pascal. Only some time-critical parts have been optimised by assembler or C language. If necessary, the microcomputers are complemented with special hardware.

The following subsystems with their main task groups have been implemented, further subsystems such as a clutter-map processor are under development. The tasks of the processors are the following:

(i)   Master processor:

  - coordination of all radar functions according to priorities
  - coordination of all control/remote inputs
  - coordination and distribution of all radar results
  - data recording for experimental or debugging purposes
  - connection to a tracking computer
  - simple $\alpha-\beta$ tracking for standalone operation without a tracking computer
  - display of tracking results

(ii) Console processor:

  - human interface with a menu dialogue, mouse and keyboard
  - command interpreter for online or batch control
  - background editor for menu files or command files
  - plan position indicator for target plots

(iii) Antenna control processor:

  - execution of low-level radar orders
  - pattern generation for beam scanning (search)

(iv)   Antenna monitoring processor:

- module testing, phase alignment and failure analysis
- display of antenna diagrams and transmit pulses
- display of beam positions

(v)   Search processor for sequential detection:

- control of noncoherent sequential detector

(vi)   Search processor for coherent processing by Doppler filter processing.
(vii)   Processor for target location and parameter estimation.
(viii)   Processor for special tasks:

- control of digital beamforming
- control of test-signal generation
- recording and processing of subarray output signals

An overview of the processors and the displays for observing the systems functions and results is given by Figure 17.17.

The individual radar functions are asynchronous and overlap in time. After the antennas have finished a certain task and the required hardware is cleared the next task is started.

First, the steering commands for all the phase shifters are computed. At the same time the signal-processing subsystems will be prepared: counters are reset and memories are cleared. Then the transmission of a predetermined or variable number of pulses is started. During the reception of echoes the signal processing starts. The final post processing by signal processors is performed with a certain time lag. So several radar tasks may be active at different stages of processing at the same time.

To the radar tasks within the multifunction operation certain priorities are assigned in the following decreasing order:

tracking–acquisition–search–monitoring or measurements

The results given by target plots with their coordinates are displayed by a monitor for plan position indication. The plot types are distinguished by colours: search results, blue; acquisition results, yellow; tracking location results, red.

All plots may be integrated during a selected time interval. The history of the observation of targets of opportunity (generally the air traffic departing and approaching Cologne airport and on several airways) may be observed for demonstrations of ELRA. The track results are displayed by a vector indicating the position and velocity of targets. With an alphanumeric display all available and measured target parameters are displayed after selection of an individual target. Figure 17.18 shows an example of target observation during 20 minutes. The partial figure above represents search results, including many false alarms from noise and clutter residues. The partial figure in the centre shows the acquisition results: most of the false alarms are not confirmed. At the bottom we see the results from track commands, generated by the tracking process. These tracking plots now form a dense series of target plots demonstrating the effective target tracking with high tenacity.

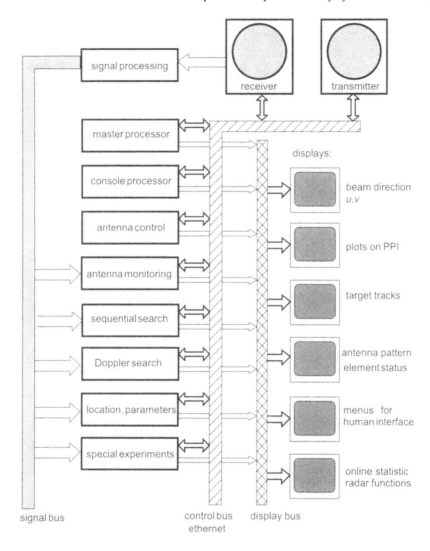

Figure 17.17   Block diagram for ELRA radar control

For a certain time period the ELRA radar system described in this chapter has been connected to a general-purpose mainframe computer for the performance and investigation of search control and tracking procedures. Track initiation, strategies for adaptation of the tracking rate to the target manoeuvres and estimation of target parameters on the basis of a Kalman-Bucy filter have been studied extensively [15–17].

In Figure 17.19 we see an enlarged view of two target tracks: the single grey points indicate search results. These give the target positions with poor accuracy because of the low range resolution and missing monopulse. These points lie besides the track path. They are followed by the acquisition plots, indicated by the larger dark points,

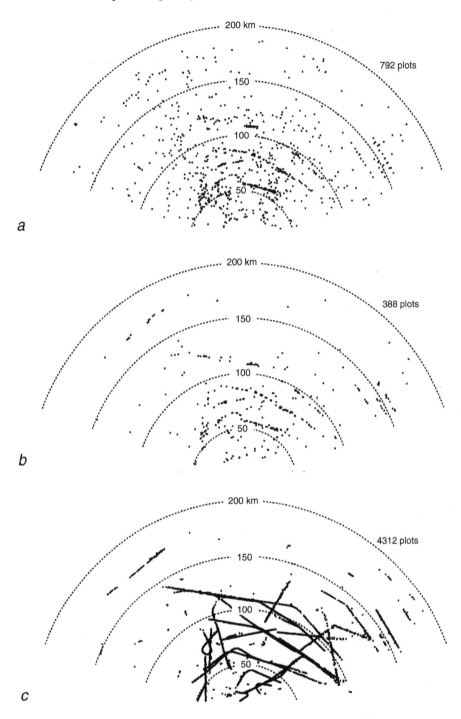

*Figure 17.18   ELRA system: search, acquisition (confirmation) and tracking plots
for observed air traffic within a time interval of 20 min
a search results; b confirmed search plots; c tracking plots*

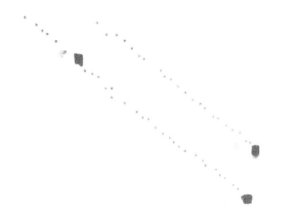

Figure 17.19   Search, acquisition and tracking plots

which are lying on the target path with good accuracy. The following dense series of tracking plots, indicated by the small points, demonstrates target tracking with high tenacity.

17.9   Experiments with ELRA

The main purpose of the system's development has been achieved by numerous experiments and demonstrations with successful multifunction operation.

Special experiments and some results have been already discussed in previous chapters, examples are:

- pulse compression with high range resolution in chapter 7
- adaptive clutter suppression in chapter 8
- sequential detection in chapter 9
- adaptive null steering in chapter 10
- monopulse correction for failure of antenna channels in chapter 11
- polarisation robustness
- some ISAR results in chapter 15
- antenna pattern measurement and beamforming in this chapter
- pulse compression: Doppler effects, pulse droop and phase variation within the transmit pulse

17.10   Conclusions

The ELRA system was developed with active array antennas, signal and data processing subsystems for the demonstration of a multifunction radar system. The antenna components have been developed and built within the labs of FGAN-FFM electronics department (EL). During all phases this project has stimulated theoretical studies, especially in the field of sequential detection and array processing. The project will be continued.

17.11 References

1 WIRTH, W. D.: 'Elektronisch gesteuertes Radar, Aufbau eines Experimentalsystems'. DGON Symposium on *Radar*, Ulm, Germany, 1972 (6), pp. 91–118
2 WIRTH, W. D.: 'Radar signal processing with an active receiving array'. Proceedings of IEE international conference on *Radar*, London, UK, 1977, pp. 218–221
3 WIRTH, W. D.: 'Array antennas for electronic scanning'. Proceedings of international conference on *Radar*, Paris, France, 1978, pp. 423–430
4 SANDER, W., and WIRTH, W. D.: 'ELRA-experimental phased array radar'. Conference Proceedings *Military microwaves*, London, UK, 1980, pp. 109–114
5 WIRTH, W. D.: 'Phased array radar with solid state transmitter'. Proceedings of international conference on *Radar*, Paris, France, 1984, pp. 141–145
6 WIRTH, W. D.: 'Solid state multifunction phased array radar'. AGARD conference proceedings, CP-381, Toulouse, France, 1986, 2.1–2.8
7 BAACK, C.: 'Der V-Dipol als Strahler für Multielement-Antennen', *AEÜ*, 1975, **29**, pp. 31–36
8 SANDER, W.: 'Monitoring and calibration of active phased arrays'. Proceedings of IEEE international conference on *Radar*, Arlington, USA, 1985, pp. 45–51
9 SANDER, W.: 'Beam forming with the ELRA system'. Proceedings of IEE international conference on *Radar*, London, UK, 1982, pp. 403–407
10 SANDER, W.: 'Experimental phased array radar ELRA: antenna system', *IEE Proc. F, Commun. Radar Signal Process.*, 1980, **127** (4), pp. 285–289
11 WIRTH, W. D.: 'Signal processing for target detection in experimental phased array radar ELRA', *IEE Proc. F, Commun. Radar Signal Process.*, 1981, **128** (5), pp. 311–331
12 SANDER, W.: 'Beamforming with phased array antennas'. Proceedings of international IEE conference on *Radar*, London, UK, 1982, pp. 403–407
13 GRÖGER, I., SANDER, W., and WIRTH, W. D.: 'Experimental phased array radar ELRA with extended flexibility'. Proceedings of IEEE international conference on *Radar*, Arlington, USA, 1990, pp. 286–290
14 HAYDL, W. H., SANDER, W., and WIRTH, W. D.: 'Precision SAW filter for a large phased array system', *IEEE Trans. Microw. Theory Tech.*, 1981, **29**, pp. 414–419
15 VAN KEUK, G.: 'Adaptive computer controlled target tracking with a phased array radar'. Proceedings of IEEE international conference on *Radar*, Washington, USA, 1975, pp. 429–432
16 MIETH, H. J.: 'Adjusting measurement error and dynamic parameters by evaluating real-time experiments with electronically steerable radar ELRA'. Proceedings of IEEE international conference on *Radar*, Washington, USA, 1990, pp. 291–294
17 VAN KEUK, G., and BLACKMAN, S. S.: 'On phased araay radar tracking and parameter control', IEEE Trans. Aerosp. Electron. Syst. 1993, **29** (1), pp. 186–194

Chapter 18

Floodlight radar concept (OLPI)

18.1 Introduction

Because a surveillance radar must transmit continually, it is easy to detect by reconais-
sance and ESM and therefore vulnerable to jamming and anti-radar missiles. In
contrast, a tracking radar is only activated to support defence actions after the detection
of penetrating targets. The intention of this chapter is to discuss a radar operational
concept which permits minimum vulnerability against these threats. It is therefore
especially suitable for a medium-range search function [1]. This concept could in
principle be implemented by a special radar system or it may also be applied to the
search mode of a multifunction phased-array radar.

For the purpose of protection against ESM and ARM, two aspects have to be
considered:

- we have to avoid or render as difficult as possible the reconnaissance of our radar
  by the opposing ESM platform, usually on an aircraft which is also carrying the
  ARMs
- an incoming ARM must be deceived effectively by decoy transmitters, without
  interrupting our own radar's search operation

Standard radar systems are exposed electronically, especially by the high pulse power
distributed by the scanning main beam and the antenna pattern sidelobes. The sidelobe radiation of radars, which may be continuously evaluated by an ARM seeker
for extracting its guidance information, may be covered by decoy transmitters operating at a suitable distance from the radar with omnidirectional radiation. The total
power radiated from this additional decoy transmitter has to be of the same order of
magnitude as the active transmitter. In contrast, it is impossible to cover the scanning main beam; this would need another complete radar transmitter as a decoy. The
energy density of the radiation within the main beam is several orders of magnitude
higher than that in the sidelobe region. Therefore, radiation from the decoy transmitter may be relatively simply neglected by the ARM platform by setting appropriate
thresholds for signal sorting. It is anticipated that future ARMs may perform their

guidance function by only using the more or less periodic information presented by our scanning main beam.

18.2   The proposed concept

From the introduction it follows logically that there is a need for a radar or radar operation without a scanning main beam for transmitting. The principal antenna diagrams are sketched in Figure 18.1. The transmitter has to illuminate continuously the observation space, e.g. a certain sector or even, with an omnidirectional pattern, the complete azimuth range. Additionally, we propose a continuous wave (CW) transmit signal to distribute the radiated energy as much as possible in space and time, to minimise the power density of our transmission in all places and all the time. In comparison to scanning pulse radars the power density is reduced by a factor $10^5 - 10^7$ relative to the peak pulse in a scanning main beam.

The CW transmit signal has to be phase coded for ranging of targets. This type of signal modulation is known as LPI (low probability of intercept) waveform. So we have chosen for this concept the abbreviation OLPI (omnidirectional LPI).

The transmitter should be separated from the receiving system by a suitable distance, e.g. some 100 m. Because the transmitter is relatively inexpensive several suitably distributed transmitters could be used, acting as mutual decoys and backup. All transmitters could operate at a different frequency and code. Because we know these frequencies and codes we can select one transmitter as active and apply the corresponding processing for receiving.

conventional radar:

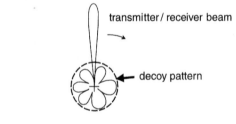

OLPI-radar:

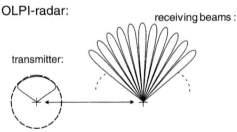

Figure 18.1   Antenna pattern of conventional and OLPI radar

Although it would of course be possible to locate an isolated OLPI transmitter with ESM techniques as discussed in Reference 2, this would be much more difficult if several OLPI systems were operating and all other radiating sources also taken into account.

These basic ideas are simple, however, interesting problems arise from a closer look at the different steps of radar operation, especially signal processing.

On the receiving side a multiple-beam antenna has to provide a continuous coverage of the illuminated space. This is necessary to avoid wasting of any transmitted energy. Additionally, the multiple beams deliver the direction information of detected targets. The received energy from targets has to be integrated for each resolution cell during an adequate time interval which may be selected, for example, as being equivalent to the usual radar scanning period.

The proposed radar concept is intended for medium-range surveillance applications. The range limitation follows from the CW operation with the resulting problems of the dynamic range of the received signals. We receive simultaneously echoes from the surrounding fixed objects with very high amplitudes and weak echoes from distant flying targets. By signal processing these weak Doppler-shifted target echoes have to be extracted out of the high-clutter signals. The limitations of this procedure determine the achievable range.

Following these general conceptual ideas we developed an experimental system for measurements and demonstrations. The critical components have been made and experiments with the complete system confirmed our ideas. In the following, descriptions for the critical subsystems are given. The block diagram is shown in Figure 18.2.

18.3 The transmitter

For transmitting a simple antenna is applied for sectoral illumination. Eight vertical dipoles in a column are combined by a microstrip feeding network resulting in a horizontal fan-beam pattern with a width of about 20° in elevation and 120° in azimuth. Phases (length of lines) within the feeding network are selected to achieve lower sidelobes looking to the ground. This will result in less backscattering from the surrounding clutter and therefore improve the conditions for signal processing by reducing the dynamic range problem. The antenna is fed by a bipolar transistor power amplifier, which is well suited for CW operation. The transmit power is 10 W. Our experimental system operates at S band (2.82 GHz, $\lambda = 11$ cm). The transmitter is connected by a cable with a length of about 100 m to the central control and receiving system and is fed from there with a coherent phase-modulated reference signal.

18.4 The receiving system

The receiving system has to perform a continuous observation of the illuminated space, to avoid any loss of radiated energy. With a multiple-beam receiving antenna

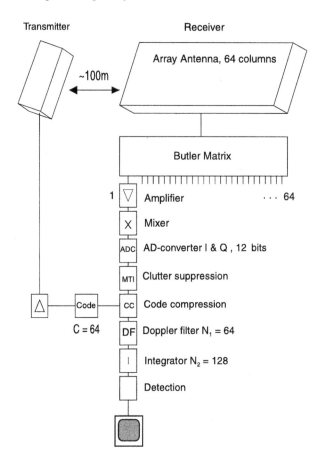

Figure 18.2   OLPI block diagram

the signals from different directions are separated. The experimental receiving antenna
has a planar array of 64 columns, each column containing eight dipoles combined by
a network which is the same as for the transmit antenna. The total number of antenna
elements is therefore 512. Figure 18.3 shows this planar-array antenna on the roof
of an FGAN building and on the right-hand side the transmit antenna on a tower.
The receiving array antenna can be adjusted in azimuth within a certain sector for
measurement purposes, for example to measure the pattern of the multiple receiving
beams.

The multiple beams in azimuth are formed by a 64-port Butler matrix which was
manufactured using microstrip technology. The Butler-matrix realisation is shown by
Figure 18.4. The half-power beamwidth of each beam is about 2°. The Butler matrix
operates like a fast Fourier transform. The antenna pattern of a single beam is therefore
in theory given by a $(\sin x)/x$ function, as presented in chapter 4 with equation 4.7.
All beams are overlapping, and are orthogonal, which means that each beam is at the
first pattern zero of its neighbour. The ensemble of beams with their pattern looks

Figure 18.3   Receiving array for OLPI radar with 64 dipole columns and transmitter

Figure 18.4   Butler matrix with 64 input/output ports in microstrip technique

like the patterns of the Doppler filter bank shown in Figure 8.3. The measured loss from input to output is about 3 dB. This loss has no negative effect on dynamic range because all signals are attenuated likewise. Beamforming has been successfully tested by mechanically rotating the array antenna while receiving a signal from a farfield source, thus measuring the antenna pattern of all beams. These patterns correspond to the above described expectation.

The Butler matrix is built into the antenna case with short connecting cables to the array columns. Each output of the Butler matrix represents one beam and is connected to a receiving chain with the following devices (Figure 18.2):

- low-noise amplifier, directly at the outputs of the Butler matrix
- mixer to the IF frequency of 30 MHz
- IF amplifier with bandpass filter which is realised with a surface acoustic wave (SAW) device
- mixer to the baseband for I and Q signals
- AD converter for I and Q components with 12 bits each and sampling at a rate of 250 kHz

Only eight receiving channels out of 64 have been included in the experimental system, for cost reasons. Therefore these channels cover only a sector of $8 \times 2° = 16°$. By switching these receiving channels to different outputs of the Butler matrix the observation sector can be changed.

The dynamic range of the receiving channels has to be as high as possible. Because of the CW operation we receive permanently as the predominant signal echoes from the surrounding clutter (ground, trees, buildings, vehicles etc.). The power of the clutter is proportional to the transmit power, which on the other hand determines the achievable range. By separating the transmitting and receiving antennas, as suggested and realised, there is almost no direct coupling between the two antennas. The predominant received clutter signals are caused by reflections from the surrounding. The problem of the clutter dynamic is distinctly decreased by separating the antennas because the common ground area which is covered by both antenna diagrams begins at a higher range. The receiving dynamic range, the ratio of the highest signal amplitude without a nonlinear distortion by saturation and the noise RMS amplitude, forms a major problem which is typical for CW radars. It has to be solved by appropriate signal processing. The higher the dynamic range of our receiving system, the higher could be the transmit power giving an increased detection range.

18.5   Resolution cell

Each resolution cell is illuminated continuously by our transmitter. If there is a target all its reflected energy has to be used for detection. The received signals for each resolution cell must be integrated during a time comparable to the search period of conventional main beam radars. We selected arbitrarily a search period of $T_s = 2$ s. The targets have to remain within one resolution cell for this time to avoid any problems with range migration. For an assumed radial velocity of 300 m/s a range resolution of 600 m is therefore adequate. The cross-range resolution b is given by the beamwidth, which for our experimental system is $2°$. For a range of e.g. 30 km it follows $b = 1000$ m. So both resolutions are approximately matched in order of magnitude. The length of the subpulse is therefore chosen $\tau = 4\,\mu$s corresponding to the required range resolution.

18.6 The phase code

The range resolution can be achieved by a suitable modulation. For continuous-wave (CW) operation a phase code is necessary. To achieve Doppler-tolerant detection only a polyphase code derived from linear or nonlinear frequency modulation can be applied. Suitable codes have already been discussed in chapter 7 section 7.8. With a subpulse length of $\tau = 4\,\mu$s and a code length of $N = 64$ we arrive at a code period of 256 μs which is equivalent to an unambiguous range of 38.4 km. Because we have to take into account a Doppler frequency shift f_d of the received signal according to maximum radial velocities we get for the product $f_d N \tau$ values greater than one. Binary phase codes which are attractive because of their simplicity are not applicable because of their Doppler sensitivity, which means that the compressed pulse would almost disappear for higher Doppler shifts. A very suitable code is given by the polyphase P3 or the Frank code. The P3 code is produced by taking samples from the linear frequency modulation LFM signal (chirp signal), as described in chapter 7. The phase follows a quadratic function with time. The phase samples, taken modulo 2π, then result in the P3 polyphase code. The Frank code is derived from a stepwise approximation of the LFM as presented in chapter 7 with equation 7.12. For our code length of $N = 64$ we have $n = 8$ and there result eight phase states, which may be generated by a three-bit phase shifter.

In our application we have to repeat the code periodically without any interval. The compression is performed by a transversal filter of a length equal to the code length and with a filter or reference function equal to the complex conjugate of one code period. This operation (convolution) is described in chapter 2 section 2.4. In the case of periodic repetition both codes show the advantageous feature of having range sidelobes equal to zero if there is no Doppler shift [3]. This applies especially to echoes from fixed or ground clutter, that is from the radar's surrounding. These signals are compressed then in theory only into the first range element for 0-600 m which is neglected for target detection.

The ambiguity function of the compressed periodic P3 code for range and Doppler shifts is shown in Figure 7.14. For the Frank code almost the same ambiguity function results. In this figure the theoretical zero sidelobes are recognised as the front curve of the waterfall presentation.

The practical achievable sidelobe level for ground clutter depends strongly on the accuracy of the generated phase-coded transmitted signal and on clutter fluctuations. For increasing Doppler frequencies the sidelobes are increasing too, as demonstrated by Figure 7.14. Weak targets at a higher range would be masked by the sidelobes of stronger targets at lower ranges. Therefore it is very important to reduce the sidelobes.

By applying a weighting function, e.g. a Hamming function (section 7.5.1), to the reference or filter function for the code compression, the range sidelobes can be suppressed very effectively as demonstrated by Figure 7.15.

18.7 Signal processing

The task of the OLPI radar is the detection of flying targets out of the clutter echoes. The decisive parameter for distinguishing the target signals from disturbing signals is the Doppler frequency shift, resulting from the radial velocity component of targets. The stages of signal processing for the OLPI radar are the following:

- suppression of clutter echoes
- code compression
- Doppler filtering
- integration of target echoes

This series of processing has to be performed for each resolution cell. With our assumed parameters we have 4096 resolution cells from 64 beams and 64 range elements.

The first step of signal processing is performed by a recursive notch filter to suppress ground clutter echoes. It acts individually for all 64 range elements. Because we have no beam scanning and therefore a stationary transmission of periodically repeated code sequences, each fixed point scatterer delivers a constant or slowly fluctuating echo signal with respect to its range element. Because we cascade stages with linear processing, the clutter-suppression filter may be placed before or after code compression. The recursive clutter-suppression filter, made with special digital hardware [4], is a relatively simple device and achieves a considerable reduction in the dynamic range of the disturbing clutter signals. Therefore it is placed at the beginning of the signal-processing chain. To save hardware complexity the first realisation with special hardware applied a transversal filter without any multiplier. Only weighting factors 2^k (with k a positive or negative integer) are applied, which means applying only binary shift operations and summing. An example of the frequency transfer function of this filter is given by Figure 18.5 [4]. Around the Doppler frequency zero the stop band serves to suppress clutter residues. This filter quality prevents clutter residues from affecting the following Doppler filters through sidelobes, and causing false alarms. For future realisations one could apply digital signal processors (DSPs) and then choose arbitrary filter-weight values for even better matching against the clutter spectrum.

The second step of signal processing is code compression. After this step the echo signals of a target appear at the corresponding range cell (one out of 64). In other words, now the target echoes appear as pulses in the same way as for a pulse radar. This code compression stage has first been realised cost effectively for the Frank code by a special digital device which uses the relationship of this code to the FFT. Clutter residues passing the notch filter are shifted by pulse compression into their corresponding range element. Residues from the surrounding are compressed into the

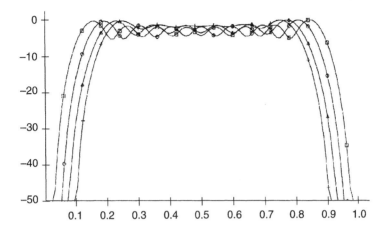

Figure 18.5   Frequency transfer filter functions of clutter filter, horizontal axis: frequency normalised to code recurrence frequency (courtesy R. Klemm)

first range element. To achieve more flexibility with respect to selecting other codes, this stage has also been implemented with a digital signal processor TMS320C40 from TI. This allows, for example, the application of Hamming weighting for sidelobe suppression for Doppler shifted signals, with a result according to Figure 7.15.

The gain of code compression is given by coherently integrating 64 signals. Because the phase is unknown, there is a loss of 1 dB compared to fully coherent integration: the net gain is $10 \log_{10} N - 1 = 17.06 \, \text{dB}$.

As the third step of signal processing a Doppler filter bank has the objective to further improve the signal-to-noise ratio by coherent integration and to estimate the target Doppler shift. The width of these filters has to be matched to the possible Doppler frequency variation of tangential flying targets. For tangential velocity v at range R the radial acceleration is:

$$\ddot{r} = \frac{v^2}{R} \qquad (18.1)$$

The change of velocity during our search period T_s then follows to $\Delta v = \ddot{r} T_s$. This results in a Doppler frequency variation of:

$$\Delta f_d = \frac{2 \Delta v}{\lambda} = \frac{2 \, v^2 T_s}{R \lambda} \qquad (18.2)$$

For $v = 250 \, \text{m/s}$ and $R = 38 \, \text{km}$ Δf_d becomes about 60 Hz. Doppler filter processing has been discussed in chapter 8 section 8.2. The filter width, given by $1/NT$, of the Doppler filter should be matched to this value of 60 Hz, resulting in a coherent integration time of about 16 ms, which corresponds on the other hand to 64 code periods of 0.256 ms each. This means that blocks of 64 compressed pulses are integrated for each range cell in a series.

The production of a Doppler filter bank with 64 channels and for 64 range elements in 64 beams is an expensive operation. It may be extremely simplified if the signal is digitised into one bit, that means the signal is only $+1$ or -1 depending on its sign. Then this processing stage can be achieved by relatively simple special hardware. The preconditions for this simplification are:

- the target signal is below the noise level, which is the case at this stage of signal integration for targets, which will be detected after final integration during 2 s
- there is no higher signal at another Doppler frequency deceiving the target signal

From this second condition it follows that clutter residues have to be suppressed by the notch filter and code compression well below the noise level at the input of the Doppler filter bank.

The integration gain by the Doppler filter bank of a factor of 64 or 18 dB is reduced only by about 2 dB by this simplification. Because Doppler frequency and phase are unknown there is an additional loss of 2.5 dB. The remaining integration gain is therefore 13.5 dB.

The Doppler filters have been realised by integrated digital correlators. In addition, the one-bit quantisation results in constant false-alarm rate (CFAR) operation and suppression of interfering single pulses.

The fourth step of signal processing is noncoherent integration of the outputs of the Doppler filter bank. To achieve the total integration time of $T_s = 2$ s a further 128 signals have to be integrated in amplitude individually for all range-Doppler cells. The noncoherent final integration may also be simplified by a one-bit quantisation of the Doppler filter output amplitude by threshold comparison before summation, which is then merely counting. The loss caused by this simplification is only 0.5 dB [5]. There are now altogether $64 \cdot 64 \cdot 64 = 262\,144$ resolution cells (beams, range cells, Doppler channels). In each cell we have to integrate by Doppler filtering and binary summation $64 \cdot 128 = 8192$ range-compressed signals.

After this summation the results for all resolution cells are compared with a detection threshold for the decision 'target present' or 'noise alone'.

The incoherent integration gain results in 12.7 dB. The loss for this binary incoherent integration compared to fully coherent integration amounts to 8.9 dB. The losses by signal processing compared to an ideal (not realisable) coherent integration sum up to 14.4 dB. The signal-processing gain from all three stages sums up to $17 + 13.5 + 12.7 = 43.2$ dB. A common value of the SNR at the threshold decision stage for reliable detection is 13 dB. The corresponding SNR at the input of signal processing results then to SNR $= -30.2$ dB. On the other hand, the clutter level may at this input point be up to 60 dB above noise.

18.8   Range performance

To check our parameter choice with respect to range performance we apply the radar equation (R for range, P_T for transmit power, G_T for transmit antenna gain, G_R

for receiving antenna gain, λ for wavelength, σ for target radar cross section, T_s for integration time, $(SNR)_o$ for signal-to-noise ratio at the output of signal processing, kTF for noise power density, L for total losses):

$$R^4 = \frac{P_T G_T G_R \lambda^2 \sigma T_s}{(SNR)_o (4\pi)^3 \, kT F L} \tag{18.3}$$

Applying our parameter values and converting them into dB we get:

$P_T = 10\,\text{W}$	$10\,\text{dB}$
$G_T = 8\pi$	$14\,\text{dB}$
$G_R = 512\pi$	$32\,\text{dB}$
$\lambda = 0.11\,\text{m}$	$-19.17\,\text{dB}$
$\sigma = 5\,\text{m}^2$	$7\,\text{dB}$
$T_s = 2\,\text{s}$	$3\,\text{dB}$
$(SNR)_o = 20$	$-13\,\text{dB}$
$(4\pi)^3$	$-32.98\,\text{dB}$
$kT = 4\cdot 10^{-21}$	$204\,\text{dB}$
L	$-20\,\text{dB}$
	$\Sigma = 184.85\,\text{dB}\ (\text{m}^4)$

The range follows to $R = 41.8\,\text{km}$. So we expect to detect also small aircraft as targets of opportunity within our unambiguous range of 38 km. The low gain of the transmitting antenna and the low power are compensated for by the continuous integration of received signals in a long time interval.

The resistance to ECM is comparable to that of any other radar. The target detection is determined by the ratio E/N_o. E is the total signal energy and N_o the noise power density. This ratio will be decreased intentionally by jammers, but this effect would be the same for all types of radar, so there is no disadvantage introduced by the OLPI concept.

The multiple receive beams establish, on the other hand, an effective basis for adaptive mutual jammer cancellation, following the concept applied to adaptive clutter suppression, discussed in chapter 8 section 8.3. The individual receive beams correspond to the individual Doppler filters. The actual beam would be adaptively combined with weighting coefficients with some neighbouring beam outputs to cancel the jammer signals.

18.9 Coherent and sequential test function with ACE

The processing loss of the experimental system is about 10 dB compared to coherent processing with a filter bank for 8192 signals. Because of the possible Doppler frequency variation of targets flying with a tangential velocity component, such a filter bank could not be applied. This processing loss resulted from the application of incoherent integration (about 8 dB) and the quantised Doppler filter bank (about 2 dB).

To increase the range without increasing the transmitting power these losses should be reduced.

In chapter 8 section 8.6.2 we presented and discussed a test function which has a detection efficiency comparable to that of coherent processing by a filter bank but has the advantage of improved Doppler-shift tolerance. This test function is based on autocorrelation estimates (ACE). With ACE we may apply coherent processing for a longer partial series, e.g. for 2048 signals. We may then perform the final integration incoherently, that is for four ACE-test values, so that all 8192 signals are finally integrated for the decision of the respective resolution cell. As a further improvement the final amplitude integration may be performed with a sequential test, as discussed in chapter 9 section 9.5.7. With this further improvement we achieve a detection efficiency which is almost the same as with a filter bank for all 8192 signals and for signals with a constant Doppler frequency shift [13], but with a sufficient Doppler tolerance. The Doppler tolerance for different processing methods is compared in Figure 9.18.

The ACE-test function is given by equation 8.30 and 8.32. It may also be computed with an FFT, to use its computational advantage for longer signal series. With N signals z_n we get the frequency coefficients, using equation 2.20 from chapter 2:

$$F_i = \sum_{n=1}^{N} z_n \exp\left(-j\frac{2\pi}{N}in\right) \quad i = 1, \ldots, N$$

The power spectrum is then given by:

$$S_i = F_i F_i^*$$

For the autocorrelation estimates follows with the Wiener Khintchine theorem, discussed in chapter 2 section 2.6 equation 2.34:

$$\rho(k) = \frac{1}{N^2} \sum_{i=1}^{N} S_i \exp\left(j\frac{2\pi}{N}ik\right) \tag{18.4}$$

The test function follows with equation 8.32. This method of processing has been successfully implemented with digital signal processors TMS320C40.

18.10   OLPI and multifunction radar

To apply the OLPI concept to a multifunction radar, for example the ELRA system, the following facilities and prerequisites are required:

- A separate transmitter with a simple floodlight antenna as for the special OLPI system.

- A cluster of receiving beams to cover the azimuth sector of about 120°; these beams have to be formed out of the subarrays of the ELRA receiving antenna, described in chapter 17 section 17.2 and Figure 17.4. For one certain phase setting of the phase shifters the possible beam cluster, formed with the subarray outputs by corresponding sets of complex weighting factors, according to chapter 5 section 5.3.6, would cover with ten beams only the subarray beam width of about 15° in azimuth. These beams could be generated using only the centre row of six subarrays, to achieve in a simple manner a broader elevation coverage of about 15°. To cover the whole required azimuth range of 120° with narrow beams in azimuth these six subarrays would need $120°/15° = 8$ sets of parallel operating phase shifters and subarray combiners. Then 80 final beams would be generated in parallel from these sets, similar to those of the OLPI system. The sets of phase shifters have to steer this centre row of subarrays in parallel into different directions to cover the whole azimuth sector with their subarray beams.
- For each beam the described signal-processing chain is required.

Instead of the CW transmission discussed above it would also be possible to apply a pulsed transmission as presented in chapter 17 for the ELRA system, but with a floodlight transmit antenna pattern. So the dynamic range problem for the received signals would be avoided and the advantage would be to achieve a higher maximum detection range. This type of transmission could be covered against the ARM threat by a corresponding pulsed decoy transmitter with the same floodlight antenna pattern. The power of the active and the decoy transmitter could in this case be similar. In the case of pulsed operation the pulses may be transmitted by the defocused array of the multifunction radar itself.

Because of the long target-illumination time, methods for target classification such as inverse synthetic aperture (ISAR), discussed in chapter 15, or spectral analysis for jet-engine modulation (JEM), discussed in chapter 16, could be applied with the OLPI concept.

18.11 Experimental results

Eight receiving and signal-processing channels have been made and used for experiments to validate the suggested principles and assumptions. While transmitting, echoes from the surrounding ground and trees have been received as the most problematic interference. The interference was sufficiently suppressed by the described signal processing, and residues appeared mainly only in the first range element. First, for target simulation with a defined signal strength, a Doppler-shifted and phase-coded signal was generated and superimposed on the received clutter signals by an auxiliary antenna radiating to the receiving array. Target detection was observed according to expectation, derived from the radar equation and the gain and loss values given above. A target signal with a level −30 dB below noise was clearly detected. The detection of targets of opportunity then was the next step. Figure 18.6 shows target plots, integrated over a time interval of about 20 s.

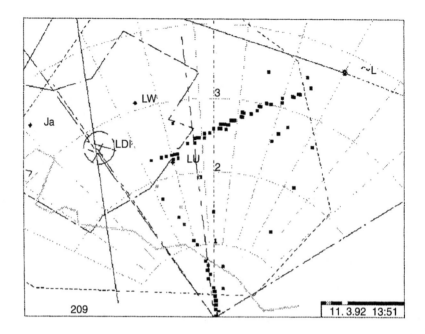

Figure 18.6   Integrated plots of aircraft detected by OLPI radar, maximum range is 40 km, © IEEE 1995 [7]

Received signals may be recorded and are then available for offline evaluations with different signal-processing methods.

18.12   Detection and classification of hovering helicopters

A special example for the application of the continuous floodlight transmission for target detection and classification is investigated and presented in the following.

The detection of hovering helicopters, which are above terrain masking only for a short time, is difficult to achieve with conventional radar. During hovering the skin echoes have no Doppler shift and are therefore suppressed by the necessary MTI filter. Only short echoes from the rotor blades or flashes allow detection during hovering, because these echoes are passed by the MTI filter. Conventional scanning radars would need a broad beamwidth, slow rotation rate and a high PRF to get a chance of detecting at least single flashes. All possible rotor flash echoes are received by the OLPI radar due to its CW and floodlight illumination [1]. This is illustrated by Figure 18.7.

The series of reflection flashes from the rotor blade, like that from a rotating mirror, is represented by the function $\sigma_H(t)$. The course of antenna gain with time $a(t)$ is constant for the OLPI radar and a pulsed function for the scanning radar. The received echo $r(t)$ is the product of $\sigma_H(t)$ and $a(t)$.

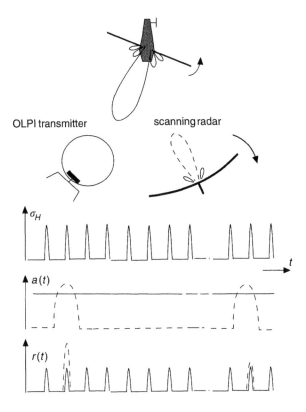

Figure 18.7   Helicopter rotor flashes and their reception by OLPI radar (solid line) or by a radar with a scanning main beam (dashed line)

In theory [8] the reflectivity or radar cross-section flash from a rotor blade with a constant reflection coefficient along its length is given by the time function, by using equation 4.7 in chapter 4 for the reflection pattern:

$$\sigma(t) = \left(\frac{\sin x}{x}\right)^2 \quad \text{with } x = \frac{2\pi L}{\lambda} \sin(\omega t) \tag{18.5}$$

with L the blade length, λ the wavelength and ω the angular rate of rotation. ωt is the aspect angle of the rotor blade with time. The duration τ of a flash between the zero points results in (with n for number of rotor blades and v_L for the velocity at the tips of the rotor blades):

$$\tau = \lambda/4v_L \quad n \text{ even}$$

$$\tau = \lambda/2v_L \quad n \text{ odd}$$

Usually v_L is chosen by the designer of the helicopter to be near 250 m/s. So we can give a simple rule of thumb for the minimal flash duration according to the simple

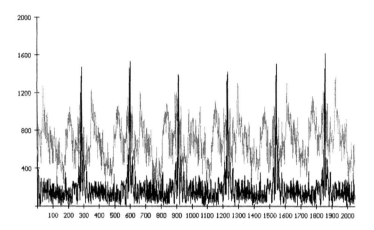

Figure 18.8   Series of echo signal amplitude before (grey) and after (black) suppression of ground clutter above time, showing the flash pulses

model assumed above:

$$\tau(\text{measured in ms}) \approx \lambda(\text{measured in m}) \quad \text{for } n \text{ even}$$

$$\tau(\text{measured in ms}) \approx 2\lambda(\text{measured in m}) \quad \text{for } n \text{ odd}$$

In practice, the flashes are longer because of inhomogenous reflectivity along the rotor blade. Now the question arose 'what complex reflection function is generated within the flashes and what are the effects on the phase-modulated OLPI signal with respect to code compression?' Fortunately for further investigations measured data from four different types of helicopter have been available. These measurements have been made by Siemens AG (UB N FR under contract to German MOD) using tracking radars looking steadily with a high PRF on hovering helicopters. So it was possible to extract the phase modulation within the flashes. Figure 18.8 shows as an example a section of the data: amplitudes before and after ground clutter suppression. The typical flash pulses with a constant period determined by the helicopter type are shown.

For the OLPI experimental system with its wavelength of 0.11 m a flash pulse length of more than 110 µs follows. From the measured data an effective pulse length of about 250 µs follows. This is just by chance the selected code period of the OLPI system. Single flashes have been extracted by a program out of the data series and analysed. The series of complex received signals is shown in Figure 18.9: a time axis starts at the upper right-hand corner and goes to the left-hand lower corner. On this time axis the signal vectors start. The resulting three-dimensional complex flash function repeats approximately with constant shape for a certain type of helicopter, but is different from type to type.

The next step was applying the complex flash function to polyphase code modulation by simply multiplying both functions. This was performed with several selected

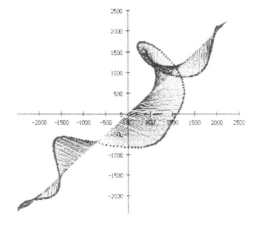

Figure 18.9   Complex helicopter flash function with time

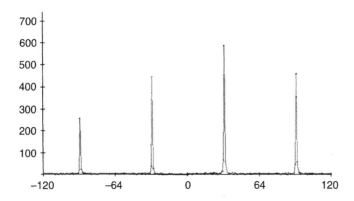

Figure 18.10   Compressed flash signal in four code periods

mutual initial phases, because this is random. Thereafter code compression was applied. Figure 18.10 shows an example for the code-compressed signal.

The result is shown for four successive unambiguous range or code periods. The compressed signals appear on the respective range elements with some residues on neighboured range elements. In other words: a single flash results in samples of the flash function compressed to the corresponding range element. Because of the continuous illumination a series of all flashes is received and appears at the corresponding range cell as long as the helicopter is visible by the OLPI radar.

The radar range with respect to the detection of rotor blades can be estimated with radar equation 18.3. We assume for example a flash period of 40 ms and an integration time of $T_s = 2$ s. The flash duration is assumed to be 0.25 ms. So we find a duty cycle of the rotor blade radar cross section of $d = 0.25/40 = 0.00625$.

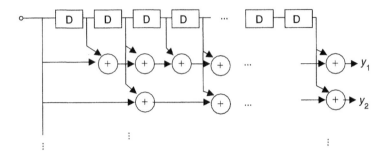

Figure 18.11   Sketch for pulse filter with delay segments, D

The assumed maximum value for the rotor blade cross section $\sigma = 10\,\text{m}^2$ has to be multiplied by d. With otherwise the same parameters we find a detection range $R = 14\,\text{km}$. This range could be prolonged by an increased transmit power, a longer wavelength or a longer integration time.

18.12.1   Classification by flash period

For the final detection of a helicopter using the flash pulses we need an incoherent filter for the series of flash echoes during a time interval of e.g. 2 s. Because the flash period depends on the helicopter type a bank of filters with the corresponding time-lag values between the taps for summation must be provided. This filter is sketched in Figure 18.11. It performs signal integration and estimation of the flash period equally well.

The delay elements D correspond to multiples of the code period 256 μs. The time lags between the taps for the individual filters have to correspond to the range of rotor periods occurring in practice, e.g. 100 to 450 code periods or 25 to 120 ms. The arithmetic elements are accumulating multipliers. By binary quantisation of the signals into the values 0 and 1 by a threshold comparison of the flash signal a special hardware device would be effectively simplified.

An alternative detection may be achieved with autocorrelation, discussed in chapter 2 section 2.5, of the flash signals. For all time-lag values corresponding to the possible rotor periods the signals have to be multiplied and summed. For the time lag which equals the rotor period of the observed helicopter, the autocorrelation value is maximum. By binary quantisation of the flash signals as above the multiplication is merely an 'and' operation, the output of which has to be counted. An example for the result of this kind of signal processing is given in Figure 18.12. The detection is then given by threshold comparison of the maximum output value.

The correlator time-lag value for this maximum output is the estimate for the flash period. This estimated flash period proves to be an effective classification feature, because all helicopter types have their individual flash period.

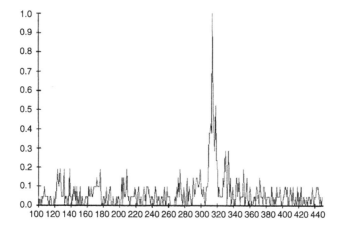

Figure 18.12   Result for binary autocorrelation of flash pulses

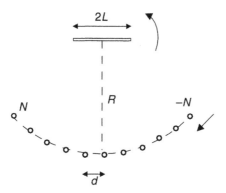

Figure 18.13   Circular synthetic aperture with respect to rotor blade of length 2 L

18.12.2   ISAR image and spectrum of rotor blades

Another interesting possibility for classification of helicopters is the ISAR image of the rotor blade. ISAR has been presented in chapter 15. This image is equal to the Doppler spectrum of single flashes.

The series of signals received during a flash can, in our imagination, be located on a circular arc, because of the continuous rotation. This is sketched in Figure 18.13.

A special focusing is not necessary, because the distance R is constant. The resolution and the unambiguous range may be derived from the expression for forming the synthetic beam which has to resolve the rotor blade as our target, according to equation 15.4:

$$S(\Theta) = \sum_{k=-N}^{N} z_k \exp\left[-j(4\pi d/\lambda)k \sin \Theta\right]$$

with $2N+1$ measured signals z_k and the distance d between synthetic array elements:

$$d = \frac{2\pi R}{T_R f_s}$$

given by the rotor rotation period T_R and the sampling frequency for the signal f_s. Because of the CW illumination the received signal may be sampled at an arbitrary frequency. From the previous detection the range is known and we know the code shift with respect to our transmitted code reference. So we may reverse the phases of the received code series of subpulses. The result is a sin signal with the flash envelope which is only phase modulated by the rotor movement. Then f_s may be selected as required.

The angular raster of the synthetic beams is:

$$\Delta\Theta = \frac{\lambda}{2d(2N+1)}$$

and the resolution at the target:

$$\Delta l = R\,\Delta\Theta = \frac{\lambda R}{2d(2N+1)} = \frac{\lambda T_R f_s}{4\pi(2N+1)}$$

The unambiguous cross range interval is:

$$l_0 = (2N+1)\Delta l = \frac{\lambda\,T_r\,f_s}{4\pi}$$

The number n of rotor blades is generally unknown and therefore the rotor period cannot be determined from the flash period. Therefore Δl also cannot be determined unambiguously.

It is nevertheless interesting to have a look on some rotor blade images, evaluated from the available measured data. In Figures 18.14 and 18.15 we have two examples for different helicopter types but with an even number of rotor blades. For the pair of blades on the left-hand side the blade is moving away, and on the right-hand side the blade is moving closer.

Despite the fact both figures are for different helicopter types, the images are alike. In Figure 18.16 a series of succeeding images is plotted together and we recognise a certain fluctuation of the image around its mean (dark curve). Figures 18.17 and 18.18 are measured for one helicopter with an odd number of blades. Only one blade is reflecting at a time, receding (Figure 18.17) and approaching (Figure 18.18) blades are alternating.

The shape of the ISAR image is equivalent to the Doppler spectrum. The maximum Doppler frequency results from echoes from the rotor blade tips. Because of the common value for v_L this Doppler shift will be approximately invariant.

A more promising classification would result from estimating the flash period, which is an unambiguous parameter of helicopter type.

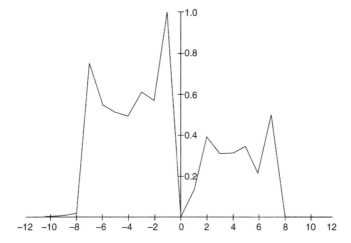

Figure 18.14   ISAR image of rotor blade, even number of blades, type 1

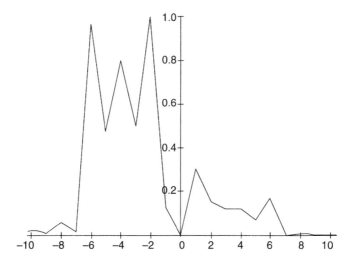

Figure 18.15   ISAR image of rotor blade, even number of blades, type 2

18.13   Spatial coding

The OLPI concept discussed and realised up to now uses a transmit signal which is radiated by an antenna with a broad pattern in azimuth, that is without a narrow main beam. The aim was to distribute the transmitted energy evenly within the observation space. By avoiding the narrow transmit beam a degraded angular resolution compared to a system with narrow beams for transmit and receive results. The reason is that by multiplication of both patterns the effective beam pattern becomes narrower.

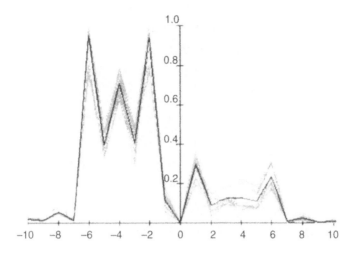

Figure 18.16   Series and mean of 20 succeeding ISAR images of rotor blade, even number of blades, type 2

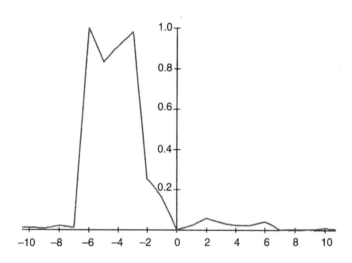

Figure 18.17   ISAR image of receding rotor blade, odd number of blades

Because the same transmit signal is distributed in all directions, the enemy, wanting to locate our radar, may apply two receiving systems at different locations and apply trilateration techniques to estimate the time of arrival (TOA). That is, find out the time difference with respect to these two receivers by a correlation analysis and derive a hyperbola as one local line. So, only one system with the additional ability for direction or azimuth estimation for the second local line is necessary for location of our position. This possibility could be countered if we

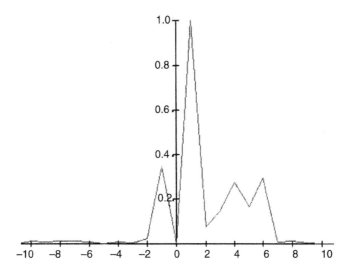

Figure 18.18   ISAR image approaching of rotor blade, odd number of blades

transmit into different directions different, and if possible orthogonal or uncorrelated, signals.

These aims, angular resolution improvement and improved deception, could be achieved by applying spatial coding for the transmit signal. This could be done with an array antenna with each antenna element transmitting individual and mutually orthogonal signals. Then the energy will be further distributed evenly in all directions. In the following we will discuss some examples of orthogonal signals and the resultant systems' behaviour. Such an idea has been discussed already in Reference 9 and applied in the French RIAS system. A circular array transmits from each antenna element a pulse with a different frequency. By superposition a different signal is generated for each direction, dependent on the relative phases. For receiving a compression is performed by correlation with the respective known superposed signal.

We assume a regular linear array according to Figure 18.19 with element distance d and element number n from $-N$ to N .

Antenna element n transmits a signal $s_n(t)$ in all directions. For $d = \lambda/2$ and direction $u = \sin \Theta$ the superposition of all partial waves results in the signal:

$$s_0(t) = \sum_{-N}^{N} s_n(t) \exp\left(\frac{2\pi j \, d u n}{\lambda}\right)$$

$$= \sum_{-N}^{N} s_n(t) \exp\left(\pi j u n\right) \tag{18.6}$$

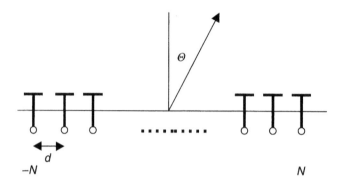

Figure 18.19   Linear array for spatial coding

We assume as usual a code-recurrence period T. The element n may transmit a signal with frequency n/T. We omit the carrier signal $\exp(j\omega_0 t)$ for simplicity:

$$s_n(t) = \exp\left(\frac{2\pi j}{T} nt\right) \quad \text{with } t = -T/2, \dots, T/2 \tag{18.7}$$

These signals are an example of an orthogonal set according to the Fourier transform. Then follows with equations 18.6 and 18.7:

$$s_0(t) = \sum_{-N}^{N} \exp\left[j\left(\frac{2\pi}{T} t + \pi u\right) n\right] \tag{18.8}$$

$$\approx (2N+1) \frac{\sin x(t)}{x(t)} \quad \text{with } x(t) = \frac{2N+1}{2}\left(\frac{2\pi}{T} t + \pi u\right) \tag{18.9}$$

Equation 18.9 represents a pulse with envelope $(\sin x)/x$ in the time domain. The time for the maximum is given simply by setting $x(t) = 0$:

$$t = -\frac{uT}{2} \tag{18.10}$$

The pulse formed by superposition in space is formed at a time t for direction u. With a higher value for N we get a shorter pulse. If we would know the target range we could derive from the receiving time, even with an omnidirectional receiving antenna, the target azimuth. With a multiple-beam receiving antenna the target range could again be estimated from the echo delay time. The high peak pulse with period T is generated by the superposition for all directions although each of the antenna elements radiates a CW signal. Protection against ARMs as intended with the OLPI concept, avoiding a high peak pulse, is not achieved with this choice for an orthogonal signal set.

To avoid the peak pulse we may distribute the signals given by equation 18.7 to the antenna element in a random manner. Antenna element n radiates the signal:

$$s_n(t) = \exp\left(j\frac{2\pi}{T} p_n t\right) \tag{18.11}$$

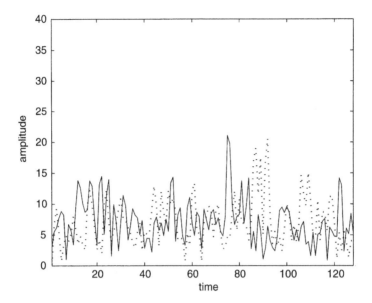

Figure 18.20 Spatial random coding, signals in direction u = 0 and 0.03, N = 32

Each p_n is selected randomly out of the numbers $-N, \ldots, N$ without repetition. Additionally, we may add a random phase α_n and get the superposed signal:

$$s_0(t) = \sum_{-N}^{N} \exp\left[j\left(\frac{2\pi}{T} p_n t + \alpha_n + \pi u \right) \right] \qquad (18.12)$$

Now we produce with equation 18.12 a pseudorandom signal different for all directions. By the addition of the random phase α_n we avoid a high peak pulse for $u = 0$. This choice for the orthogonal signal set was evaluated by a MATLAB program. In Figure 18.20 the resultant signals for $u_1 = 0$ and $u_2 = 0.03$ and $N = 32$ are displayed. This angular difference corresponds to the mutual first pattern zeros of the linear array antenna. Both signals are already uncorrelated for this angular difference.

The superposed signals transmitted in a certain direction may be computed with our known parameters p_n and α_n. These resultant functions are used as the reference for a matched filter or a correlation for the compression and detection of the received signals. Thereby a selection with respect to the direction is achieved which corresponds to the virtual directivity of the transmit array. This directivity multiplies to the directivity of the receiving beams, thus improving the angular resolution as desired. The compression result is shown as an example in Figure 18.21.

The compression result for a signal from a different neighbouring direction u_2 with the reference for u_1 is shown in Figure 18.22. There is no compression peak because of the orthogonal signals superposed into different directions.

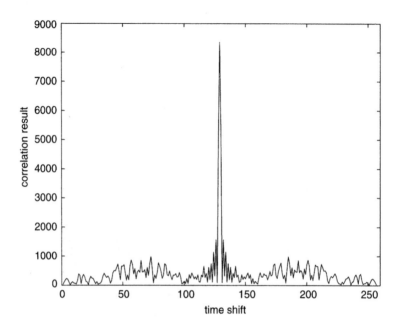

Figure 18.21   Spatial random coding, correlation or compression result, $u_1 = u_2$, N = 32

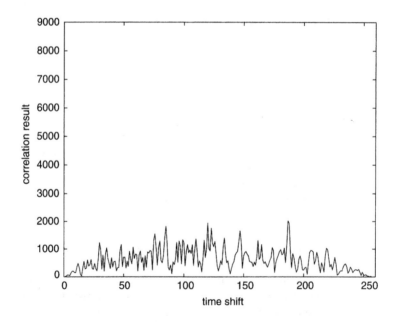

Figure 18.22   Compression for signal from direction u_2 and the reference for u_1 above time

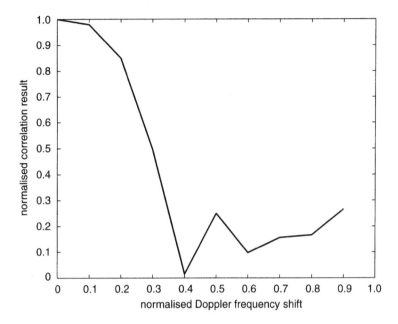

Figure 18.23   Decay of compression peak dependent on Doppler factor d

Therefore by this spatial coding the superposed transmit signal also shows a directivity and contributes to the angular resolution.

But we have also to consider the possible Doppler frequency shift of aircraft. In the next step of investigation a Doppler phase rotation was added to the simulated target signal. The factor d measures the phase in multiples of 2π within the period T. In Figure 18.23 the decay of the compressed signal maximum dependent on d is shown. Efficient detection is possible only up to $d = 0.3$. This behaviour corresponds to the random binary phase code discussed in chapter 7 section 7.3. For the OLPI parameters for a radial velocity of 200 m/s a Doppler factor $d = 0.93$ results. To apply spatial coding would require three different reference functions for Doppler frequencies at $d = 0, 0.5$ and 1.0 which are applied in parallel to cover the Doppler range of interest.

We can expect for other choices for the signal set, for example pseudorandom phase codes, the same lack of Doppler tolerance. This Doppler tolerance, if required, is achieved only with the LFM or NLFM modulation, as discussed in chapter 7.

18.14   Summary and conclusions

Starting from the aim of countering or reducing the threat to a search radar by antiradar missiles (ARMs) and electronic warfare intercept receivers we developed the concept of a floodlight and CW radar with an LPI waveform, which distributes its radiated energy in space and time as much as possible. In comparison to a conventional radar the peak power density for a certain location is reduced by the product of the usual transmit gain and the inverse duty factor, that is together about 10^5–10^7. The protection would be especially effective if several systems operated in the same area at the same time, acting as mutual decoys.

The principle has been demonstrated with an experimental system, the OLPI radar. It constitutes a ubiquitous radar with some new problems for signal processing but also with new possibilities for target observation. First we need a receiving multibeam facility, realised with an array antenna and multiple beamforming, in future achieved by digital techniques. The signal processing has to perform long-term integration; solutions have been discussed. A reasonable realisation of signal processing with few DSPs for each receiving beam and for online operation has been demonstrated.

To avoid range-walk problems during the integration time the range resolution has been chosen to be modest. For an approaching target the necessary integration time will reduce dramatically according to the radar equation. Then the size of the range cells can also be reduced if necessary for target tracking.

Because there is no antenna modulation of clutter echoes the clutter-suppression filter can be recursive and needs no initialisation. The multiple receiving beams allow cancelling of jammers by adaptive processing in beam space. The OLPI radar experimental system has been realised for a fixed frequency, but there is no problem in principle to applying several frequencies together or in agility modes.

The possibilities for imaging and classification by continuous target observation seem interesting. The results for helicopter detection and classification are an example. ISAR and JEM evaluation could also be applied.

The application of spatial coding has been discussed. This could be a means of countering an opposing electronic reconnaissance. The problem of sufficient Doppler tolerance has been considered.

18.15   References

1 WIRTH, W. D.: 'Omnidirectional low probability of intercept radar'. Proceedings of international conference on *Radar*, Paris, France, 1989, pp. 25–30
2 SCHRICK, G., and WILEY, R. G.: 'Interception of LPI radar signals'. IEEE international conference on *Radar*, Arlington, USA, 1990, pp. 108–111
3 COOK, C. E., and BERNFELD, M.: 'Radar signals' (Academic Press, New York, 1967) p. 256
4 KLEMM, R.: 'Multiplier-free filters for ground clutter suppression', *IEE Proc. F, Commun. Radar Signal Process.*, 1986, **133** (1), pp. 12–15
5 STORZ, W., and WIRTH, W. D.: 'Automatische Auswertung digitalisierter Radarsignale', *NTZ*, 1963, (12), pp. 643–656
6 WIRTH, W. D.: 'Zur Entdeckung aufschwebender Hubschrauber mit Radar'. Actes du Colloque franco-allemand ISL-Rapport R 104187, 'Detection des helicoptres', Saint-Louis, France, 1987, IV-7-1..13

7 WIRTH, W. D.: 'Long term coherent integration for a floodlight radar'. IEEE international conference on *Radar*, Alexandria, USA, 1995, pp. 698–703
8 RETZER, G.: 'Zur Detektion von Zielen im Clutter unter Ausnutzung von Zielechomodulationen durch rotierende Strukturen'. Forschungsbericht FGAN-FHP, Wachtberg, Germany, 1986 (256)
9 DOREY, J., BLANCHARD, Y., and CHRISTOPHE, F.: 'Le Projet "RIAS": une approche nouvelle du radar de surveillance aérienne'. International conference on *Radar*, Paris, France, 1984, pp. 505–510

Chapter 19

System and parameter considerations

In this final chapter we will discuss some aspects of the choice of some main systems' parameters and their relationships for multifunction operation. We consider the observation of the air space to detect, locate and track flying targets, the classical task of a surveillance radar. The aim is to establish a track for each target as early as possible. We have discussed in several preceding chapters the arbitrary movement of the agile beam, which is generally a pencil beam. There is the desire to use this freedom in a most intelligent and effective way, which means to achieve reliable tracking of incoming targets with adequate location accuracy. The tracks should be established at a maximum range with a minimum of mean power. Thereafter, the tracks should be maintained with an adequate tracking rate and power.

19.1 Parameter selection for search and tracking

We may imagine ourselves, for a moment, sitting somewhere scanning the sky for flying objects, like insects, birds or aircraft. We look around in an intuitive regular or random pattern. If something excites our attention we will immediately look more closely or concentrate more to recognise the type of the detected object and the direction of its movement. Then we proceed looking for other objects, but we will, at a suitable time interval, look back at our already recognised objects so as not to lose them and hold them under our observation or tracking.

A similar observation function will be performed by a multifunction phased-array radar for air surveillance. After the search and detection of targets follows an acquisition or confirmation procedure as the starting operation for establishing a track. A fundamental quality given by the agile beam is decoupling of the search and tracking functions, allowing both types of function to operate with individual parameters.

This is in contrast to classical mechanical rotating radars, where the rotation period, usually several seconds, is a compromise between the search and tracking functions. It could be worthwhile to use longer search periods and shorter track intervals. But the tracking must be performed with the rotation period with which the targets are scanned, and it is therefore named track while scan (TWS).

The tracking process in a multifunction radar starts with the first detection of the target. With a short time delay after this detection the resolution cell is revisited for a confirmation that a target is present, or a false alarm is deleted. After detection confirmation the tracking in this beginning phase may be performed at such a high data rate and increased pulse energy as to establish and maintain the target track under observation with a very high tenacity. In other words, once detected targets will almost never be lost. That is, we have established with a high degree of confidence the existence of the target and estimated its position and velocity vector.

19.1.1   Search period

We will first discuss the choice for the search period. The search period is denoted as the time until revisiting a certain resolution cell again with our beam. The search period will generally be different for certain areas of the search volume. In the following we discuss some considerations of Billam [1], in simplified form, for some general conclusions.

Our prime interest may be for the case of approaching targets. A target may fly with a closing velocity v from infinity to our radar.

We ask for the probability of detecting a target for the first time as a function of range. The detection probability P_d increases during the target's approach from 0 to near 1. We have to consider together all chances for the target's detection and this leads to the cumulative detection probability P_c. We can expect a higher P_c for a short search period T_f, because there are more chances for a detection, and *vice versa*. We introduce a range-dependent $P_d(i)$, with i as the range cell number. Then we have for detection possibilities at the ranges i_n, with $n = 1, \ldots, N$ as indicated in Figure 19.1, the probabilities $P_d(i_n) = P_d(n) = P_n$, and for the cumulative detection probability P_c:

$$P_c = 1 - \prod_{n=1}^{N}(1 - P_n) \tag{19.1}$$

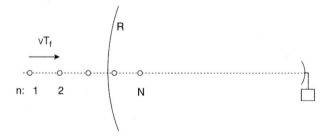

Figure 19.1   Detection sequence for approaching target

The product term alone is the probability for 'not detecting the target in any search period n'. P_c is then the complement to one, the probability of detecting a target at least at one position n.

To get the values for P_n we have to look as a first step to the radar equation, given in chapter 4 with equation 4.48. C may be a constant comprising all other parameters apart from the range R. For a certain range R the received signal-to-noise power ratio is then:

$$\text{SNR} = \frac{C}{R^4}$$

and for the design range R_0, characterised by a $P_d = 0.5$ we may have:

$$(\text{SNR})_{R_0} = \frac{C}{R_0^4}$$

or together by eliminating C:

$$(\text{SNR})_R = \left(\frac{R_0}{R}\right)^4 (\text{SNR})_{R_0} \tag{19.2}$$

For a target with closing velocity v and a search period T_f we try to detect the target at range intervals $\Delta R = vT_f$, starting at a range R_m well above R_0 but within the unambiguous range, given by the pulse recurrence frequency PRF. With equation 19.2 we can compute the SNR values for these range steps ΔR:

$$(\text{SNR})_n = \left(\frac{R_0}{R_m - n\Delta R}\right)^4 (\text{SNR})_0 \tag{19.3}$$

Then we need the operating characteristic for target detection, that is the function of the detection probability dependent on the SNR. This depends on the target fluctuation model and the type of integration. As an example we have for a nonfluctuating target model the curves in chapter 8 in Figure 8.19. All curves have approximately the same form and we can take the SNR value for $P_d = 0.5$ as the reference. With the SNR steps n from above the corresponding values for $P_d(n)$ follow. For different values of ΔR the curves for the cumulative detection have been computed [1], an example is given in Figure 19.2. The ΔR are normalised to R_0, the dashed curve is for a single detection with probability $P_d(R)$. We recognise a substantially improved range by considering the cumulative detection probability.

One can define a cumulative surveillance range R_c for $P_c = 0.9$. The ratio R_c/R_0 can be considered as a gain by looking to the target or sampling it at a period T_f, so this gain has been named the sampling gain. This may be expressed in dB by $40 \log (R_c/R_0)$, equivalent to a power gain for the SNR.

During searching, the mean power spent in one beam position i is, with P for the mean radar power and T_d for the dwell time in one beam position, given by:

$$P_i = P \frac{T_d}{T_f} \tag{19.4}$$

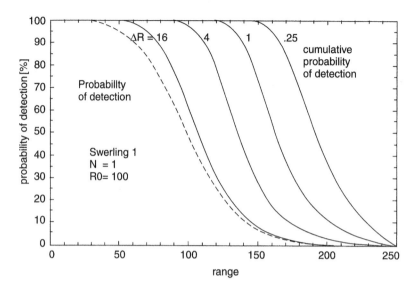

Figure 19.2   Probability of detection and cumulative probability of detection curves for values of $\Delta R / R_c$ of 0.25, 1, 4 and 16%, $P_{fa} = 10^{-4}$, $N = 1$, $R_0 = 100$, after Billam [1] © IEE 1982

We may trade T_d and T_f for a selected or given mean power P_i for a certain beam position i. An increase in search period T_f is accompanied by an increase in ΔR and therefore a decrease in sampling gain. On the other hand, T_d then increases likewise causing an increase of the integration gain.

The product of integration and sampling gain gives information about if and where an advantage can be achieved. The integration gain may be also expressed in dB, e.g. for coherent detection by $10 \log N$, with N for the number of integrated pulses. The product of both gains in the dB scale is then given by the sum. At the point of maximum of this product the required mean power will be a minimum. The corresponding curves for the required mean power, shown in Figure 19.3, have been given by Billam in Reference 1. The ordinate indicates the relative required mean power to achieve a certain range. The range of the optimum is quite broad and depends on the type of target fluctuation and integration and is found at $\Delta R/R_0$ equal to four to 20 per cent. A high value for $\Delta R/R_0$ results especially for coherent detection, which is the most effective approach and should be applied if possible, corresponding to the suggestions in chapters 8 and 9.

For multifunction operation we also need system availability or time for the tracking task. The amount of resource availabe for tracking is highly dependent on the scenario, it may amount to e.g. 50 per cent; this reduces the time available for search. Compensation is possible by a reduction of the dwell time, causing a corresponding reduction of R_0 and R_c. This reduction has also been computed as a function of the tracking load [1] and is given in Figure 19.4. Surprisingly, a tracking load of 50 per cent results only in a ten per cent reduction of R_c. The dwell time for the

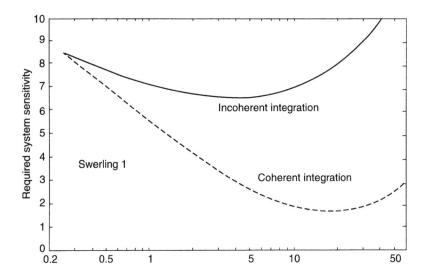

Figure 19.3   Relative system sensitivity (mean power) required to achieve a given R_c as a function of $\Delta R / R_c$ (%), $P_{fa} = 10^{-4}$, after Billam [1] © IEE 1982

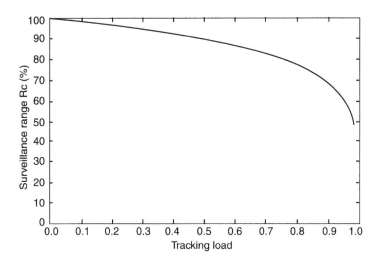

Figure 19.4   Reduction of surveillance range R_c versus tracking load, after Billam [1] © IEE 1982

tracking tasks will be fairly high to achieve a high probability of detection and it should be proportional to R^4 for an adaptation to range according to the radar equation. So the long-range targets predominantly determine the load. This in turn means that a reduction of range for search reduces the tracking load more than proportionally.

A parameter example may illustrate the above result. With $R_c = 200\,\text{km}$, $v = 300\,\text{m/s}$ and $\Delta R/R_c = 10$ per cent as a mean value we get an optimum search period of 66 s. This is surprisingly long, unusual for our conventional rotating surveillance radars. This fact, on the other hand, makes it easier to scan the high number of beam positions within the required search volume of a high-resolution radar with a narrow pencil beam.

19.1.2 False-alarm probability

Another parameter to be selected is the false-alarm probability P_{fa}. Each false alarm activates the confirmation mode and consumes system power. As discussed in chapter 9 for sequential detection with its two decision thresholds, the value for P_{fa} can be selected to be very low without a decrease in P_d and a penalty of increased mean power. But for all detection processing modes with a fixed pulse number N and one decision threshold, P_d and P_{fa} are coupled. Again following Billam [2] one can define a false-alarm dismissal loss L_{pfa} describing the wasted power necessary to cope with the false confirmation load. We have further to consider the SNR variation to maintain a $P_d = 0.5$ for a varying P_{fa}. Then we have additionally a confirmation loss L_c, which describes the additional power for the detection to compensate for the detection probability of confirmation being less than one. All losses together are shown in Figure 19.5.

The resulting net loss shows a flat minimum for $P_{fa} = 10^{-5}$, so the choice is not critical, but one should not go with a false-alarm probability above about $5 \cdot 10^{-5}$.

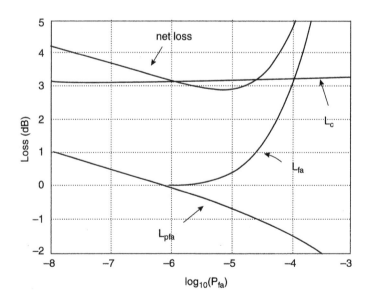

Figure 19.5 Losses in the track initiation process versus P_{fa}, after Billam [2] © IEE 1992

For the confirmation mode we can increase the P_{fa} to a value of about 10^{-2} with a corresponding increase in detection probability, because there are only a few resolution cells in this mode. This reduces the confirmation loss and it is not necessary to increase the power for this mode from the viewpoint of efficiency. There may be, on the other hand, the desire to have some surplus energy in favour of accurate target location and for the case of target cross-section fluctuations. It is important to avoid any interruption of tracks by target missing because otherwise a local search with confirmation would be necessary to reestablish the track.

19.1.3   Beam position separation

The beam positions (BP) will be, within the (u, v) coordinates, on a triangular grid. The question arises of what is the optimal distance D, measured in 3 dB beamwidth (one way), for this grid. We have to find a compromise between the beam shape loss, increasing with D, and an increase in mean power with decreasing D. The relative power for achieving a certain track initiation range has also been computed by Billam [2], assuming a uniform distribution probability for target directions. The result is seen in Figure 19.6.

The optimum value is found to be about $D = 0.8$. A further reduction in beam shape loss for $D \geq 0.9$ can be achieved by applying a three-scan cycle with interlaced beam positions to fill the holes within the simple triangular grid [3].

19.1.4   Range resolution

A high range resolution is not required for the search mode, because it is only preparation for the confirmation or acquisition mode, which follows directly. As discussed in

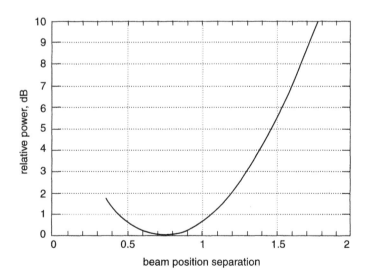

Figure 19.6   Relative required power to achieve a specified track initiation range versus beam position separation, after Billam [2] © IEE 1992

chapter 9, the efficiency of sequential detection decreases with an increasing number of range cells M within a beam position. The processing load increases also with M. For search in heavy clutter a high resolution may be advisable however, because the size of the resolution cells determines the clutter power level.

In any case we have to avoid losses from range walk. The target should not leave the actual resolution cell by a radial movement during the dwell time. Otherwise the processing complexity would increase dramatically because the signal-integration process would be distributed over several range cells. With conventional processing, assuming only one range cell, detection losses would occur. Therefore with v = radial velocity, T = pulse period, N = number of integrated pulses and Δr for the size of the range cell the dwell time should be limited by:

$$vT_d = vTN \ll \Delta r$$

Equation 19.4 relates, for a required mean radar power, to a long dwell time T_d, a proportionally long search period T_f. Because T_d is limited by $\Delta r/v$ we have also a limit for T_f.

For example, with $v = 300 \, \text{m/s}$, $T = 2 \, \text{ms}$, $N = 64$ we have $vT_d = 38.4 \, \text{m}$. A range cell in the search mode should be, for example, not less then 150 m, according to a subpulse length of 1 μs.

19.1.5   Tracking parameters

A fundamental advantage of the agile beam affects target tracking. The tracking process may be arranged completely independently of the search function. The freedom in beam steering may be used to control the track update rate and the target dwell time. The target dwell time offers control of the effective signal-to-noise ratio SNR by determining the length of signal integration or pulse number N. Further, beam agility offers tracking of multiple targets at the same time by multiplex operation. The tracking capacity or the number of targets tracked at the same time is limited only by the radar's resource in time and power. Therefore, it is worthwhile trying to optimise the resource allocation to individual tracks with the aim of not wasting any energy. In critical situations, as at the beginning of a target track, during target manoeuvres or with a locally increased false-alarm rate by interference signals, we can apply an increased update rate as a means of adaptation to the actual situation.

These problems and possibilities have been studied and discussed especially by Blackman and van Keuk [5–11]. We will restrict ourselves here only to some general remarks concerning the relevant parameters.

After detection of a target there immediately follows the acquisition with an accurate measurement of target location and Doppler frequency shift, indicating the radial velocity. Very soon a second measurement will follow to initiate a track. The standard processing method for tracking is the Kalman filter, described for example in Reference 11. It estimates the target state for the parameters (r, u, v), that is range, azimuth, elevation, and the first and second derivatives in time. As a measure of track performance the covariance matrix of the estimated errors is computed and updated.

The track process predicts the position of a target for steering the beam and a range gate to measure the target again for the next update. The error P for the predicted position (u and v), given by the Kalman filter, should be less than the beamwidth, otherwise the SNR at the next measurement decreases or the target even gets lost. The error P is increasing, as expected, with the update interval T. The necessary update interval results from the condition:

$$P(T) = V_0 \Theta \tag{19.5}$$

Θ is the 3 dB beamwidth of the array antenna (chapter 4), V_0 is a dimensionless constant to be selected appropriately. The update interval T is determined just to meet the above condition. Blackman and van Keuk have found [8] a good compromise value for V_0:

$$V_0 = 0.3$$

The implication of equation 19.5 is also an adaptive update rate: if in the case of target manoeuvres the error increases then a shorter update interval results. Recently Hong and Jung [9] have also confirmed this choice for V_0 for nonstationary situations. The track update interval results from equation 19.5 to the following range of values:

track initiation phase and during manoeuvres: $T = 0, 2–1, 3\,\text{s}$
track maintenance phase: $T = 1, 2–4\,\text{s}$

The other important parameter for the tracking function is the SNR. It determines the measurement accuracy for target locations according to chapter 11, section 11.3, equation 11.19. A generally optimum value has been found [9] with :

$$\text{SNR} = 15\,\text{dB}$$

The two parameters T and SNR determine the radar design for tracking. For the search function generally a value SNR $= 13\,\text{dB}$ is assumed sufficient. For tracking a slightly increased value ($+2\,\text{dB}$) is suggested.

Tracking becomes more difficult in the case of multiple neighbouring targets and with locally multiple false detections produced by clutter echoes or false alarms by jamming residues. All these target or false target reports then may be connected to an existing track. To avoid the generation of new and separated tracks with an increasing number, with the consequent explosion effects, some refined tracking methods have been developed. They are all discussed in detail in Reference 11. Only a very simplified idea for some different methods will be given here.

The simplest method for the allocation of multiple observations to an existing track is to choose the nearest neighbour (NN) to the predicted point. If this observation is just a false alarm the track will be confused.

With probabilistic data association (PDA) a single track will also be continued in the presence of multiple observations, which are all lying within the track gate. The track gate forms a possible area for new and valid observations to an existing track. For all the new observations the association probabilities belonging to the track are computed. In other words, for each observation exists a separate hypothesis to belong to the track with a certain probability. With these probabilities a weighted sum of all observed locations is formed. The resultant location is then used in the Kalman filter to update the single track.

Joint probabilistic data association (JPDA) is a similar method. Now multiple target tracks are considered. The association probabilities are computed by using separate tracks. For each individual track the association probabilities are computed and the weighted sum of observations are derived. So each track or hypothesis is continued with its corresponding resultant observation.

A further refinement, with increased complexity, is multiple hypothesis tracking (MHT). The decision for a certain hypothesis is deferred, whenever there are association conflicts between observed locations and existing tracks. All possible hypotheses are propagated with the new observations until future observations lead to a decision and resolving of uncertainties. To avoid an explosion in the number of hypotheses, several techniques have been developed. They include clustering or merging of tracks and also deletion (pruning), dependent on the respective hypothesis probability.

In the preceding sections we have given some rules for the choice of parameters for the search and tracking functions. These parameters may be used for system design and for an initial setting. But the phased-array radar system offers the challenging advantage of changing the parameters to adapt the system to special situations given by the interference and target behaviour. After acquisition the target cross section is estimated by the received amplitude and the necessary energy to achieve the required SNR for future measurements may be selected. The clutter distribution will be known by an adaptive clutter map. Interference by jamming will also be measured and mapped. Therefore, the condition for achieving an overall adaptive radar function will be given.

19.2   Search procedure

The volume for a target search may be scanned by the pencil beam with any pattern. Examples are a series of azimuth scans for a series of elevation levels, or *vice versa*. Also, a pseudorandom pattern may be applied as a kind of ECCM measure. A spiral search may be applied as a super search after the loss of a tracked target. In each beam position a test for a target present has to be performed, for example by sequential detection, as discussed in chapter 9.

It would be desirable to favour for the search function those beam positions with a high probability of containing a target. If *a priori* knowledge in the form of probabilities for targets present in each individual beam position was available, then the beam positions could be ranked in a sequence according to descending target probabilities. To some extent this might be the case in air traffic control using the information given by flight plans and location of airports and air lanes.

If no *a priori* information exists, there is a possibility for an adaptive search strategy. With all detection procedures, a pulse series with a fixed or variable number of pulses is applied for each beam position. Following a proposal given by Posner [12] as a first step, only one pulse is transmitted into each beam position. The maximum echo signal out of each beam position is stored and used for establishing a ranking order according to descending amplitudes. The amplitudes are used as substitutes for the target probabilities which are not available. In a second step a final test with the decision 'target present' or 'no target' is performed. By this strategy some advantages are obtained:

- The transmit energy is distributed according to actual requirements and short reaction times against targets appearing out of shadowed areas are achieved.
- If the search has to be interrupted in favour of other radar functions then those beam positions with the highest probability for containing targets have already been tested, independent of their distribution in space contrary to a systematic search pattern. This serves as a graceful degradation of the system in case of overload by tracking.
- The period for the repetition of the ranking procedure may be chosen according to the requirements for the reaction time for different parts of the search volume.

If the number of targets is known, e.g. only one as in the case of a lost tracked target [12], the local search may be terminated after detection of these targets. Energy and time saving would result by avoiding the testing of empty beam positions. If the target number is unknown it is likewise not useful to perform the final tests up to the end of the rank order, because the last part of this rank order will be determined by noise only.

An improved strategy would be the following. After forming a rank order with the first transmitted pulse per beam position, a certain part of this order starting from its beginning is tested for targets. This part would contain stronger targets. Then this first test is stopped, a second pulse is transmitted into the remaining beam positions and the echoes added to the first ones, improving the SNR. A new and improved rank order based on the echo sum is then established. In this manner learning periods with the ranking procedure and periods with final testing may alternately be performed, establishing a cyclic search procedure.

This version has been studied by Dannemann with simulations [13]. It was shown for the general case of an unknown target number that the costs of the search, expressed by the overall number of transmitted pulses, were not higher for this adaptive search compared to those for a systematic search, despite the fact of necessary additional pulses for the learning function. A saving of energy seems to be possible only in the case of a known target number.

Of course, clutter echoes have to be suppressed before the ranking procedure, which should therefore be applied only for long-range applications or in combination with an MTI filter. In this last case at least a pair of pulses has to be transmitted to form a difference of subsequent signals for each range cell for fixed clutter suppression. This cyclic search procedure has not been realised and tested up to now.

19.3 References

1 BILLAM, E. R.: 'Design and performance considerations in modern phased array radar'. IEE international conference on *Radar*, London, UK, 1982, pp. 15–19

2 BILLAM, E. R.: 'Parameter optimisation in phased array radar'. IEE international conference on *Radar*, Brighton, UK, 1992, pp. 34–37

3 HANLE, E.: 'Suchraumabtastung mit einem elektronisch gesteuerten Radarsystem', *NTZ*, 1977, **30**, p. 161

4 BILLAM, E. R.: 'The problem of time in phased array radar'. IEE international conference on *Radar*, Edinburgh, UK, 1997, pp. 563–567

5 VAN KEUK, G.: 'Adaptive computer controlled target tracking with a phased array radar'. Proceedings of international conference on *Radar*, Washington, USA, 1975, pp. 429–433

6 FLESKES, W., and VAN KEUK, G.: 'Adaptive control and tracking with the ELRA-phased array experimental radar system'. Proceedings of international conference on *Radar*, Washington, USA, 1980, pp. 8–13

7 FLESKES, W., and VAN KEUK, G.: 'On single target tracking in dense clutter environment'. IEE international conference on *Radar*, London, UK, 1987, pp. 130–134

8 BLACKMAN, S. S., and VAN KEUK, G.: 'On phased array radar tracking and parameter control', *IEEE Trans. Aerosp. Electron. Syst.*, 1993, **29** (1), pp. 186–194

9 HONG, S.-M., and JUNG, Y.-H.: 'Optimal scheduling of track updates in phased array radars', *IEEE Trans. Aerosp. Electron. Syst.*, 1998, **34** (3), pp. 1016–1022

10 VAN KEUK, G.: 'Multihypothesis tracking with electronically scanned radar', *IEEE Trans. Aerosp. Electron. Syst.*, 1995, **31** (3), pp. 916–927

11 BLACKMAN, S., and POPOLI, R.: 'Design and analysis of modern tracking systems' (Artech House, Boston, London, 1999)

12 POSNER, E.C.: 'Optimal search procedures', *IEEE Trans. Information Theory*, 1963, **9** (2), pp. 157–160

13 DANNEMANN, H.: 'Optimale Suchverfahren für Elektronisches Radar'. Nachrichtentechnische Zeitschrift, **26**, 1973, pp. 118–122

Index

Glossary

a	signal amplitude
A	high decision threshold for sequential test
$\mathbf{A}[N,K]$	matrix with K antenna vectors as columns of length N
a_0	design signal for sequential test function
ABF	adaptive beamforming
ACE	autocorrelation estimate
ADC	analogue-to-digital converter
AER	airborne experimental array radar of FGAN
a_i	design signal for sequential test function at range i
ARM	antiradar missile
ASR	air surveillance radar
α	false-alarm probability
B	bandwidth
B	low decision threshold for sequential test
$\mathbf{b}$	signal vector
BB	baseband
BP	number of beam positions
BW	beam width
β	abbreviation for $2\pi/\lambda$, target-missing probability
CNR	clutter-to-noise power ratio
CPI	coherent processing interval
CRI	cross-range repetition interval
CW	continuous wave signal
$\mathbf{C}_w(\omega)$	cross-spectral matrix of clutter
D	dynamic range
d	Doppler factor
$\mathbf{d}$	Doppler vector, difference weighting vector
D	beam separation
$D(u(t))$	directional two-way antenna pattern
DOA	direction of arrival
DSP	digital signal processor

Δ	Jacobi determinant
ΔR	range interval according frame time
$E\{x\}$	expectation of variable x
ECCM	electronic counter countermeasures
ECM	electronic countermeasure
ELRA	experimental electronic steerable radar
η	decision threshold
F	noise factor
f	frequency, sweep factor
$f(\theta/\Theta_0)$,	array factor for steering direction Θ_0 or u_0
f_d	Doppler frequency shift
FET	field effect transistor
FFT	fast Fourier transform
FIR	finite impulse response
FSST, FST	fixed sample size test, fixed sample test
G	antenna gain
$g, \mathbf{g}$	weighting coefficient or vector
g_a	amplitude-fluctuation estimator
g_p	phase-fluctuation estimator
$\boldsymbol{\Gamma}$	projection matrix
h	filter function in time domain
H	filter function in frequency domain
HTP	Hung-Turner projection
HPRF	high pulse recurrence frequency
$\mathbf{I}$	identity matrix
I, Q	in-phase and quadrature signal components
IF	intermediate frequency
$\text{Im}\{z\}$	imaginary part of z
$I_0(x)$	modified zero-order Bessel function
ISAR	inverse synthetic aperture radar
j	$\sqrt{-1}$
K	number of jammers, possible notches, data snapshots
kTF	noise-power density
L	aperture length, loss factor, number of subarrays, necessary degrees of freedom
LFM	linear frequency modulation
LMI	lean matrix inversion
LMS	least-mean-square
LNA	low-noise amplifier
LOS	line-of-sight
LRT	likelihood ratio test
LS	least square
LSMI	loaded sample matrix inversion
λ	wavelength
$\lambda(z)$	test function for signal z

mod	*n mod N*, the residue of *n* by adding or subtracting integer multiplers of *N*
M	number of signals in a time series, number of range elements, target number
MF	matched filter
MMIC	miniaturised microwave integrated circuit
MSAR	multichannel SAR
MTBF	mean time between failure
MTI	moving target indication
MTP	matrix transform projection
MUSIC	multiple signal classification
μ	recursion constant
n	noise vector
N	number of array elements or signals in time or space
NF	noise figure
NLFM	nonlinear frequency modulation
OLPI	omnidirectional low probability of intercept
ω	parameter set, frequency
P	projection matrix into signal space
P	tracking prediction error
$p(x)$	probability density function for variable *x*
$p(x, y)$	joint probability density function for variables *x* and *y*
$p(x/y)$	conditional probability density function for variable *x* given *y*
$p(z; \theta)$	probability density of signal *z* for a signal parameter θ
PC	personal computer
P_c	cumulative detection probability
PDF	probability density function
P_D	detection probability
P_F	false-alarm probability
PNL	phase code from NLFM
PTMF	parametric target model fitting
Q	covariance matrix of clutter or other interference
Q̂	estimated covariance matrix
q	correlation coefficient
Q	expression for estimation error
R	target range
R_0	range with $P_d = 0.5$
R_c	cumulative surveillance range
Re $\{z\}$	real part of *z*
RF	radio frequency
RSL	relative sidelobe level
ρ	autocorrelation coefficient
s	signal vector
S	number of snapshots, sum beam output

SAR	synthetic aperture radar
SAW	surface acoustic wave
SLL	sidelobe level
SMI	sample matrix inversion
SNCR	signal-to-noise-plus-clutter ratio
SNIR	signal-to-noise-plus-interference power ratio
SNR	signal-to-noise power ratio
STAP	space-time adaptive processing
σ	target cross section, standard deviation of noise
σ^2	variance
T	time period, especially radar pulse period
$\mathbf{T}$	transform matrix
t	time
T_d	target dwell time
T_f	frame time for search
T_K	test function with autocorrelation estimates up to K lag values
TOI	third-order intercept point
trace $(\mathbf{Q})$	sum of all main diagonal elements of matrix $\mathbf{Q}$
TRM	transmit/receive module
T_{sa}	time for generating a synthetic array
τ	length of subpulse
Θ_B	half-power beamwidth of antenna main beam
Θ_s	beamwidth of synthetic array
$\mathbf{u}$	directional unity vector
u, v	projected directions
v	target velocity
VGA	variable-gain amplifier
$\mathbf{w}$	filter vector
WVD	Wigner-Ville distribution
y	output signal
$\mathbf{z}$	signal vector
$\mathbf{z}^*$	conjugate transpose vector
z, z_k	complex signal
$\mathbf{Z}[N, M]$	signal matrix with N rows and M columns